Restoring Tropical Forests

A practical guide

This book is dedicated to the memory of Surat Plukam. A talented artist and illustrator, his clear and simple artwork has made forest restoration more accessible to both children and adults, from villagers to government officers, throughout SE Asia.

This publication has been made possible with funding from the Darwin Initiative, supported by the RBG Kew Programme in Restoration Ecology, John Ellerman Foundation, Man Group plc, and Kew's Millennium Seed Bank Partnership.

For bibliographic purposes, this book should be referred to as:

ELLIOTT, S. D., D. BLAKESLEY AND K. HARDWICK, 2013. Restoring Tropical Forests: a practical guide. Royal Botanic Gardens, Kew; 344 pp.

RESTORING TROPICAL FORESTS

A PRACTICAL GUIDE

By Stephen D. Elliott,
David Blakesley and Kate Hardwick

Artwork by Surat Plukham and Damrongchai Saengkam
Joseph Nkongho Agbor

Sponsored by Britain's Darwin Initiative

Kew Publishing
Royal Botanic Gardens, Kew

First published in 2013 by
Royal Botanic Gardens, Kew,
Richmond, Surrey, TW9 3AB, UK
www.kew.org

Distributed on behalf of the Royal Botanic Gardens, Kew in North America by the University of Chicago Press, 1427 East 60th Street, Chicago, IL 60637, USA

ISBN 978-1-84246-442-7

British Library Cataloguing in Publication Data
A catalogue record for this book is available from the British Library

Production editor: Sharon Whitehead
Design, typesetting and page layout: Margaret Newman
Publishing, Design & Photography, Royal Botanic Gardens, Kew
Cover design: Lyn Davies Design

Printed and bound in Italy by Printer Trento S.r.l.

For information or to purchase additional copies of this book (in English, French or Spanish) and other Kew titles please visit www.kewbooks.com or email publishing@kew.org

Kew's mission is to inspire and deliver science-based plant conservation worldwide, enhancing the quality of life.

Kew receives half of its running costs from Government through the Department for Environment, Food and Rural Affairs (Defra). All other funding needed to support Kew's vital work comes from members, foundations, donors and commercial activities including book sales.

Contents

CLARENCE HOUSE

As Patron of the Foundation and Friends of the Royal Botanic Gardens, Kew, I was delighted to have been asked to contribute a foreword to this wonderful book, 'Restoring Tropical Forests: A Practical Guide'. I can only congratulate the authors for their achievement and wish all those who implement its clear practical measures on how to restore tropical forests around the world - in South and Central America, Africa and Asia – every possible success in their vitally important endeavours.

Nature has a remarkable capacity, if given half a chance, to recover and renew herself and it is for this reason, above all, that I believe this book to be so very welcome. I am particularly drawn to its emphasis on the need to restore rich tropical forests, wherever possible with native species; its description of how best to engage local communities in restoration efforts and its focus on the need for landscape and silvopastoral approaches to forest restoration, all of which seem to me absolutely crucial.

I am also intrigued by the book's account of 'rainforestation', the technique pioneered in the Philippines by which indigenous tree species are planted in order to restore ecological integrity and biodiversity, whilst at the same time producing a diverse range of timbers and other forest products for local people.

For many decades, I have been deeply concerned about the plight of the world's tropical forests, inspired both by their timeless grandeur, by the extraordinary biological and cultural diversity to which they are home and by the profound knowledge that neither humanity nor the Earth itself can survive without them, in particular in the face of global climate change. With this in mind, a few years ago I set up my own Rainforests Project, in the hope of drawing attention to the urgent need to establish an international agreement to protect the forests, coupled with a financial mechanism – R.E.D.D.+ – intended to contribute to this protection at the scale required. I have been heartened by progress made in many countries, including Brazil, since then, but note that the global pressures on our remaining forests remain acute. Restoration has a fundamental role in advancing these efforts in the years ahead.

Wangaari Maathai, whose death we all continue to mourn, said, 'We owe it to ourselves and to the next generation to conserve the environment so that we can bequeath our children a sustainable world that benefits all'. What better place to start than with the sound recommendations and practical steps identified in this book?

PREFACE

"One touch of nature makes the whole world kin."
William Shakespeare, from *Troilus and Cressida*, 1601–1603.

Twenty years ago, when our Forest Restoration Research Unit at Chiang Mai University (FORRU-CMU) was no more than a few wishful bullet points scribbled on the back of an envelope, the decline of the world's tropical forests was seen as an inevitable and irreversible consequence of economic development. The idea that tropical forest ecosystems could actually be restored was viewed by many as naïve idealism. Scientists saw tropical forests as being far too complex to be re-constructed, while conservation NGOs regarded the idea as an unnecessary distraction from the vital task of funding the protection of remaining primary forest. Even one of our unit's early funders candidly remarked that he considered the concept to be 'armchair conservation'.

Now, thankfully, attitudes have undergone a paradigm shift. Restoration is seen as complementary to protecting primary forest, especially where protected areas have failed to prevent deforestation. Two decades of research have yielded tried and tested methods that have transformed forest restoration from a romantic 'pipedream' into a readily achievable goal. By combining nature's regenerative capacity with tree planting and other management practices, it is now possible to restore rapidly both the structure and ecological functioning of tropical forests and thus to achieve substantial biodiversity recovery, within 10 years of starting restoration activities. Conservation organisations now recognise restoration as vital to reviving degraded landscapes and improving rural livelihoods, by providing a diverse range of forest products and by developing Payments for Environmental Services (PES) programmes. The inclusion of restoration in the UN's REDD+ scheme[1], to 'enhance carbon stocks' and mitigate global warming, has resulted in unprecedented demand for restoration knowledge, skills and training. Such knowledge is vital to enable developing tropical countries to cash in on the global trade in carbon credits, while reducing biodiversity losses and meeting the needs of local communities. But very little practical advice has been published to satisfy this demand.

This book seeks to provide such advice. It presents scientifically tested techniques for the restoration of diverse climax tropical forest ecosystems that are resilient to climate change, using indigenous forest tree species, for biodiversity conservation and environmental protection and to support the livelihoods of rural communities. It is based on more than 20 years of research, carried out by FORRU-CMU, as well as on local knowledge and experiences, exchanged over the past 20 years at hundreds of workshops, conferences and project consultations. Plant names in the book generally follow those listed as 'accepted' on Theplantlist.org website, as of June 2013.

Our book presents generic concepts and practices that can be applied to revive forest ecosystems on all tropical continents, in an accessible format and initially in three languages (English, French and Spanish). It includes case studies that illustrate a diversity of successful restoration projects from around the world. It is aimed at the full range of stakeholders, whose collaboration is vital to the success of restoration projects. It provides planners, policy makers and funding agencies with viable and

---

[1] 'Reducing emissions from deforestation and forest degradation' — a set of policies and incentives being developed under the UN Framework Convention on Climate Change (UNFCCC) to reduce CO_2 emissions derived from clearing and burning tropical forests. www.scribd.com/doc/23533826/Decoding-REDD-RESTORATION-IN-REDD-Forest-Restoration-for-Enhancing-Carbon-Stocks

practicable alternatives to conventional mono-culture plantations that can be used to attain their reforestation goals. For protected area managers, communities, and the NGOs that work with them, the book provides some solid advice on planning restoration projects, as well as scientifically tested instructions for growing, planting and caring for native forest tree species. And for scientists, the book suggests dozens of research project ideas and provides details of standardised research protocols, which can be used to develop new restoration systems that meet local needs. There's even an appendix of proforma for data collection, so that researchers can collect data sets that are comparable with those now being replicated in FORRUs in several countries.

The continued destruction of tropical forests is probably the greatest threat to our planet's biodiversity. Although both awareness of the problem and a willingness to solve it have never been higher, they are ineffective without sound, scientifically based practical advice. We therefore hope that this book will not only inspire more people to get involved in saving Earth's tropical forests, but also provide them with effective tools to do so.

Stephen Elliott
Email: stephen_elliott1@yahoo.com
Website: www.forru.org
 Facebook Page: Forest Restoration Research Unit

David Blakesley
Email: David.Blakesley@btinternet.com
Website: www.autismandnature.org.uk

Kate Hardwick
Email: k.hardwick@kew.org
Website: www.kew.org

ACKNOWLEDGEMENTS

This book is the main output of the project entitled 'Restoring Tropical Forests: a Practical Guide', sponsored by the United Kingdom's Darwin Initiative. We are very grateful for the Darwin Initiative's support of the production costs of this manual, to the Royal Botanic Gardens, Kew, who provided in-house services, to John Ellerman Foundation for funding Kate Hardwick, to Kew Publishing, especially Sharon Whitehead for copyediting and Margaret Newman for layout, and to Kew's Millennium Seed Bank Partnership and Man Group plc, who covered additional costs.

The book is based substantially on the work of the Forest Restoration Research Unit of Chiang Mai University, northern Thailand, and the authors would like to take this opportunity to thank all staff members of the unit, past and present, whose dedicated research has contributed the book's contents, currently Sutthathorn Chairuangsri, Jatupoom Meesana, Khwankhao Sinhaseni and Suracheat Wongtaewon. Australian Youth Amabassador, Robyn Sakara, funded by Biotropica Australia Plc, and FORRU's research officer Panitnard Tunjai contributed significantly to Chapters 2 and 5, respectively.

The authors also thank all those who contributed text, photos or information: Dominique Andriambahiny, Sutthathorn Chairuangsri, Hazel Consunji, Elmo Drilling, Patrick Durst, Simon Gardner, Kate Gold, Daniel Janzen, Cherdsak Kuaraksa, Roger Leakey, Paciencia Milan, William Milliken, David Neidel, Peter Nsiimire, Andrew Powling, Johny Rabenantoandro, Tawatchai Ratanasorn, Khwankhao Sinhaseni, Torunn Stangeland, John Tabuti and Manon Vincelette.

Photographs are mostly by Stephen Elliott and FORRU-CMU staff, unless otherwise credited. Line drawings are by Damrongchai Saengkham and the late Surat Plukam. However, we thank the many others who have provided photographs and illustrations including: Andrew McRobb and other contributors to the Kew photo library, NASA, IUCN for the maps, Tidarach Toktang, Kazue Fujiwara, Cherdsak Kuaraksa and Khwankhao Sinhaseni.

We also thank all reviewers of sections or chapters of the manuscript for their helpful feedback: Peter Ashton, Peter Buckley, Carla Catterall, John Dickie, Mike Dudley, Kazue Fujiwara, Kate Gold, David Lamb, Andrew Lowe, David Neidel, Bruce Pavlik, Andrew Powling, Moctar Sacande, Charlotte Seal, Roger Steinhardt, Nigel Tucker, Prasit Wangpakapatanawong and Oliver Whaley.

We are especially grateful to Val Kapos and Corinna Ravilious (WCMC) for the maps reproduced in Chapter 2.

We thank Joseph Agbor and Claudia Luthi for translating the French and Spanish editions, respectively, and Norbert Sonne and Maite Conde-Prendes for copyediting and proof-reading the translations.

All opinions expressed in this book are those of the authors and not necessarily those of the sponsors or reviewers. The compilers would like to take this opportunity to thank anyone not already mentioned above who has contributed in any way towards the work of FORRU-CMU and the production of this book. Finally, we are grateful to the Biology Department, Science Faculty, Chiang Mai University, for institutional support of FORRU-CMU since its inception, and East Malling Research, Wildlife Landscapes and Royal Botanic Gardens Kew for institutional support of the Darwin Initiative research and the capacity- building programme over a number of years.

Chapter 1

Tropical deforestation: a threat to life on Earth

Tropical forests, which are home to around half of the Earth's terrestrial plant and animal species, are being destroyed at rates unprecedented in geological history. The result is a wave of species extinctions that is leaving our planet both biologically impoverished and ecologically less stable. Although this is widely accepted by scientists, putting precise global figures on rates of tropical deforestation and species losses is not straightforward.

1.1 Rate and causes of tropical deforestation

How fast are tropical forests being destroyed?

Since pre-industrial times, the Earth's tropical forests have shrunk in area by 35–50% (Wright & Muller-Landau, 2006). If losses continue at current rates, the last remnants of primary tropical forest will probably disappear sometime between 2100 and 2150, although global climate change (if unchecked) would undoubtedly accelerate the process.

The United Nations' Food and Agriculture Organisation (FAO) provides the most comprehensive global estimates of tropical forest cover, collating statistics reported by the forest agencies of individual countries (FAO, 2009). Such estimates are, however, far from perfect and are often revised as survey methods become more reliable. Furthermore, definitions of 'forest' vary (e.g. plantations are sometimes included, sometimes not), there is often debate over where the 'edge' of a forest lies, and geographical information technologies are constantly changing. A review of FAO estimates by Grainger (2008) reported that between 1980 and 2005, the area of natural tropical forests[1] worldwide declined from 19.7 to 17.7 million km^2 (**Table 1.1**), an average loss of about 0.37% per year.

The loss of original primary forests[2] is of particular concern for the conservation of biodiversity[3]. Globally[4], FAO (2006) estimates that an average of 60,000 km^2 of primary forest has been destroyed or substantially modified each year since 1990, with just two tropical countries, Brazil and Indonesia, accounting for 82% of this global loss. In terms of percentage losses, both Nigeria and Vietnam lost more than half of their remaining primary forest between 2000 and 2005, while Cambodia lost 29% and Sri Lanka and Malawi each lost 15% (FAO, 2006).

Table 1.1. Natural tropical forest[1] cover (million km^2), 1980–2005 (adapted from Grainger (2008)).

Region	1980[a]	1990[b]	2000[b]	2005[b]
Africa	7.03	6.72	6.28	6.07
Asia-Pacific	3.37	3.42	3.12	2.96
Latin America	9.31	9.34	8.89	8.65
Totals	**19.71**	**19.48**	**18.29**	**17.68**

Sources: *Food and Agriculture Organisation Global Forest Resource Assessments*, [a]1981 and [b]2006. Adapted from Grainger (2008).

[1] 'All naturally occurring woody vegetation with >10% canopy cover, excluding timber plantations, shrub-land etc.'
[2] Forests of native species, with undisturbed ecological processes and not seriously impacted by human activity.
[3] Biodiversity is the variety of life forms, including genes, species and ecosystems (Wilson, 1992). In this book we use the term to refer to all species that naturally comprise the flora and fauna of tropical forests, excluding exotic or domesticated species.
[4] Excluding Russia.

The front line of tropical deforestation — in this case for establishment of oil palm plantations in Southeast Asia. This wholesale destruction is the main cause of the biodiversity crisis and is contributing substantially to global warming. (Photo: A. McRobb).

Although global estimates of tropical forest loss may be problematic, there are many well-documented examples of severe and rapid deforestation at the regional level. For example, between 1990 and 2000, the Indonesian island of Sumatra lost 25.6% of its forest cover (at least 50,078 km^2 of forest). The scale of the destruction is well illustrated on Google Earth (www.sumatranforest.org/sumatranWide.php)

In Brazilian Amazonia, forest cover has been reduced by 10% (377,108 km^2), since 1988. About 80% of forest loss has been caused by clearance for cattle ranches, with much of the rest following highway construction. However, up to 30% of deforested areas may be undergoing natural regeneration (Lucas *et al.*, 2000).

Loss of primary tropical forests and their replacement with secondary forests are likely to continue, despite greater awareness of forest biodiversity and the impact of forest destruction on the environment and climate change. Therefore, whereas conservation of primary forest remains important, management of regenerating secondary tropical forests is fast becoming a major global issue in minimizing biodiversity losses.

Deforestation in the Brazilian state of Mato Grosso, following paving of highway BR 364 (forest in red): left 1992, right 2006 (from NASA Earth Laboratory).

Why are tropical forests destroyed?

The ultimate cause of tropical forest destruction is too many people making too many demands on too little land. The United Nations (2009) predicts that the global human population will surpass 9 billion by 2050 (up from 7 billion at the time of writing); well on the way to exceeding Earth's estimated carrying capacity of about 10 billion (United Nations, 2001). The fate of tropical forests, and that of most other natural ecosystems, ultimately depends on controlling human population growth and consumption.

In most tropical countries, forest destruction usually begins with logging. Logging opens up forest areas by introducing roads and, as the supply of timber trees becomes exhausted, the loggers are followed by landless rural people looking for farmland. The remaining trees are cleared and replaced with small-scale agriculture. Small-holders may initially practice low-intensity, slash-and-burn agriculture, but as a growing population increases pressure on the land, more intensive agricultural systems are typically adopted. As the land value increases, small-scale farmers often sell out to large agro-companies, moving on to clear forest elsewhere.

However, logging is now declining as a primary cause of tropical forest loss as more timber is produced from plantations. Asia-Pacific leads the way in plantation forestry, having a total of 90 million ha of plantations for wood production in 2005. So, although logging has historically been a major cause of tropical deforestation, it has now been overtaken by the exponential surge in demand for farmland, driven by global markets (Butler, 2009).

In Africa, more than half (59%) of deforestation is carried out by families establishing small-scale farms, whereas in Latin America deforestation is mostly (47%) the result of industrial agriculture, caused by global demand for agricultural products. In Asia, conversion of forest to small-scale farms and replacement of shifting agriculture with more intensive agricultural practices account for 13% and 23% of deforestation, respectively, whereas industrial agriculture, particularly oil palm and rubber plantations, account for 29% (FAO, 2009).

Tropical deforestion often begins with logging for the timber industry, but many other factors are involved.

Charcoal making in Brazil. The reliance of more than 80% of people in developing countries on wood or charcoal to cook their food contributes significantly to forest degradation. (Photo: A. McRobb).

Montane forest has been destroyed to make way for tea plantations in Likombe, Cameroon. (Photo: A. McRobb).

An over-grazed landscape in northeast Brazil. (Photo: A. McRobb).

The development of infrastructure, especially roads and dams, can also have a very destructive effect on tropical forests. Although such development impacts relatively small areas of forest, it opens up forest areas for settlement and fragments them, isolating small wildlife populations in ever-shrinking forest fragments.

Finally, weak governance is a major factor that enables deforestation to occur. Although most countries have laws to control forest exploitation, forest departments often lack the authority and funding needed to enforce them. Consequently, in many tropical countries, more than half of the timber produced is extracted illegally (Environmental Investigation Agency, 2008). Forest officials are often poorly paid and are therefore easily corrupted. Local communities are marginalised in decision making and therefore lose their sense of forest stewardship. Consequently, strengthening governing institutions, as well as empowering local communities is fundamental to the survival of Earth's tropical forests.

1.2 Consequences of tropical deforestation

The disastrous effects of tropical forest destruction have been well-documented for decades (Myers, 1992). Of most concern is the greatest extinction event in our planet's geological history.

How much biodiversity is being lost?

Although tropical forests now cover only about 13.5% of Earth's land area, they are home to more than half of the planet's terrestrial plant and animal species. So, it is not surprising that their destruction is causing a substantial proportion of Earth's biota to go extinct. It is difficult to put a precise figure on exactly how many species are likely to die out as a result of tropical deforestation, however, because there is no definitive list of all tropical forest species. Vertebrates and vascular plants have been fairly well counted and named, though new species discoveries are not uncommon so this task is certainly not complete. But it is the smaller animals, particularly insects and other arthropods,

In the 1980s, fogging insects in the canopies of tropical forests began to show that Earth's biodiversity was much higher than anyone expected and that tropical forest destruction was a major threat to it.

that contribute most to tropical biodiversity and there are not enough taxonomists working in the tropics to identify and count all of these species.

Back in the 1980s, the work of Terry Erwin began to reveal just how many arthropod species there might be in tropical forests. Erwin (1982) studied beetle communities in tropical tree crowns. He used an insecticidal fogging machine, hoisted into the crowns, to knock down insects. In the crowns of trees of just one species (*Luehea seemannii*), he found 1,100 beetle species, of which about 160 lived exclusively in that tree species. Since beetles account for about 40% of insect species, we can estimate that the crowns of *L. seemannii* trees probably support around 400 specialist insect species, with a further 200 species living on other parts of the tree. The number of tropical tree species known to science is around 50,000. Were each of these to support a number of specialist insect species similar to that of *L. seemannii*, then the world's tropical forests could support around 30 million insect species.

Even though this calculation makes many (still largely untested) assumptions, and relies on work that is 30 years old, it remains one of the most widely quoted estimates of tropical biodiversity; a sad reflection on the progress of taxonomy in tropical forests over the past three decades. A more recent study by Ødegaard (2008), which tested some of Erwin's assumptions, suggested that the global arthropod fauna may be nearer 5–10 million species.

If counting surviving species is problematic, then counting extinct ones is even more so. The continued existence of a species is verified from a single observation, but it is impossible to be certain that a species is extinct, as it may persist where biologists have not yet looked. Rediscovery of 'extinct' species still happens, so, we must rely on biological theory instead of direct species counts to estimate extinction rates.

The most widely applied model is the species-area curve, which is derived from counting species in consecutive, equal-sized, sample plots. As the number of sampled plots increases, the cumulative number of species discovered increases. At first, the increase is steep but the curve levels off, as more sample plots are added because fewer species remain to be discovered. The number of new species in each subsequent sample plot eventually declines to zero when all species have been discovered, and thus the species-area curve reaches an upper asymptote.

To estimate extinction rates, species-area curves are used in reverse to address the question: "how many species will disappear as the area of a habitat is reduced?" Using this logic, Wilson (1992) estimated that about 27,000 tropical forest species go extinct each year on the basis of published rates of forest destruction and a species-area curve that predicts an eventual 50% decline in species numbers when a forest is reduced in area by 90% (**Figure 1.1**).

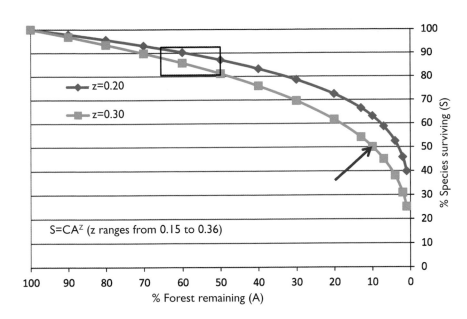

Figure 1.1. Despite their flaws, species-area models still contribute to predictions of extinction rates. For tropical forests, values of the parameter 'z' vary from 0.2 to 0.35 (from empirical studies). A value of 0.3 predicts a 50% decline in biodiversity with 90% forest loss (arrow). The rectangle shows an 8–20% loss of tropical species since pre-industrial times (assuming a 35–50% reduction in tropical forest cover).

Wright and Muller-Landau (2006) also incorporated species-area relationships into their analysis of tropical species extinctions. They also demonstrated a negative relationship between human population density, especially in rural areas, and forest cover. These authors predicted the continued loss of primary forests for timber exploitation, but expected a fall in rural population density in tropical countries by 2030, resulting in the regeneration of secondary forests on abandoned land. Consequently, they predicted little change in overall forest cover over the next 20 years, although most primary forest will be replaced by secondary forests, with the latter providing a refuge for most tropical forest species[5]. Applying species-area relationships to this scenario, the authors projected species extinctions of 21–24% in Asia, 16–35% in Africa and 'significantly less' in Latin America by 2030.

There are several problems with these projections. One is that species-area relationships are based on the total area of remaining forest, rather than on the size of individual forest fragments. If a country's total forest cover is high but that forest is highly fragmented, each fragment may not be large enough to support viable plant and animal populations. In this situation, inbreeding will gradually kill off each small population, fragment by fragment, and as species start to disappear, the web of species relationships that is vital for the maintenance of tropical forest biodiversity will unravel. As plants lose their pollinators or seed dispersers, they will fail to reproduce, and as key species die out, a cascade of extinctions will reduce the rich biodiversity of tropical forests to a few, common weedy species that dominate the landscape. Thus, it is not the overall rate of deforestation that drives extinction, but also the degree to which the remaining forest is fragmented.

Another problem is Wright and Muller-Landau's assumption that secondary tropical forests will provide refuges for primary forest species (Gardner *et al.*, 2007), especially if such areas are separated by vast expanses of agricultural land, over which most primary forest species cannot move. That is, the problem may have more to do with fragmentation of forests rather than simply whether a forest is 'secondary' or 'primary'. And last, their analysis does not consider the effects of hunting and global climate change on species extinctions.

[5] In Asia, fragmented secondary forest already covers a greater area than primary forest (Silk, 2005).

Just a few of the many tropical animal species threatened with extinction as a direct result of deforestation.

The spectacular black and white Roloway monkey (*Cercopithecus diana*) has been brought perilously close to extinction by conversion of West African forests to agricultural land. Hunting now endangers the few remaining animals.

The spineless forest lizard (*Calotes liocephalus*) is endemic to tropical moist montane forest in Sri Lanka. It is threatened by habitat destruction and fragmentation due to cardamom cultivation, grazing livestock and logging.

The flat-headed cat (*Prionailurus planiceps*) is endangered in Indonesia and Malaysia, mostly due to conversion of its tropical lowland forest habitat into oil palm plantations.

The golden lion tamarin (*Leontopithecus rosalia*) is endemic to lowland coastal forests of Rio de Janeiro, one of the most endangered of tropical forest types. Now reduced to fewer than 1,000 individuals, the species continues to teeter on the brink of extinction despite a re-introduction program.

The Alagoas currassow (*Mitu mitu*) is extinct in the wild because of destruction of lowland primary forest in Brazil. Consequently, this forest ecosystem has lost an important seed disperser. Two captive populations remain the only hope for the survival of this species.

Gurney's Pitta (*Pitta gurneyi*) has already been declared extinct because of the conversion of lowland evergreen tropical forest in Thailand and Burma to rubber and oil palm plantations. Its rediscovery in 1986 was followed by frantic efforts to protect, restore and 'unfragment' the tiny forest patches at the rediscovery site.
www.birdlife.org/news/features/2003/06/gurneys_pitta_stronghold.html

Although a loss of between a quarter and one-third of tropical biodiversity over the next 20 years is serious, many scientists argue that Wright and Muller-Landau actually underestimated tropical extinctions. The rise of industrial agriculture and plantations as the main drivers of tropical deforestation may render the relationship between human population and deforestation invalid. Cattle ranches, tree plantations and bio-fuel production often increase deforestation, while simultaneously reducing human population density.

Clearly a better model is needed for estimations of extinction rate, but developing ever more precise predictions of species extinctions will not solve the problem. In a world where secondary tropical forests will largely replace primary forest, the survival of most species will depend on ensuring that secondary forests grow well, support rapid biodiversity recovery and are well-connected, so that they become ecologically similar to primary forests as quickly as possible. The science of tropical forest restoration can certainly help with that.

Contribution of tropical deforestation to global climate change

Deforestation makes a significant contribution to global climate change. Carbon dioxide (CO_2), released by clearing or burning tropical forests currently contributes about 15% of the total CO_2 emitted into the atmosphere from human activities (Union of Concerned Scientists, 2009). The rest comes from burning fossil fuels. Several countries have deforestation and degradation as their largest source of CO_2 emissions, with Brazil and Indonesia jointly accounting for almost half of global CO_2 emissions from tropical deforestation (Boucher, 2008).

Tropical forests store 17% of the total of carbon contained in all of Earth's terrestrial vegetation. The pan-tropical average works out at about 240 tonnes of carbon stored per hectare of forest, split more or less equally between the trees and soil (IPCC, 2000). Drier tropical forests store less than this average, whereas rain forests store more. By contrast, crop lands store, on average, only 80 tonnes of carbon per hectare (almost all of it in the soil). So, on average, clearing 1 ha of tropical forest for agriculture emits approximately a net 160 tonnes of carbon, while reducing future carbon absorption by diminishing the global carbon sink. Furthermore, agriculture (particularly rice cultivation and cattle ranching) often releases substantial quantities of methane, which is 20 times more efficient at trapping heat in the atmosphere than CO_2.

These facts show that although tropical forest destruction contributes significantly to global climate change, forest restoration could be a significant part of the solution.

Deforestation and water resources

Tropical forests produce huge quantities of leaf litter, resulting in organic-matter-rich soils that are capable of storing large amounts of water per unit volume. These soils soak up water during the rainy season, helping to replenish groundwater and thus ensuring that water is released slowly during the dry season. Deforestation results in an increase in overall water yield from a catchment (as trees, which transpire water through their leaves, are removed), but that increased yield often becomes more seasonal. Without input of leaf litter into the soil and tree roots to reduce soil erosion, the absorptive top soil is rapidly washed away. Soil compaction (resulting from exposure

Deforestation can cause water sources to dry up in the dry season, as pictured here in northeast Brazil. (Photo: A. McRobb).

to intense rainfall), disappearance of soil fauna, overgrazing and road construction all reduce infiltration of rain water into the soil and groundwater replenishment. So in the rainy season, storms result in rapid surges of water from the catchment, sometimes causing floods. Conversely in the dry season, insufficient water is retained in the catchment to sustain stream flow. Streams dry up and agricultural production in the dry season declines (Bruijnzeel, 2004).

Deforestation dramatically increases soil erosion, especially where the understorey and soil litter layer are damaged (Douglas, 1996; Wiersum, 1984). This in turn causes siltation of streams, rivers and reservoirs, which reduces the life span of irrigation systems that are vital for agriculture downstream.

Effects of deforestation on communities

People living near forests are the first to be affected by deforestation, losing the environmental benefits described above, as well as foods, medicines, fuels and construction materials.

Millions of forest-dwelling people depend on forest products for subsistence. In times of necessity, gathering or selling such products provides a safety net for the rural poor (Ros-Tonen & Wiersum, 2003). For a few, trade in forest products provides significant regular cash income, although problems with marketing and changing lifestyles have limited the commercial development of this trade (Pfund & Robinson, 2005).

However, because most forest products are not bought or sold in markets, their value does not contribute to economic development indices, such as gross domestic product (GDP). Hence, their importance is often ignored by policy makers, who sacrifice forests for conversion to other uses. Consequently, poverty worsens when local people are forced to spend cash to buy substitutes for lost forest products. Paradoxically, such transactions do count towards GDP, giving a false impression of economic growth.

1.3 What is forest restoration?

Reforestation and forest restoration are not always the same

'Reforestation' means different things to different people (Lamb, 2011) and the term can refer to actions that return any kind of tree cover to deforested land. Agro-forestry, community forestry, plantation forestry etc. are all kinds of 'reforestation'. In the tropics, tree plantations are the most common form of reforestation. Even-aged plantations of single species (often exotics) may be needed to meet economic demand for wood products and to take the pressure off natural forests. They cannot, however, supply local people with the diversity of forest products and ecological services they need, nor can they provide the range of habitats for all the plant and animal species that once inhabited the forest ecosystems they replace.

Forest restoration is a specialised form of reforestation but, unlike industrial plantations, its goals are biodiversity recovery[6] and environmental protection. The definition of forest restoration used for this book is:

… "actions to re-instate ecological processes, which accelerate recovery of forest structure, ecological functioning and biodiversity levels towards those typical of climax forest" …

… i.e. the end-stage of natural forest succession — relatively stable ecosystems that have developed the maximum biomass, structural complexity and species diversity possible within the limits imposed by climate and soil and without continued disturbance from humans (see **Section 2.2**). This represents the *target ecosystem* aimed for by forest restoration.

Since the climate is a major factor in determining the composition of the climax forest, changes in climate may alter the climax forest type in some areas and thus might change the aim of restoration (see **Sections 2.3** and **4.2**).

Forest restoration may include passive protection of remnant vegetation (see **Section 5.1**) or more active interventions to accelerate natural regeneration (ANR, see **Section 5.2**), as well as planting trees (see **Chapter 7**) and/or sowing seeds (direct seeding) of species that are representative of the target ecosystem. Tree species that are planted (or encouraged to establish) should be those typical of, or providing a critical ecological function in, the target ecosystem. Wherever people live in or near the restoration site, economic species can be included amongst those planted in order to yield subsistence or cash-generating products.

Forest restoration is an inclusive process that encourages collaboration among a wide range of stakeholders including local people, government officials, non-government organisations, scientists and funding agencies. Its success is measured in terms of increased biological diversity, biomass, primary productivity, soil organic matter and water-holding capacity, as well as by the return of rare and keystone species that are

---

[6] Throughout this book, 'biodiversity recovery' refers to the re-colonisation of a site by the plant and animal species that originally inhabited the climax forest ecosystem. It excludes exotic species and domesticated species.

characteristic of the target ecosystem (Elliott, 2000). Economic indices of success can include the value of forest products and the ecological services generated (e.g. watershed protection, carbon storage etc.), which ultimately contribute towards poverty reduction.

Where is forest restoration appropriate?

Forest restoration is appropriate wherever biodiversity recovery is one of the main goals of reforestation, whether it be for wildlife conservation, environmental protection, ecotourism or to supply a wide variety of forest products to local communities. Forests can be restored in a wide range of circumstances, but degraded sites within protected areas are a high priority, especially where some climax forest remains as a seed source. Even in protected areas, there are often large deforested sites: logged over areas or sites formerly cleared for agriculture. If protected areas are to fulfil their role as Earth's last wildlife refuges, restoration of such areas must be routinely included in their management plans.

But wildlife is not the only consideration. Many restoration projects are now being implemented under the umbrella of 'forest landscape restoration' (FLR; see **Section 4.3**), defined as a "planned process to regain ecological integrity and enhance human well-being in deforested or degraded landscapes". FLR recognises that forest restoration may also provide social and economic functions. It aims to achieve the best possible compromise between meeting both conservation goals and the needs of rural communities. As human pressure on landscapes increases, forest restoration will most commonly be practiced within a mosaic of other forms of forest management, to meet the economic needs of local people.

Is tree planting essential to restore forest ecosystems?

Not always. A lot can be achieved by studying how forests regenerate (see **Section 2.2**), identifying the factors that limit regeneration and devising methods to overcome them. These can include weeding and adding fertiliser around natural tree seedlings, preventing fire, removing cattle and so on. This is 'accelerated' or 'assisted' natural regeneration (ANR; see **Section 5.2**). This strategy is simple and cost-effective, but it can only operate where trees, mostly pioneer species, are already present. Such trees represent only a small fraction of the total tree species that comprise climax tropical forests. Therefore, for full biodiversity recovery, some tree planting is often required, especially of poorly dispersed species with large seeds. It is not feasible to plant all of the many hundreds of tree species that may have formerly grown in the original primary tropical forest and, fortunately, it is usually unnecessary if the framework species method can be used.

The framework species method

Planting a few, carefully selected tree species can rapidly re-establish forest ecosystems that have high biodiversity. First developed in Queensland, Australia (Goosem & Tucker, 1995; Lamb et al., 1997; Tucker & Murphy, 1997; Tucker, 2000; see **Box 3.1**, p. 80), the framework species method involves planting mixtures of 20–30 indigenous forest

In the 1980s, conservation organisations warned that, once destroyed, tropical forests could never be recovered. Thirty years of restoration research is beginning to challenge this long-accepted truth.

(a) This site in Doi Suthep-Pui National Park, northern Thailand, was deforested, over-cultivated and then burnt, but local people subsequently teamed up with Chiang Mai University to repair their watershed.

(b) Fire prevention, nurturing existing regeneration and planting framework tree species began to produce results within a year.

(c) Nine years later, the blackened tree stump is dwarfed by the restored forest.

tree species that rapidly re-establish forest structure and ecosystem functioning (see **Section 5.3**). Wild animals, attracted by the planted trees, disperse the seeds of additional tree species into planted areas, while the cooler, more humid and weed-free conditions, created by the planted trees, favour seed germination and seedling establishment. Excellent results have been achieved with this technique in Australia (Tucker & Murphy, 1997) and in Thailand (FORRU, 2006).

Limits to forest restoration

"Tropical forests, once destroyed, can never be recovered" — this was the clarion call of conservation organisations 30 years ago when raising funds for tropical forest protection projects. Although restoration science has achieved much in the intervening years, protecting remaining areas of primary tropical forest, as the "cradles of evolution", must remain the top global conservation priority when seeking to reduce biodiversity loss. Although some attributes of primary forests can now be restored, their long, unbroken history of species evolution cannot. Once species that are most sensitive to forest disturbance become extinct, no amount of habitat restoration can bring them back. Furthermore, restoration is expensive and laborious, and the outcome cannot be guaranteed, so advances in restoration techniques cannot be used to support a "destroy now — restore later" policy of forest management.

1.4 The benefits of forest restoration

Reliable techniques are essential for the success of forest restoration, but they are of little consequence without the support, motivation and hard work of local communities. Local people benefit most from the environmental services and forest products that result from forest restoration, but they also incur the highest cost in terms of giving up potentially productive land. Their participation is assured only when they are fully aware of all benefits and confident that they will receive their fair share of them.

Numerous studies have quantified the values of tropical forests (www.teebweb.org/), but such values are realised only when someone is prepared to pay for them. Politicians, policy makers and businessmen will continue to ignore the value of tropical forests unless such values contribute to indices of economic growth. Various valuation mechanisms are now being developed that may fairly reward all those who invest their effort in forest restoration. Carbon trading is probably the most advanced of these mechanisms, but payments for water supplies, biodiversity-offset schemes and income-generation from ecotourism and trade in forest products are now also growing in acceptance.

The market value of biodiversity

Forest products.

One of the most obvious ways to value a tropical forest is to calculate the total substitution value of the products extracted from forests by local people. For example, if villagers lose their firewood supply because of deforestation and buy gas canisters in the market, the substitution value of the firewood is the price paid for the gas. This then is one measure of the value of the forest. Interestingly, the loss of the firewood has no effect on GDP (as it is not typically bought or sold), but the purchase of gas does contribute to GDP. In this way, deforestation appears to increase national prosperity, although the villagers become poorer. Forest restoration reverses this paradox. Restoring the supply of forest products to communities provides a powerful motive for local people to plant trees. It is a directly measurable value of forest restoration.

The value of tropical forest products can be calculated from market prices and traded volumes. At least 150 different forest products, including rattan, bamboo, nuts, essential oils and pharmaceuticals, are traded internationally, contributing at least US$ 4.7 billion/ year to the global economy. Forest restoration could

Preparing for ecotourists. At the Himmapaan Project, a tree nursery and an exhibition centre have been constructed specifically to involve eco-tour clients in forest restoration activities. Eco-tour guides are thoroughly trained in restoration techniques, ready to guide their clients through nursery and field techniques.

play an important role in meeting the increasing demand for such products, while generating income for local communities. The provision of such products can be included in the design of forest restoration projects, either by planting the appropriate economic species or by creating conditions that enhance their natural colonisation of the restored forest. Of course, income from the extraction of forest products can only be maintained if such products are harvested sustainably[7] and the benefits shared fairly amongst community members. However, this is more likely to occur in forests that villagers have worked to restore than in natural forests, where such resources are considered to be 'free'. Rainforestation, a restoration method developed in the Philippines, is perhaps the best-known approach for incorporating forest products into forest restoration projects (www.rainforestation.ph) (see **Box 5.3**, p. 135).

Income from ecotourism is another way to value the return of biodiversity resulting from forest restoration. For example, the Harapan[8] Rainforest Initiative in Indonesia, run by a coalition of conservation organisations[9], aims to restore more than 1,000 km² of Sumatran rain forest for wildlife conservation and plans to generate funding for the project by creating a unique ecotourism destination. By contrast, eco-tour companies[10] and villagers in northern Thailand have set up a forest restoration project, The Himmapaan[11] Foundation, to involve their clients in tree seed collection, working in the project's tree nursery and planting and caring for trees in restored sites.

International markets that put a value on biodiversity as a whole are also being developed. In some countries, the destruction of biodiversity by development must be amended by restoring equivalent biodiversity elsewhere. This is called 'biodiversity-offset' or 'bio-banking'. Developers purchase biodiversity credits that are generated by conservation projects that restore or enhance biodiversity. For example, a mining company, which destroys 100 hectares of tropical forest in one location, pays the full cost of restoring an equal area with the same biodiversity elsewhere. Such schemes could pay for forest restoration, but they are highly controversial. Buying the 'right to destroy biodiversity'

[7] i.e. the amount harvested per year does not exceed the annual productivity.
[8] 'hope' in Indonesian.
[9] Burung Indonesia, Birdlife International, Royal Society for the Protection of Birds and others (www.birdlife.org./action/ground/sumatra/harapan_vision.html).
[10] East West Siam Travel, Asian Oasis, Gebeco and Travel Indochina.
[11] A mythical forest in oriental cultures, equivalent to the Garden of Eden (himmapaan.com).

is morally questionable. By its very nature, biodiversity is not a uniform commodity (like carbon). For highly diverse tropical forests, the restoration of *all* of the species impacted by a development at another site is impossible to guarantee, no matter how much money is spent. So, whilst corporate sponsorship of forest restoration is laudable, biodiversity 'offset' in its current form remains of questionable conservation value.

The value of carbon storage

Tropical forests absorb more CO_2 through photosynthesis than they emit by respiration. Recent research has quantified this 'sink' at about 1.3 gigatonnes of carbon (GtC) per year (Lewis *et al*., 2009), equivalent to 16.6% of carbon emissions from the cement industry and burning fossil fuels[12] and contributing 60% of the sink provided by all of the terrestrial vegetation on Earth. In Africa, tropical forests actually absorb more carbon than is released by fossil fuel emissions (Lewis *et al*., 2009). As atmospheric CO_2 concentration increases, tropical forests could become even more efficient at mopping up CO_2, as high CO_2 concentrations stimulate photosynthesis. Tropical forests cannot be relied upon to solve the problem of global climate change, but they may help to slow it down sufficiently to provide the time needed for the seismic shift from a carbon-based global economy to a carbon-neutral one.

Trading in carbon credits could turn the carbon storage potential of forest restoration projects into cash. The idea seems simple. Carbon dioxide is the most important greenhouse gas. Power stations that burn coal or oil release CO_2 into the atmosphere, while tropical forests absorb it. So if a power company pays for forest restoration, they could continue to emit CO_2 without actually increasing the atmospheric CO_2 concentration. A company that buys carbon credits buys the right to emit a certain amount of CO_2. The money paid for those carbon credits could then used to finance forest restoration thereby increasing the capacity of the global carbon sink. Carbon credits are traded, like stocks and shares. So their prices can go up or down according to demand. There are two kinds:

- Compliance credits are bought by corporations and governments in order to meet their international obligations under the Kyoto Protocol, thereby offsetting some of the carbon they emit. The protocol's Clean Development Mechanism (CDM) channels the credited money into projects that absorb CO_2 or reduce emissions.
- Voluntary credits are bought by individuals or organisations seeking to reduce their 'carbon footprints'. The 'voluntary market' is much smaller than the compliance market and the credits are cheaper because the projects supported by it don't have to meet the stringent requirements of the CDM.

At present, few forest restoration projects have been approved for support under the CDM because it is difficult to measure the amount of carbon stored in forests, which have very variable growth rates and which could easily burn or become degraded. Furthermore, credits could encourage the establishment of plantations of fast-growing trees over large areas, which displace local people. So, several obstacles must be overcome before compliance credits could generate income for forest restoration projects.

The voluntary principle, however, is proving to be much more successful. All over the world, corporations are sponsoring tree planting, partly to off-set their carbon footprints, but also to promote a cleaner, greener image. The challenge is to ensure

[12] 7.8 GtC per year, as of 2005, increasing by 3% per year (Marland *et al*., 2006).

that such projects result in more than just carbon storage by restoring biodiversity-rich forest ecosystems that will provide the full range of products and environmental services to both local people and wildlife.

Another international scheme worthy of mention here is REDD+, which stands for 'reducing emissions from deforestation and forest degradation'. This is a set of policies and incentives being developed under the UN Framework Convention on Climate Change (UNFCCC) to reduce CO_2 emissions derived from clearing and burning tropical forests. The concept was recently expanded to include the 'enhancement of carbon stocks', i.e. forest restoration to actually increase CO_2 absorption[13]. Once established, this international framework will provide approved funding and monitoring mechanisms for both forest conservation and forest restoration projects that enhance the net global forest 'sink' for CO_2, while also conserving biodiversity and benefiting local people. Funding would come from both established carbon credit markets and specially created international funds, but as yet no formal international agreement has been reached. The success of REDD+ will also depend on considerable improvements in forest governance as well as capacity-building at all levels, from villagers to policy makers. Despite these challenges, several pilot REDD+ projects are already underway, which will doubtless provide valuable lessons for the future development of the program.

Forest stream in Thailand.

What about water?

In many tropical countries, clean water supplies depend on the conservation of forested catchments. The organic-matter-rich soil beneath forests provides a natural storage mechanism and a natural filter, which maintain dry-season water flows and prevent siltation of water infrastructure (Bruijnzeel, 2004). Maintaining forest cover incurs a cost to the people that live in the catchments (i.e. agricultural land foregone) but benefits farmers and city dwellers downstream. To guarantee clean water supplies, therefore, some water companies have come up with novel mechanisms to pay for forest conservation. For example, the Public Utilities Company of Heredia, Costa Rica, charges customers an extra 10 US cents per cubic meter of water consumed. This money is paid to state forest parks and landowners to protect or restore forests at a rate of US$ 110/ha/yr (Gamez, undated). In fact, Costa Rica leads the world in payments for environmental services (PES). The country's National PES Program, funded mostly from a fuel tax, pays forest owners for four bundled environmental services (watershed protection, carbon storage, landscape beauty and biodiversity). Over 9 years, it paid out US$ 110 million to 6,000 owners of more than 5,000 km^2 of forest (Rodriguez, 2005).

[13] www.scribd.com/doc/23533826/Decoding-REDD-RESTORATION-IN-REDD-Forest-Restoration-for-Enhancing-Carbon-Stocks

The value of tropical forests

If *all* forest values were marketed and paid for, forest restoration could become more profitable than other land uses. The Economics of Ecosystems and Biodiversity study (TEEB)[14] has estimated the average total value of all ecosystem services from tropical forests at more than US$ 6,000/ha/yr (**Table 1.2**), which is more profitable than palm oil. The elegance of the forest restoration business model is that it generates several different revenue streams that are shared amongst many stakeholders. So, if the market price of one service or product falls, another one can be developed to maintain overall profitability. Forest restoration is no longer just a pipedream of conservationists; it could very well become a highly lucrative global industry.

Table 1.2. Average values of ecosystem services from tropical forest.

	Average value (US$/ha/y)	No. of studies
Provisioning services		
Food	75	19
Water	143	3
Other raw materials	431	26
Genetic resources	483	4
Medicinal resources	181	4
Regulating services		
Air quality	230	2
Climate regulation	1,965	10
Water flow regulation	1,360	6
Waste treatment/water purification	177	6
Erosion prevention	694	9
Cultural services		
Recreation and tourism	381	20
Total	**6,120**	**109**

Source: TEEB (2009)

[14] www/teebweb.org/

CASE STUDY 1   Cristalino

Country: Brazil

Forest type: Lowland tropical evergreen forest, seasonally flooded forest, lowland tropical dry forest and white sand formations.

Ownership: State and private protected areas, smallholdings and cattle ranches.

Management and community use: Conservation management, cattle ranching and swidden (slash-and-burn) agriculture.

Level of degradation: Substantial areas of degraded pasture and secondary vegetation.

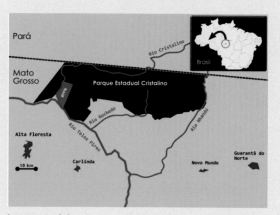

Location of the study area.

Background

Cristalino State Park in Mato Grosso, lies at the frontier of the northward spread of deforestation into the southern Brazilian Amazon. It forms part of a proposed conservation corridor designed to block this process. Even though the area is officially protected, it has lost substantial areas of natural vegetation to cattle ranching since its establishment in 2000. Its southern and eastern boundaries have been severely deforested as a result of both legal and illegal land occupation by ranchers and smallholders.

Building the baseline: biodiversity research in the Cristalino region

In close collaboration, the Royal Botanic Gardens, Kew, the Cristalino Ecological Foundation (FEC) and the State University of Mato Grosso (UNEMAT) have carried out species inventories, vegetation mapping and quantitative analyses of species composition to provide baseline data for management planning and restoration. The work has generated a checklist of approximately 1,500 species, linked to vegetation types and ecology (Zappi *et al.*, 2011). This basic understanding of forest composition and diversity is recognised as a fundamental starting point for the development of restoration activities in the region, where the flora had not previously been studied in any significant depth.

Degraded areas in the Cristalino State Park. The red/white hatched areas were deforested before the establishment of the reserve, the solid red areas subsequently.

Discussions with local governmental and non-governmental organisations highlighted the need for the strategic recuperation of degraded areas and the development and dissemination of locally appropriate methodologies and incentives for reforestation.

Opportunities, approaches and methods for restoration

Opportunities for restoration were identified in areas of abandoned cattle pasture within the reserve, in degraded land occupied by smallholders, and along the margins of water courses in the buffer zone around the park. The selection of appropriate framework tree species for restoration will depend on both the ecological and human context. The demand for relatively short-term economic benefits within smallholdings, dictates the inclusion of species with economic value, either direct (e.g. food plants, timber trees etc.) or indirect (shade trees to nurture understorey cash crops). Data on local plant uses, collected during the baseline studies, were supplemented with published information on the uses of the same species elsewhere in the Amazon.

Fourteen native species of *Inga* (Leguminosae), a nitrogen-fixing genus capable of fast growth on poor or highly degraded soils, have been recorded in the area. They include species that are adapted to flooded forest, riversides and terra firme (dry land) forest. *Inga* seeds are surrounded by sweet white arils, which attract wildlife and are widely eaten by indigenous communities across the Amazon. *Inga edulis*, a cultivated species that also occurs in the wild at Cristalino, has been used successfully in alley-cropping trials on degraded land elsewhere in the Neotropics (Pennington & Fernandes, 1998). It enriches the soil with nutrients and organic matter (assisted by periodic pruning in the alleys) and rapidly shades out exotic *Brachiaria* grass, which inhibits tree regeneration. This system is equally appropriate for establishing forest trees, which can be planted in corridors between the managed rows (T. D. Pennington, personal communication).

In the Cristalino region, successful reforestation will inevitably take place at the interface between agro-forestry, forestry and ecological restoration. A local NGO, Instituto Ouro Verde (IOV), has developed a prototype web-based database to provide data on locally appropriate species for agro-forestry systems. This will enable the selection of framework species for forest restoration in the region and will provide guidance for their management. In response to a growing water-shortage problem, IOV, with input from Kew, is also engaging local communities in the restoration of gallery forest in smallholdings and is providing fencing materials. Drawing on the baseline botanical diversity data, the opportunity now exists to develop a proactive tree-planting programme that uses species adapted to the local situation.

Undisturbed evergreen forest in the Cristalino State Park.

Inga marginata, one of several native species found in the region.

In the Cristalino region, the native vegetation is highly variable and strongly influenced by edaphic and hydrological factors. Soils vary from almost pure white nutrient-poor sand to more fertile clayey latosols, the former commonly associated with water stress in the five-month dry season and, in places, water-logging during the wet season. This complexity necessitates careful matching of selected species with site conditions and thus underlines the importance of detailed baseline vegetation studies. For example, the terra firme (dry land) forest on clayey soils at Cristalino is dominated by species of the Burseraceae family, with *Tetragastris altissima* abundant. This large canopy tree is well adapted to the region, attracts wildlife with the sweet arils that surround its seeds and

Brachiaria pasture in the Cristalino State Park.

Dry (deciduous) forest on a granite hill, with evergreen forest on lower ground.

has several popular uses. In the semi-deciduous forest on sandy soils, however, Leguminosae is the dominant family, with abundant *Dialium guianense* and *Dipteryx odorata*. Both are commercial timber species. The latter attracts bats, which are important seed-dispersers. These important tree species are therefore promising framework species candidates. Similarly, observations on secondary vegetation have also been useful for the identification of potential framework pioneers. Both *Acacia polyphylla* (Leguminosae) and *Cecropia* spp. (Urticaceae) are excellent local candidate species; the latter also being bat-dispersed.

The future impact of climatic change will also influence the choice of species for reforestation. Preliminary models for the southern Amazon predict a shift from evergreen to dry-adapted vegetation types (Malhi *et al.*, 2009) because of a drier climate. Given that dry habitats already occur in the Cristalino region, where water availability is restricted during the dry season, it may prove beneficial to incorporate dry-adapted species such as *Tabebuia* spp. (Bignoniaceae) into experimental plantings in localities where they would not naturally occur under current conditions.

By William Milliken

CHAPTER 2

UNDERSTANDING TROPICAL FORESTS

Think of a 'tropical forest' and images of equatorial rain forest probably spring to mind — evergreen forest, teaming with wildlife and drenched in rain — but many other forest types grow in the tropics. In seasonally dry climates, evergreen forest types in wetter areas alternate abruptly with deciduous forest types in drier sites, grading into grassy savannahs in the driest areas. Likewise, on mountains, forest structure changes dramatically with elevation. In more limiting environments, there are swampy peat forests, salty mangrove forests and acidic heath forests. Different forest types function differently, and each has distinctive characteristics that present restoration projects with different challenges. In evergreen forests, the major challenge is ensuring the rapid recovery of the high levels of biodiversity that characterize such ecosystems; whereas in drier forests simply getting planted trees to survive the first dry season is a major achievement. The climax forest type defines the goal of restoration (i.e. the 'target' see Section 1.2), so it is important to know which forest type you are dealing with.

2.1 Tropical forest types

Many different schemes to classify tropical forest types have been proposed. These are based on various criteria including climate, soil, species composition, structure, function and successional stage (Montagnini & Jordan, 2005). Commonly used schemes include Whitmore's system (1998) **Box 2.1**, which is based on climate and elevation, and UNEP–WCMC's forest category classification (UNEP–WCMC, 2000), which also includes disturbed forests and plantations (see **Box 2.2**).

Evergreen tropical forests (including rain forests)

Tropical rain forests are the most highly developed of evergreen tropical forests. They mostly grow within 7° latitude from the equator, where mean annual temperatures exceed 23°C and mean monthly temperatures are higher than 18°C (i.e. there is no frost). Annual rainfall exceeds 4,000 mm, with monthly rainfall averaging more than 100 mm all year round (i.e. there is no significant dry season). Other types of evergreen tropical forest grow wherever rainfall exceeds evapotranspiration (usually where mean annual rainfall is more than 2,000 mm) and the dry season is no longer than 2 months. They extend up to 10° latitude from the equator. The greatest expanses of evergreen tropical forests are in the lowland Amazon Basin, the Congo Basin, the Malay Peninsula, and the Southeast Asian islands of Indonesia and New Guinea.

Tropical rain forest at Maliau Basin Conservation Area, Sabah, East Malaysia

Box 2.1. Whitmore's simple classification of tropical forest types.

Whitmore's (1998) simple classification of tropical forest types proposes that on moving away from the equator, climax tropical forests can be broadly grouped into two main categories: seasonally dry and everwet. Superimposed upon the effects of latitude and the prevailing climate are the effects of elevation (i.e. montane or lowland forests) and substrate (e.g. forests growing on limestone or peat etc.).

Climate	Elevation	Tropical forest types
Seasonally dry		Monsoon (deciduous) forests of various types Semi-evergreen rain forest
Everwet	Lowlands	Lowland evergreen rain forest
	Montane 1,200–1,500 m	Lower montane rain forest
	Montane 1,500–3,000 m	Upper montane rain forest or cloud forest
	Montane > 3,000 m	Subalpine forest to climatic tree limit
	Usually lowland	Heath forest Limestone forest Ultrabasic forest Mangrove forest Peat swamp forest Freshwater swamp forest Freshwater periodic swamp forest

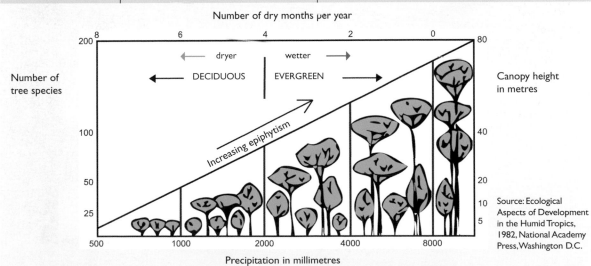

The relationship between moisture and plant life in a tropical lowland forest. The diagonal line from left to right represents a gradient of mean annual precipitation, demonstrating that as the amount of moisture increases, forests become more complex, with greater biological diversity and ecological stratification. (Source: Assembly of Life Sciences (U.S.A.), 1982.)

Box 2.2. UNEP–WCMC's forest category classification.

UNEP–WCMC's forest category classification, developed in 1990, divides the world's forests into 26 major types (on the basis of climate zone and characteristic tree species) of which the 15 listed below are tropical (UNEP–WCMC, 2000). For each tropical forest type, The International Tropical Timber Organization (ITTO) proposes another layer of classification, based on successional stage, i.e. primary, managed primary, modified natural, degraded, secondary or planted.

TROPICAL FOREST TYPES

- Mangrove
- Freshwater swamp forest
- Upper montane forest
- Lowland evergreen broadleaf rainforest
- Lower montane forest
- Semi-evergreen moist broadleaf forest
- Exotic species plantations
- Native species plantations
- Mixed needleleaf/broadleaf forest

- Needleleaf forest
- Sclerophyllous dry forest
- Deciduous/semi-deciduous broadleaf forest
- Thorn forest
- Sparse trees/parkland
- Disturbed natural forest

OTHER TYPES

- Temperate and boreal forests
- Water bodies
- No data

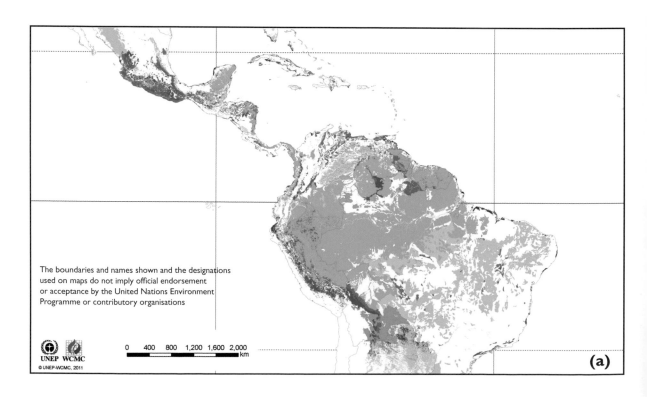

The boundaries and names shown and the designations used on maps do not imply official endorsement or acceptance by the United Nations Environment Programme or contributory organisations

UNEP WCMC
© UNEP-WCMC, 2011

0   400   800   1,200 1,600 2,000
                                    km

(a)

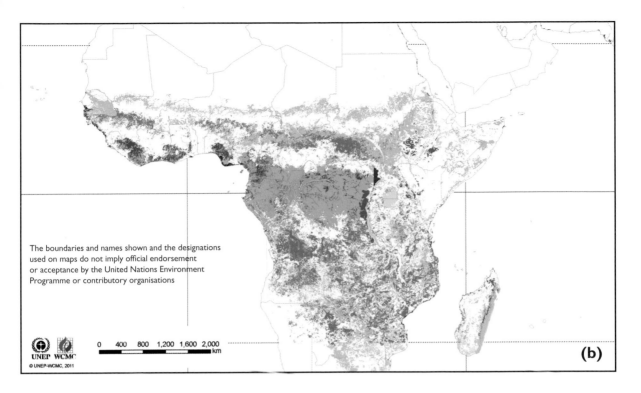

The boundaries and names shown and the designations used on maps do not imply official endorsement or acceptance by the United Nations Environment Programme or contributory organisations

0 400 800 1,200 1,600 2,000 km

UNEP WCMC
© UNEP-WCMC, 2011

(b)

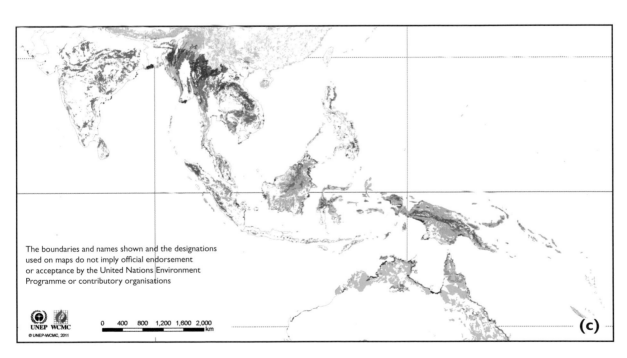

The boundaries and names shown and the designations used on maps do not imply official endorsement or acceptance by the United Nations Environment Programme or contributory organisations

0 400 800 1,200 1,600 2,000 km

UNEP WCMC
© UNEP-WCMC, 2011

(c)

Extent of major tropical forest types of a) Central/South America b) Africa and c) Asia, based on UNEP–WCMC's classification of 1990, derived from a number of different national and international sources. The scales and dates vary between sources, and this synthesis can be considered to show global cover in approximately 1995. The forest classification was designed to reflect the characteristics of forests that are relevant to conservation and to facilitate harmonisation between various national and international classification systems. © UNEP–WCMC, 2011.

Evergreen tropical forests are the most luxuriant of all tropical forests, with structural complexity and biodiversity that usually exceed those of other tropical forest types, although there is immense variability. In sample plots in Ecuador, for example, Whitmore (1998) cited extremes of 370 tree species per hectare, compared with just 23 tree species per hectare at a Nigerian site. At least five canopy strata can usually be distinguished (i.e. ground flora, shrubs (including tree saplings), understory trees, main canopy trees and emergent trees), with the main canopy occurring up to 45 m above the ground and some emergent trees soaring up to 60 m. Most light is captured by the main canopy,

Buttresses are a characteristic feature of many evergreen tropical forest tree species. The Waorani Indians make use of them for communication. The low frequency boom, produced when the buttresses are thumped carries over considerable distances.

so the shade-tolerant ground and shrub layers tend to be less dense than those in drier tropical forests. Buttressed trees are common, particularly on shallow soils. Cauliflory (i.e. the growth of flowers and fruits on tree trunks) is also characteristic, particularly for understory trees, whose leaves also tend to have 'drip tips' (i.e. elongated acumen) that allow them to shed water rapidly. Some canopy trees may be briefly deciduous, but the canopy as a whole is evergreen. Woody climbers (including rattans in Asia and Africa), fig trees (*Ficus* spp.) and dense communities of epiphytic ferns and orchids (along with bromeliads in South America and Apocynaceae and Rubiaceae species in Asia) are also characteristic of tropical rain forests.

Theobroma cacao pods are an example of a cauliflorus fruit.

Most food resources provided by evergreen forests (i.e. leaves, fruits, insects etc.) are in the canopy, so most of the animals live there and provide the trees with services vital for reproduction. The most important pollinators are bees and wasps, but nectar-eating birds and bats also pollinate many tree species. Seed dispersal is mostly carried out by frugivorous birds, along with fruit bats and primates and, when fruit falls to the ground, ungulates and rodents. Seed dispersal by wind is rare except for the tallest trees (the Dipterocarps of tropical Asia being an obvious exception). The heavy dependence of evergreen tropical forest trees on animals for reproduction is critical when considering forest restoration.

Double-eyed fig parrots feast on figs and disperse their seeds. This crucial ecological service is vital to the forest's survival and encouragement of it is essential for successful forest restoration.

Buttressed tree in a lowland evergreen forest, Cameroon. (Photo: A. McRobb)

Many tree species in evergreen tropical forests (like this *Baccaurea ramiflora* from Southeast Asia) produce flowers and fruits directly from trunks or branches. The flowers are more visible to pollinators and the fruits to seed dispersers as they are not hidden by foliage.

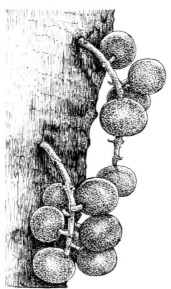

In evergreen tropical forests, the leaves of many tree species have elongated acumen or 'drip tips' which help them to shed rain water more efficiently, thereby preventing the smothering growth of mosses and lichens on the leaf surface.

Box 2.3. The essential nature of fig trees (*Ficus* spp.).

Fig trees are keystone species in tropical forest ecosystems, so restoration projects should always include them. The pantropical genus *Ficus* comprises more than 1,000 species of vines, woody climbers, shrubs and large trees and it is their unique reproductive mechanism that makes them keystone species. Sometimes mistaken for fruit, the parts of the fig tree that are eaten (called 'syconia' in botanical language) often grow on short stalks on the trunk or large branches and are a vital food for forest animals. Syconia are actually the swollen stalks of inflorescences (receptacles), which have become inverted to enclose many tiny flowers or fruits inside.

The flowers of each *Ficus* species are pollinated by one (or very few) fig wasp species. Figs provide the only means for wasp reproduction, and the wasps are the only way that fig flowers can be pollinated. Fig wasps complete their life-cycle in just a few weeks so, somewhere in the forest, figs of all species must be available all year round so that the wasps do not die out, leaving the fig trees unable to reproduce. The loss of *Ficus* species from a tropical forest is disastrous because it causes arboreal birds and mammals that rely on figs in times of food shortage to die out gradually. Much later, the tree species that rely on these animals for seed dispersal are also lost.

Planting fig trees restores ecological balance by attracting seed-dispersing animals into restoration plots. In addition, fig trees grow very dense root systems, which enable them to grow well under the harshest conditions and to grow back rapidly after burning or slashing. *Ficus* species are therefore excellent for preventing soil erosion and stabilising river banks.

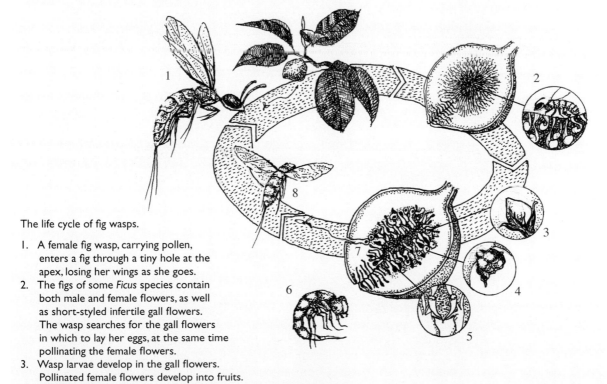

The life cycle of fig wasps.

1. A female fig wasp, carrying pollen, enters a fig through a tiny hole at the apex, losing her wings as she goes.
2. The figs of some *Ficus* species contain both male and female flowers, as well as short-styled infertile gall flowers. The wasp searches for the gall flowers in which to lay her eggs, at the same time pollinating the female flowers.
3. Wasp larvae develop in the gall flowers. Pollinated female flowers develop into fruits.
4. Wingless male wasps emerge from their gall nurseries first. They mate with female wasps shortly before the females emerge from their galls.
5. By the time the females emerge, the male flowers are producing pollen.
6. The males chew a hole through the wall of the fig.
7. Females escape through the hole, collecting pollen as they go.
8. Female wasps, laden with pollen, then fly to another fig tree and the cycle continues.

Challenges to restoring evergreen tropical forests

Achieving high biodiversity and structural complexity is the greatest challenge when restoring evergreen tropical forests. Recovery of full biodiversity is difficult to achieve when so many species are involved in such complex ecological relationships, especially because the ecology, reproductive biology and propagation of most tropical tree species are poorly understood.

Selectively logged forests, or even some clear-cut sites that have not been disturbed further, may respond well to accelerated natural regeneration (ANR; see **Section 5.2**); whereas tree planting is usually necessary in degraded sites that are dominated by grasses and herbs. The great richness of tree species in evergreen tropical forests presents a huge choice from which to select high-performing trees for planting. Focusing first on the small minority of deciduous tree species that grow in evergreen forests can often achieve rapid results because such species resist desiccation in exposed, dry, deforested sites by shedding their leaves during the driest months of the year.

One consequence of tree species richness is that trees of the same species are usually spaced far apart from each other. This makes it difficult to locate enough seed trees to ensure high genetic diversity amongst trees grown in nurseries. Furthermore, fruiting can be irregular and many tree species have recalcitrant seeds that cannot be stored easily. Many evergreen forest tree species have large seeds that can only be dispersed by large animals, many of which (rhinos, elephants, tapirs etc.) have been extirpated from large parts of their former ranges. Therefore, including large-seeded tree species amongst those chosen for planting can help to conserve them (Vanthomme *et al.*, 2010). Even small-seeded tree species are mostly dispersed by birds, bats and small mammals, so preventing the hunting of such animals is vital to enable the recruitment of non-planted tree species into planted sites.

Year-round, abundant water, warmth and light in the wet tropics mean that trees can be planted at any time of the year, and getting them to survive and grow well is less of a problem than in drier regions. However, these conditions are also optimal for weed growth, so frequent weeding is necessary and weeding costs may be high. Fire is usually less of a problem than in drier areas, but it is more likely in degraded forest and climate change will exacerbate the problem. Hence, fire-prevention measures may still be necessary.

Seasonal tropical forests

Seasonally dry tropical forests or 'monsoon' forests are most prevalent at 5–15° latitude from the equator, where rainfall and day length vary annually. Such forests grow where annual rainfall averages 1,000–2,000 mm and there is a short cool season. During the longer dry season (3–6 months), many trees shed some or all their leaves, resulting in fluctuations in canopy density. This allows more sunlight to reach the forest floor and consequently the development of dense ground and shrub layers, features that distinguish these forests from evergreen tropical forests. Diurnal and monthly fluctuations in temperature are far greater than those in evergreen forests. Mean monthly minimum temperatures can drop to 15°C and mean monthly maxima may exceed 35°C. The greatest expanses of seasonal tropical forest grow in eastern Brazil (cerrado), India (monsoon forests), the Zaire Basin and East Africa.

Seasonally dry tropical forest in northern Thailand. Around half the tree species are deciduous and half are evergreen. The stream runs dry in the hot season.

Both evergreen and deciduous tree species grow closely together, forming a continuous main canopy of up to 35 m tall. Structural features shared by seasonal and evergreen tropical forests include emergent trees, buttressed trees, woody climbers and epiphytes, although they are all less prevalent in seasonal than in evergreen forests. The presence of bamboos distinguishes seasonal forests from evergreen forests. Seasonal tropical forests retain a high degree of structural complexity, although the stratification of the canopy is usually not as highly developed as that in evergreen forests. They are usually less diverse than evergreen forests, although their tree species richness can match that of evergreen forests in some places (Elliott *et al.*, 1989). Although animals remain the main pollinators and seed dispersers in seasonal tropical forests, wind pollination and seed dispersal are more common than in evergreen tropical forests. Seasonal tropical forests may be more resilient to global warming than evergreen forests because their flora and fauna have evolved to cope with seasonal drought.

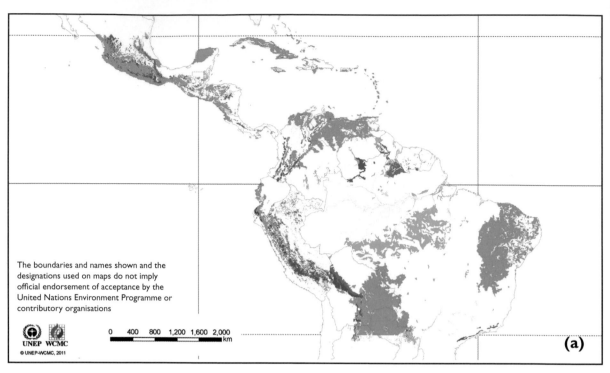

The boundaries and names shown and the designations used on maps do not imply official endorsement of acceptance by the United Nations Environment Programme or contributory organisations

UNEP WCMC
© UNEP-WCMC, 2011

0   400   800   1,200 1,600 2,000
km

(a)

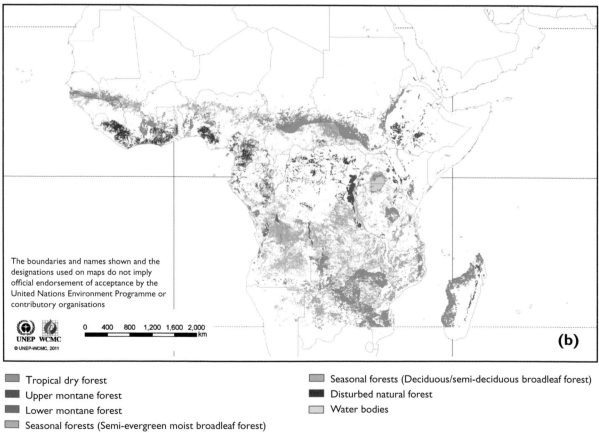

The boundaries and names shown and the designations used on maps do not imply official endorsement of acceptance by the United Nations Environment Programme or contributory organisations

UNEP WCMC
© UNEP-WCMC, 2011

0   400   800   1,200 1,600 2,000
km

(b)

- ☐ Tropical dry forest
- ☐ Upper montane forest
- ☐ Lower montane forest
- ☐ Seasonal forests (Semi-evergreen moist broadleaf forest)

- ☐ Seasonal forests (Deciduous/semi-deciduous broadleaf forest)
- ☐ Disturbed natural forest
- ☐ Water bodies

Extent of tropical dry, seasonal and montane forests of a) Central/South America b) Africa and c) Asia, based on UNEP–WCMC's classification of 1990, derived from a number of different national and international sources. The scales and dates vary between sources, and this synthesis can be considered to show global cover in approximately 1995. © UNEP–WCMC, 2011.

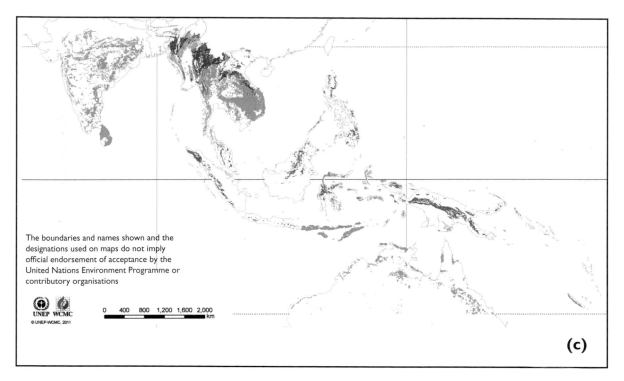

The boundaries and names shown and the designations used on maps do not imply official endorsement of acceptance by the United Nations Environment Programme or contributory organisations

UNEP WCMC
© UNEP-WCMC, 2011

0 400 800 1,200 1,600 2,000 km

(c)

Challenges to restoring seasonal tropical forests

Very little is known about the phenology, propagation and silviculture of the vast majority of tree species in these forests: clearly a problem when planning tree planting. In seasonally dry climates, trees can only be planted at the beginning of the rainy season because sufficient time must be allowed for their roots to grow deep enough to enable them to survive the first dry season. So, nursery work schedules must be devised to grow trees to a plantable size by the start of the rainy season, regardless of when seeds are produced or how fast seedlings grow. This requires a lot of research on tree phenology, seed germination and seedling growth.

Bamboos present one of the greatest challenges to the restoration of seasonal tropical forests because they suppress the growth of trees that are planted nearby. Their dense root systems fully exploit the soil, they cast dense shade and, in the dry season, they smother nearby tree seedlings with a dense layer of leaf litter. Therefore, controlling (but not eliminating) bamboos is essential for successful restoration of seasonal tropical forests. Luckily, bamboo canes and bamboo shoots are useful products, so local people usually need no encouragement to harvest them.

In some degraded seasonal tropical forests, the rich soils will have been severely depleted and are consequently low in organic matter and minerals such as phosphorous. These soils may require the addition of organic matter and/or inorganic fertiliser for tree seedlings to establish and flourish.

Invasive plants and browsing by domestic livestock are both big problems in seasonal tropical forests that must be addressed by working with local people. Seasonal tropical forests are more fire-prone than evergreen forests, so weeding, firebreak construction and an effective fire prevention program are all particularly important when restoring these forest types.

Box 2.4. Bamboos.

As aggressive giant grasses, bamboos can suppress tree establishment, but they are also a natural component of seasonally dry tropical forests and a source of several forest products. Many species exhibit mass flowering at intervals of years or decades, after which the plants die back.

Bamboos are giant 'woody' grasses in the family Poaceae, with more than 1,400 species growing mostly in the tropics and sub-tropics. They are pan-tropical with the Asia–Pacific region having the most species (1,012, with 626 in China alone) and Africa the fewest. The largest bamboos grow up to 15 m tall and have stems that reach 30 cm in diameter. They are the world's fastest-growing woody plants and are among the most useful. Bamboo canes are used for all kinds of temporary construction and furniture and are split and woven to make mats and baskets, whereas young culm buds ('bamboo shoots') are a popular vegetable in oriental cuisine.

Bamboos are classified into two types: 'clumping' and 'running'. Running bamboos produce very long, rhizomes that can spread considerable distances underground. Each rhizome node can produce a new shoot, from which a new rhizome system can develop. This characteristic is sometimes beneficial, e.g. for controlling soil erosion, but it also enables these plants to become invasive and to suppress tree establishment and growth. If forest restoration is threatened by invasive bamboos, the bamboos must be controlled. Cutting back the shoots may be effective, but if it is not followed up rigorously, it actually stimulates the spread of the rhizomes underground. Therefore, a systemic herbicide such as glyphosate (Roundup) can be applied to the cut culm stumps to kill the rhizomes. Bamboos are characteristic features of some seasonal tropical forest types so, although it may be necessary to suppress them during initial tree establishment, they should be allowed to re-grow afterwards.

Dry tropical forests

Dry tropical forests are most common at 12–20° latitude from the equator, where annual rainfall is 300–1,500 mm and the dry season lasts 5–8 months. Such forests often grow closely inter-mingled with seasonal forest types. Abrupt transitions between the two are usually the result of fire history or variations in soil moisture. The most extensive dry tropical forests are the drier type of miombo and Sudanian woodlands in Africa, caatinga and chaco in South America, and deciduous dipterocarp forest in Asia. Structurally, dry tropical forests are simpler than wetter tropical forests. They are predominantly deciduous, with an irregular and sometimes discontinuous canopy, up to 25 m high, that allows the development of a rich and varied ground layer, which is sometimes dominated by grasses. Large emergent trees, buttresses and bamboos are absent. Woody climbers and epiphytes occur infrequently but vines are more common. Dry tropical forests share many of the families and genera of plant species found in wetter tropical regions, but most of the species are different. They are less species-rich than wetter tropical forests but are home to many species that live in no other forest type (habitat endemics); this is especially true of coastal dry forests.

Dry tropical forests have a prevalence of conspicuously flowered trees (which often flower when leafless) that are pollinated by specialist bees, hawk-moths and birds (hummingbirds in the Neotropics and, to a lesser extent, sunbirds and flower-peckers in the Old World tropics). Seeds are wind-dispersed for up to one third of trees and up to 80% of woody climbers (Gentry, 1995).

Challenges to restoring dry tropical forests

Dry tropical forests are perhaps the most endangered of tropical forest types (Janzen, 1988; Vieira & Scariot, 2006) with only 1–2% of their original area remaining intact (Aronson *et al.*, 2005). They are a lot easier to clear than evergreen forests, so they have been subjected to longer and more intense degradation, including logging, chopping for fire wood, burning and cattle browsing.

Tree planting is possible only during a short window of opportunity at the beginning of the rainy season and the growing season for root development is short (usually less than 6 months before the onset of the dry season). It is important, therefore, that only high-performing tree species are planted, and these may be more difficult to find than in other tropical forests because there are fewer tree species from which to choose. Fruiting

Few epiphytes grow in dry forests and those that do are highly drought tolerant; examples include *Dischidia major* (top) and *D. nummularia* (bottom) (Apocynaceae), pictured here growing on *Shorea roxburghii* (Dipterocarpaceae) in northern Thailand. *Dischidia major* harvests nutrients that are released by the activities of the ants that it hosts its hollow leaves.

Acacia-dominated
dry woodland,
Kenya
(Photo: A. McRobb)

is more seasonal than in wetter forest types, seed dormancy is more common, and seedlings may grow more slowly in the nursery. All of these factors present challenges to nursery tree production in the dry tropics and require considerable research.

The greatest impediments to the restoration of dry tropical forests, however, are the hot, dry climate, poor soils and fire. The sites that are available for restoration are mostly those too infertile for agriculture (Aronson *et al.*, 2005). Soils are often lateritic and hard, so digging holes for tree planting is hard work and expensive. In the dry season, the upper soil layers dry out quickly. In the rainy season, they become waterlogged because of poor drainage, suffocating the roots and killing planted trees. Such problems may be overcome by soil amelioration before planting trees, e.g. use of green manure, adding water-absorbing polymer gels to planting holes, watering trees immediately after planting and applying organic mulch. All such measures may reduce post-planting mortality but they also add to the costs. Weeds grow relatively slowly at dry sites, so weeding may be needed less frequently than at wetter sites, but frequent and liberal application of fertiliser is essential throughout the first 2–3 growing seasons.

Dried grasses and leaf litter provide ideal fuel for fire. Therefore, fire prevention measures are particularly important when restoring dry tropical forests. Other intense human pressures include the introduction of invasive plant species and cattle browsing. Outreach programs for local people are essential in tackling these problems. Nonetheless, in some places, the resilience of disturbed dry forests can be high enough for forest recovery to be initiated simply by preventing fires and removing cattle (see **Section 5.1**).

Tropical forests on mountains

With increasing elevation in the tropics, rainfall usually increases while mean temperatures fall (on average by 0.6°C for every 100 m ascended), resulting in lower evapotranspiration rates and slower decomposition rates. Organic matter therefore accumulates in soils at higher elevations, increasing their water-holding capacity. Consequently, forests on mountains are cooler and wetter than those on adjacent lowlands, and their structure, stature, species composition and leafing phenology can all change abruptly over short distances. In drier parts of the tropics, deciduous forests at the foot of mountains give way to mixed deciduous forests higher up, with evergreen forests being confined to upper slopes and summits. Floristically, ascending a mountain in the tropics is analogous to travelling away from the equator: the tree genera typical of tropical lowlands are gradually replaced by those more usually associated with temperate forests.

Forests on mountains have traditionally been divided into 'lower' and 'upper' montane forests, although the transition between the two is often floristically indistinct and the elevation at which they occur is highly variable, depending on latitude, topography and prevailing climate. The most extensive montane tropical forest ecosystems grow in the Eastern tropics and on the Andes in South America. Montane forests are least extensive in Africa, where they can be found in Cameroon and along the eastern fringe of the Zaire Basin.

Lower montane tropical forests

The transition from lowland to lower montane forest is gradual and can occur anywhere between 800 and 1,300 m elevation. Lower montane forest is largely evergreen in the wetter tropics or mixed evergreen and deciduous at more seasonal latitudes. The trees tend to be shorter than those of lowland forests (15–33 m tall) with few or no emergents. Buttresses, cauliflory and lianas are less evident, whereas epiphytes are more common. Species diversity is generally high because variations in elevation, aspect and slope can result in sharp changes in rainfall, wind direction and temperature.

Upper montane tropical forests and cloud forests

The most dramatic change in montane forests occurs where the mountains meet the clouds: above 1,000 m on coastal and insular mountains or above 2,000–3,500 m inland. Drenched in persistent or frequent mist, 'cloud forests' (also referred to as 'elfin' or 'mossy' woodlands) are characterised by stunted, crooked trees with gnarled trunks and branches (usually smothered in epiphytes) and compact

Lower montane forest, northern Thailand

Cloud forest, Irian Jaya (Photo: A. McRobb)

crowns, composed of small, thick leaves. Although species diversity is generally lower here than in lower montane forests, levels of endemism are high because habitat-specific plant and animal populations evolve in genetic isolation.

Organic matter accumulates in the soils (because decomposition occurs slowly in the cold montane climate) making them highly acidic. Rainfall is high, but up to 60% of the water reaching the soil may come from mist droplets that are captured by the tree crowns (termed 'fog drip' or 'cloud stripping'). Furthermore, the organic-matter-rich soils of upper montane forests have very high water-storage potential, making such forests the most important catchment areas for the water supplies of many tropical countries. Despite this, cloud forests are now among the world's most threatened terrestrial ecosystems (Scatena *et al.*, 2010). Across Central America and the South American Andes, cloud forests are being cleared for subsistence farming and horticulture, despite their poor soils and rugged terrain. In the Americas and Africa, cloud forests continue to be cleared for rearing livestock. Other threats include logging, fuel-wood harvesting, fire, mining, road construction and hunting.

Table 2.1. General characteristics of forests on mountains in the humid tropics (adapted from Whitmore (1998)).

Characteristic	Lowland	Lower montane	Upper montane
Canopy height	25–45 m	15–33 m	1.5–18 m
Emergent trees	Characteristic (to 60 m tall)	Often absent (to 37 m tall)	Usually absent (to 26 m tall)
Pinnate leaves	Frequent	Rare	Very rare
Leaf size (woody plants)	Mesophyll	Mesophyll	Microphyll
Buttresses	Frequent, large	Rare, small	Usually absent
Cauliflory	Frequent	Rare	Absent
Large woody climbers	Abundant	Less abundant	Rare or absent
Vascular epiphytes	Frequent	Abundant	Frequent
Bryophyte epiphytes	Occasional	Common	Abundant

Challenges to restoring montane tropical forests

Working on steep, wet tropical mountains is beset with logistical problems. Access is often the greatest obstacle. Poor roads and the need for 4-wheel-drive vehicles can greatly increase restoration costs. Periodic landslides block roads and smother restoration sites, and soil erosion is a continuous problem. Nothing short of major engineering works can prevent landslides, but soil erosion may be reduced (on a small scale) by applying mulch.

Low temperatures slow the growth of planted trees, and in depressions and gullies, frost can kill them in the winter. The closer tree crowns are to the ground, the greater is the risk of frost damage. Fast-growing trees, which elevate their crowns above the danger zone, are therefore less prone to frost damage. Cutting back weeds from around planted trees reduces the height at which cold air collects. Pulling mulch away from the trunks of very young trees and wrapping them with newspaper may also help to reduce the risk of frost damage.

The exposure of planted trees to strong winds is also a particular problem on mountains. A long-term solution can be to plant the first trees as strategically placed wind breaks. The wind breaks then protect subsequently planted trees from exposure and can act as corridors for seed dispersers (especially if they are connected to remnant forest), thereby enhancing tree species recruitment (Harvey, 2000).

The extirpation of seed-dispersing animals from isolated or highly fragmented montane forests can seriously reduce the rate of seedling recruitment of new (non-planted) tree species into restored plots, and thus delay biodiversity recovery. Attracting seed-dispersing birds by planting rapidly maturing, fleshy-fruited trees (the framework species method (see **Section 5.3**)), or by placing artificial bird perches across restoration sites can help to alleviate the problem (Kappelle & Wilms, 1998; Scott *et al.*, 2000).

It has been predicted that large areas of agricultural land that formerly was cloud forest may be abandoned in Latin America as people move to urban areas, creating considerable opportunities for restoration (Aide *et al.*, 2011). Such areas can, however, develop into fire-climax grasslands, which prevent natural succession. Consequently, clearance of the grass vegetation may be necessary before the land can be planted with tree seedlings or directly seeded. Planting native cloud forest trees has been constrained by a lack of basic knowledge on the reproductive biology, seed handling, propagation and silviculture of most species (Alvarez-Aquino *et al.*, 2004).

Effects of substrate

The soil type and underlying rock can greatly affect the structure and species composition of tropical forest. For example, highly acidic (pH <4) and nutrient-poor podzols in South America and Southeast Asia support heath forest. Here, small, closely spaced, evergreen trees, often of a few dominant species, form a low, non-layered canopy of mostly microphyll species over dense woody undergrowth. Restoring such forests can be impeded by the highly acidic and erosion-prone sandy soils, which cause high mortality among planted trees and corrode metal tags attached to them for monitoring.

Heath forest,
Irian Jaya
(Photo: A. McRobb)

Limestone also supports unique and often species-rich vegetation, with many endemic species, mostly in the seasonal tropics of Southeast Asia and the Caribbean. The porous nature of limestone leads to year-round water shortages, resulting in stunted, xeromorphic, semi-deciduous forest and scrub, with low tree density. Precipitous terrain, shallow soils and high levels of endemism all present challenges to restoration. Water-logging of the substrate or inundation by fresh water, either seasonally or permanently, also generates unique forest types.

Confined to Southeast Asia, peat swamp forests (or 'moor forests') grow in low-lying flat areas, where the decomposition of dead organic matter is slowed by water-logging. This results in the accumulation of acidic peat, eventually forming 'domes' of up to 20 km across and 13 m deep (Whitmore, 1998). Up to six forest communities are distinguishable growing in more or less concentric bands from the centre of the dome to its edge (Anderson, 1961). Each community has only a few tree species, but several are habitat-specific and sensitive to the water levels within the peat. This, along with the semi-fluid nature of the substratum, complicates restoration. When dry, peat is highly flammable and peat fires are notoriously difficult to extinguish. Therefore, hydrological recovery (i.e. 're-wetting' the peat by damming drainage channels) is often the first step to restoring peat swamp forests (Page *et al.*, 2009). It prevents fires, preserves carbon stocks and creates better conditions for tree establishment.

Forest clinging to limestone crags, southern Thailand. Water shortage is the challenge to plants growing in this habitat.

Sago swamp forest,
Irian Jaya
(Photo: A. McRobb)

Fresh water swamp forests (or marsh forests) are a diverse range of forest types that are flooded periodically, in places up to 9 months of each year, growing most extensively alongside the world's largest tropical rivers (the Amazon, Congo and Mekong). In these forests, palms and dicotyledonous trees grow up to 30 m tall, often forming two canopy layers. The longer such forests are inundated each year, the lower is their tree species richness. Swamp forests rely on the accumulation of dead herbaceous vegetation before they can take root. Shrubs establish first, often followed by palms and later by larger trees. This results in a gradient of different forest types on moving away from open water. Taking such zonation into account by manipulating natural succession and/or by planting trees on flooded sites is highly problematic, but thanks to the high-nutrient soils, restoration can progress rapidly once tree establishment has been achieved.

In tropical estuaries and along coast lines, freshwater swamp forests give way to mangroves in the inter-tidal zone. Mangroves are dominated by a few species of salt-tolerant trees, often with characteristic pneumatophores (exposed roots for gaseous exchange) that allow the plants to overcome the anaerobic conditions in the sediment in which they grow. Like other swamp forests, mangroves are zoned into different forest types along a wet-to-dry gradient. Most produce water-dispersed seeds annually in large numbers and some are viviparous (i.e. the seeds germinate on the tree before dispersal). Restoration projects on tidal mudflats are both difficult and dangerous. Planting propagules or small seedlings has a very low success rate. Planting larger saplings is more expensive but more successful. Desiccation, high salinity and attacks by herbivorous insects being the most common problems (Elster, 2000).

Mangrove forest,
Irian Jaya
(Photo: A. McRobb)

Succession proceeds rapidly in tree-fall gaps within intact forest. (A) Nearby fruiting trees provide (B) a dense seed rain. The surrounding forest provides habitat for (C) seed-dispersing animals. (D) Damaged trees and (E) tree stumps re-grow. (F) Seedlings and (G) saplings, which were formerly suppressed by the dense forest canopy, now grow rapidly. (H) Seeds in the soil seed bank germinate. In large deforested areas, many of these natural mechanisms of forest regeneration are reduced or blocked entirely by human activities.

Regional variations

The earlier account merely outlines the broadest tropical forest types. Within each of these, the forest classification schemes of individual countries distinguish many sub-types, often with inconsistent terminologies.

2.2 Understanding forest regeneration

Forest restoration is all about accelerating natural forest succession, so its success depends on understanding and enhancing the natural mechanisms of forest succession.

What is succession?

Succession is a series of predictable changes in ecosystem structure and composition that occurs after disturbance. If allowed to run its course, succession eventually results in a final, climax ecosystem with maximum biomass, structural complexity and biodiversity within the limitations imposed by the local soil and climatic conditions.

A climax tropical forest is not a stable unchanging system but rather a dynamic equilibrium undergoing constant disturbances and renewal. Gaps are formed when large trees die, but they are rapidly filled as saplings and seedlings grow up to exploit the light. Thus a climax forest is an ever-changing mosaic of differently sized tree-fall gaps, regenerating patches and old growth, with species composition varying according to micro-habitat, disturbance history, seed dispersal limitations and chance events. All of these factors contribute to the high species diversity that is characteristic of most climax tropical forests.

More widespread disturbance of climax forest causes it to revert to an earlier, temporary ecosystem or 'seral stage' in the successional sequence. The nature of the seral stage depends on the severity of the disturbance. A major disturbance, such as a volcanic eruption, completely destroys the plant community and soil, causing the land to revert to the earliest seral stage: bare rock. Less severe disturbances, such as logging, cultivation and fire, turn forests into grass- or shrub-lands. Once disturbance ceases, sequential changes in species composition occur due to interactions among plants, animals and their surrounding environment. Bare rock becomes colonised by lichens and mosses, a process known as 'primary succession'. Grass- and shrub-lands undergo 'secondary succession', whereby shrubs shade out herbs, light-demanding pioneer trees shade out shrubs, and much later, pioneer trees are themselves shaded out by shade-tolerant climax trees. Thus, the forest becomes progressively denser, more structurally complex and more species-rich as succession propels it towards the climax condition.

Even under the best of conditions, this process may take 80–150 years to complete, and more often than not, continued human disturbance completely prevents the attainment of climax forest. Therefore, active restoration is necessary wherever more speedy return to climax forest than would happen naturally is desirable.

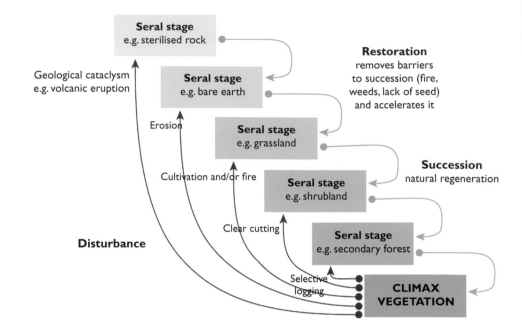

Restoration removes barriers to succession (fire, weeds, lack of seed) and accelerates it

Geological cataclysm e.g. volcanic eruption

Seral stage e.g. sterilised rock

Seral stage e.g. bare earth

Erosion

Seral stage e.g. grassland

Cultivation and/or fire

Succession natural regeneration

Seral stage e.g. shrubland

Clear cutting

Disturbance

Seral stage e.g. secondary forest

Selective logging

CLIMAX VEGETATION

Understanding forest succession is essential for designing effective forest restoration methods. Forest restoration seeks to remove those factors that prevent natural forest succession from progressing.

Pioneer and climax tree species

Tree species can be divided into two broad groups, depending on when they appear in the sequence of forest succession. Pioneer tree species are the first to colonise deforested sites, whereas climax forest tree species establish later, only after the pioneers have created shadier, cooler and wetter conditions. The main distinctions between the two groups are that seeds of pioneers can germinate only in full sunlight and their seedlings cannot grow in shade, whereas the seeds of climax trees can germinate in shade and their seedlings are shade-tolerant.

Cecropia, the largest genus of pioneer trees in the neotropics

The seeds of pioneer trees can lie dormant in the soil, germinating when a gap is formed and light intensity increases. Once the forest canopy closes, however, no more seedlings of pioneer species can grow to maturity. Therefore, pioneer trees grow rapidly and usually produce large numbers of small fruits and seeds at a young age. These are dispersed over long distances by wind or small birds, thereby finding new disturbed areas to colonise. Pioneer species can be divided into two groups: early pioneers (e.g. *Cecropia*, *Macaranga*, *Trema*, *Ochroma*, *Musanga*, *Acronychia* and *Melochia*) and late or persistent pioneers (e.g. *Acacia*, *Alstonia*, *Octomeles*, *Neolamarckia*, *Terminalia* and *Ceiba*). The former are the first species to colonise open areas but seldom live longer than 20 years, whereas the latter grow for 60–80 years and may persist even after climax tree species have begun to reach the canopy (although their seedlings are absent from the ground layer).

Climax tree species grow slowly over many years, gradually consolidating their position in the forest before flowering and fruiting. They tend to produce large, animal-dispersed seeds with low (or no) capacity for dormancy and large food reserves that can sustain seedlings in shaded conditions. Hence, climax tree species can regenerate beneath their own shade, giving rise to the relatively stable species composition of climax forest. They may live for hundreds of years.

In reality, the division between pioneer and climax tree species may be too simplistic. Many climax tree species grow very well when planted in deforested sites. Their absence from such areas is often not because they are limited by the hot, dry, sunny conditions of deforested sites but because their large seeds fail to be naturally dispersed into such areas. Most climax forest tree species are shade-*tolerant* but not shade-*dependent*. This means that tree-planting programs need not be restricted to pioneer species. Planting carefully selected climax tree species alongside pioneers 'short-circuits' succession and achieves a climax forest more rapidly than would happen naturally.

Ashton *et al.* (2001) provide a more refined view of the successional status of tree species by recognising six tree guilds. Short-lived 'pioneers of initiation' (i.e. early pioneers) are the first trees to form a canopy that shades out weeds. 'Pioneers of stem exclusion' (i.e. late or persistent pioneers) rise to dominate the canopy later. They live on while late successional (i.e. climax) canopy tree species grow up alongside them, forest biomass increases, and the tree species composition and structure of the forest become more diverse. The seedlings of pioneers disappear with the development of an understorey, marking a crucial milestone in the progress of succession. Ashton *et al.* (2001) subdivide late successional tree species into four groups, depending on the position of their crowns in the canopy: dominant (abundant in the main canopy or as emergent trees), non-dominant (less abundant in the main canopy), sub-canopy and understory. All six guilds may be present as seedlings early in succession (if seed dispersal is not limiting). If practicable, forest restoration should, therefore, attempt to mimic this by including representative species from all six tree guilds amongst those tree species planted or encouraged to regenerate.

Phases of stand development (adapted from Ashton *et al.* (2001))

PHASES OF STAND DEVELOPMENT

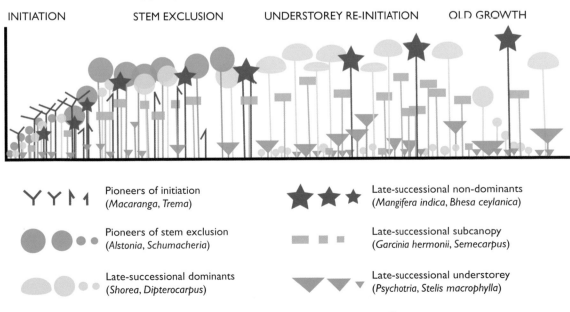

INITIATION   STEM EXCLUSION   UNDERSTOREY RE-INITIATION   OLD GROWTH

Pioneers of initiation
(*Macaranga, Trema*)

Pioneers of stem exclusion
(*Alstonia, Schumacheria*)

Late-successional dominants
(*Shorea, Dipterocarpus*)

Late-successional non-dominants
(*Mangifera indica, Bhesa ceylanica*)

Late-successional subcanopy
(*Garcinia hermonii, Semecarpus*)

Late-successional understorey
(*Psychotria, Stelis macrophylla*)

Limitations to natural reforestation

Disturbance beyond a certain 'threshold' can disrupt the usually efficient ecological mechanisms of forest recovery, causing the vegetation to enter into an 'alternate state'. A good analogy is provided by the elastic band. After moderate stretching, the band can easily return to its original ring shape. However, stretch the band too much and it breaks, becoming a short strip of stretchy rubber, i.e. an alternate state. The properties that enabled it to revert to a ring have been destroyed. It will never recover its original ring shape without human intervention to tie the two ends back together and restore the ring.

By analogy, deforestation on a large scale, followed by continuous disturbance, destroys the natural mechanisms of succession that enable forest recovery. Large deforested areas often become occupied by a persistent pre-climax seral stage (termed 'plagioclimax'), such as grassland, or by a completely new community dominated by invasive exotic species. Where human actions block succession, human intervention is needed to reinstate the mechanisms of forest recovery and allow succession to proceed towards the climax state.

Models of 'threshold dynamics' seek to explain and predict such irreversible changes. They show how 'positive feedback' mechanisms restrain the ecosystem in its degraded state, even after the disturbance has ceased. For example, cutting the trees in a tropical forest increases light levels, leading to an increase in grass cover. Hot, dry grasslands burn more easily than cool moist forest, giving rise to more fires that destroy establishing tree seedlings. This new fire regime prevents the site from reverting to forest, even if tree cutting is stopped.

Understanding such thresholds and the feedback mechanisms that cause forest ecosystems to remain in a persistent degraded state is very helpful in devising appropriate forest restoration strategies.

Regeneration in large deforested areas

In large deforested areas, the establishment of forest trees usually depends on the availability of local seed sources and the dispersal of seeds into deforested sites. Seeds must land where conditions are suitable for germination and they must escape the attention of seed-eating animals, the so called 'seed predators'. After germination, tree seedlings must win an intense competition with weeds for light, moisture and nutrients. Growing trees must also avoid being burnt by wildfires or eaten by cattle.

Limitations to regeneration from the seed bank

When a forest is cut down, large numbers of seeds remain in the soil (the seed bank). However, the vast majority of tropical tree species produce seeds that are viable only for short periods. So, if a forest is cleared and the site is then burnt and cultivated for more than a year or so, most seeds from the original climax forest seed bank die, because they have either no capacity for dormancy (Baskin & Baskin, 2005) or capacity for only a very short period of dormancy (i.e. they must germinate within 12 weeks (Forest Restoration Research Unit, 2010; Garwood, 1983; Ng, 1980)). Consequently, forest regeneration depends almost entirely on seeds that are dispersed into deforested sites from surviving forest remnants or from isolated trees in the surrounding landscape.

Coppicing

Some tree species can re-grow from old tree stumps or root fragments years after the original tree was chopped down (Hardwick *et al.*, 2000). Dormant buds around the root collar of a tree stump can sprout spontaneously, often generating several new shoots. This is called coppicing. Examples of both climax and pioneer tree species can re-grow in this way. Drawing on food reserves that are stored in the roots, coppicing sprouts can rapidly grow above surrounding weeds and have greater resilience to fire and browsing than seedlings. Larger stumps tend to produce more vigorous shoots in greater numbers than smaller stumps. Furthermore, taller stumps survive fire, browsing and weed competition better than shorter ones because the shoots are usually above the height of the disturbance. Protecting tree stumps, therefore, gives forest regeneration a head start.

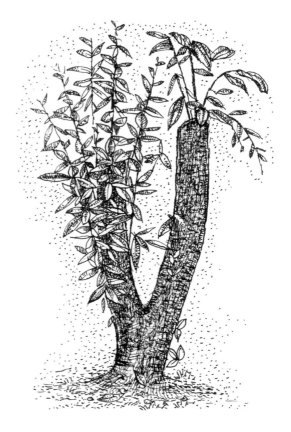

However, tree species that regenerate from stumps usually represent only a small proportion of the climax forest tree community. Although such trees can accelerate the recovery of forest structure, incoming seed is still essential to restore the full tree species richness of the climax forest.

Tree stumps are a major source of natural regeneration, particularly in recently logged over forests.

The importance of seed trees

All trees start life as seeds, so forest succession ultimately depends on the presence of fruiting trees nearby. In a largely deforested landscape, some tree species might be represented by a few scattered, isolated, individuals that somehow escaped the chainsaw, or there may be remnant forest patches producing seeds of a wide range of tree species. Fruiting trees not only provide seeds for forest regeneration, they also attract frugivorous seed-dispersing animals. In a deforested landscape, therefore, the protection of any remaining mature trees greatly enhances natural forest regeneration.

Seed rain

The seed rain consists of all seeds falling on to any particular site, either blown there by the wind or deposited by animals. The density and species composition of the tree seed rain on any deforested site depends on the nearness of fruiting trees and on the efficiency of seed-dispersal mechanisms. The seed rain is most dense and contains most tree species near to intact forest and is sparse in the centre of large deforested areas. A depleted seed rain is one of the most important causes of lack of forest regeneration or low species richness among the tree communities colonising deforested sites. Encouraging seed dispersal is, therefore, a vital element of forest restoration.

Wind-dispersal of tree seeds

In the wet tropics, relatively few tree species produce wind-dispersed seeds. Those that do are usually the tallest trees in the forest, often emergents (e.g. dipterocarps). Wind-dispersal is more common in the seasonal or dry tropics, but even there, less than half of tree species are wind-dispersed (although these species may account for up to 60% of individual trees (FORRU, 2006)).

Wind-dispersed seeds tend to be small and light, and they often have wings or other structures that slow their fall, enabling them to drift on air currents. Most are deposited within a few hundred metres of the parent tree, but some are uplifted by gales and transported over many kilometres. To maximise their dispersal distances, many wind-dispersed tropical tree species fruit at the end of the dry season, when mean maximum wind-gust speeds are at their highest. Consequently, wind-dispersed tree species are capable of colonising deforested sites up to 5–10 km from seed sources. If site conditions allow such species to become naturally established, there may be little need to include them in tree planting programs.

Seed dispersal by animals

With species living in all tropical regions, fruit-eating pigeons are the 'work horses' of natural forest regeneration, due to their seed dispersal capability. Here, wedge-tailed green pigeons feast on the fruits of *Hovenia dulcis*.

Most tropical tree species depend on animals for seed dispersal. Animals that eat fruits either discard the seeds or swallow them, later regurgitating or defecating them some distance from the parent tree (termed 'endozoochorous' dispersal). Fruits that contain animal-dispersed seeds tend to be brightly coloured to attract animals and fleshy, providing a food reward to their animal dispersers.

The dispersal of seeds from forest trees into deforested sites depends on animals that regularly move between the two habitats. Unfortunately, rather few forest animals venture out into open areas for fear of exposing themselves to predators. Apart from birds and bats, few animals travel far between eating a fruit and depositing the seed. Furthermore, many seeds are crushed by teeth or destroyed by digestive juices.

The maximum size of seeds that can be dispersed by an animal depends on the size of the animal's mouth. Small animal species are still relatively common in the tropics, but most of the larger ones, capable of dispersing large seeds, are now rare or have been hunted out. In the past, large herbivores were undoubtedly the most important dispersers of seeds from forest into deforested areas. Elephants, rhinos, tapirs, wild cattle and some deer often consume fruit in the forest, before emerging into open areas at night to graze or browse. With their large mouths, long retention times and long roaming distances, such animals can swallow the largest of seeds and transport them over long distances. The elimination of most of these large mammals over much of their former ranges in recent decades is now preventing the dispersal of many tree species with large seeds (Stoner & Lambert, 2007).

Because birds and bats can fly, they can disperse seeds over long distances. Forest birds such as macaws, parrots, hornbills, pigeons, fruit doves, fruit crows, jays, tityras and bulbuls are particularly important because some species in these groups are equally at home in both forest and deforested sites and can disperse seeds between the two. Fruit bats are also important seed dispersers because they fly over long distances and drop seeds in flight. Unlike most birds, however, bats are nocturnal and cannot be identified using binoculars. Consequently, little research has been done on their role in forest regeneration. Research on bats is, therefore, a high priority for the improvement of forest restoration techniques. Non-flying mammal species that remain relatively common and are likely to disperse seeds between forest and degraded areas include wild pigs, monkeys, deer, civets and badgers, but again, largely because of their nocturnal habits, very little information is available on the seed-dispersing capabilities of these animals.

How far are seeds dispersed?

Most tree seeds fall within a few metres of the parent tree, and the density of a single tree's 'seed shadow' declines steeply with distance away from the tree. Nevertheless, according to Clark (1998), approximately 10% of tree seeds are dispersed over much longer distances of 1 to 10 km. Little is known about long-distance seed dispersal because it is very difficult to measure, but it is vital for biodiversity recovery in any restoration site that is more than a few hundred metres away from intact forest. In the absence of natural long-distance seed dispersal, humans may have to collect seeds from forest and 'disperse' them into sites targeted for restoration in order to restore the climax forest tree community. Human-assisted seed dispersal may be the only way to ensure that large-seeded tree species are represented in restored forests.

Seed predation

A single tree produces vast numbers of seeds during its lifetime, although to replace itself, it need produce only one seed that eventually grows into a reproductively mature adult. The need for such excessive seed crops is because most seeds either fall where conditions are unfavourable for germination or are destroyed by animals. The rich food reserves contained within seeds make them nutritious meals for animals. Some seeds might pass through the digestive tracts of animals intact, but many others are crushed by teeth and digested.

Seed predation is the destruction of a seed's potential to germinate when an animal crushes or digests its embryo. It can occur when seeds are still attached to the parent tree (pre-dispersal predation), but it has most impact on forest regeneration when seeds that have already been dispersed into deforested areas are eaten (post-dispersal predation).

Levels of seed predation

Seed predation can seriously limit natural forest regeneration. Levels of seed predation are highly unpredictable, varying from 0% to 100%, depending on tree species, vegetation, location, season and so on. In deforested sites, seed predation is usually severe enough to reduce significantly the seed survival of most tree species (Hau, 1999), but levels decline as canopy closure is achieved and forest regeneration progresses. Seed predation significantly affects both the distribution and abundance of tree species. It is also a potent evolutionary force, compelling trees to evolve various morphological and chemical mechanisms to defend their seeds against attack, e.g. poisons, tough seed coats and so on.

Animals that eat seeds in regenerating forests

Small rodents and insects, particularly ants, are the most abundant seed predators, capable of affecting forest regeneration (Nepstad et al., 1996; Sanchez-Cordero & Martínez-Gallardo, 1998). Rodents thrive in the weedy, herbaceous vegetation that dominates most deforested sites, but populations decline as soon as canopy closure begins to shade out the weeds (Pena-Claros & De Boo, 2002). Younger successional stages also support higher densities of ants than more advanced regeneration (Vasconcelos & Cherret, 1995).

Susceptibility of seeds to predation

Ecological theory suggests that the susceptibility of any particular tree species to seed predation depends on the food value of its seed. Animals should consume seeds that provide them with maximum nutriment while requiring the least effort to find them. Most attention has been paid to the influence of seed size on vulnerability to predation. Large seeds provide large food rewards to those seed predators that are capable of processing them. Animals may be able to locate large seeds easily, because they are more visible and emit more odour than small seeds, but small rodents have difficulty handling very large seeds. By contrast, small seeds have low food value and are easily overlooked (Vongkamjan, 2003; Mendoza & Dirzo, 2007; Forget et al., 1998). The longer a seed lies on the ground before germinating, the greater is the probability that a predator will discover it. Consequently, seeds that have longer periods of dormancy usually suffer greater predation rates.

The nature of the seed coat is important in protecting seeds from predation. A tough, thick and smooth seed coat makes it very difficult for rodents to reach the nutritious seed contents. Low predation rates amongst seeds that have thick or hard seed coats have been reported for many forest tree species (e.g. Hau, 1999; Vongkamjan, 2003). There may be a trade-off, however, between the effects of seed coat thickness and length of dormancy on seed predation. A thick seed coat often causes prolonged dormancy, which lengthens the period during which seeds are available for attack by predators. But even the toughest seed coat must soften just before germination, presenting a window of opportunity for seed predators. Vongkamjan (2003) observed that several hard-coated tree seed species are attacked during this vulnerable period.

The teeth of rats make short work of large seeds, but rats can act as dispersers of tiny seeds.

Dispersal pattern may also affect the likelihood of predation. Seeds that are scattered thinly over a large area (a pattern that often results from wind-dispersal) are hard for predators to find, whereas a clumped dispersal pattern (characteristic of animal-dispersal) means that once one seed has been discovered, the whole clump will probably be predated. Sporadic large fruit crops can surmount this problem by satiating seed predator populations: seed predators cannot possibly eat all the seeds in such large crops, so many seeds escape predation.

When it comes to seed predation, the literature is full of contradictory statements and opposing viewpoints. The effects of seed predation undoubtedly depend on complex interactions among many variables, including the nature of the environment, the availability of alternative food sources and the individual preferences and seed handling capabilities of the particular seed predator species present. But seed predation is certainly a factor that must be considered in forest restoration projects, particularly those that include direct seeding. Models that can accurately predict the overall effects of seed predation have yet to be made; therefore, the effects of seed predation must be evaluated for each individual site.

Seed dormancy

After being deposited in a deforested site, a seed might not germinate immediately. The dormant period is the length of time during which a mature seed fails to germinate under favourable conditions. It enables seeds to be dispersed at the optimal time, survive the rigours of dispersal (such as being swallowed by an animal) and then germinate when conditions are favourable for seedling establishment.

In general, tree species that grow in cooler, drier climates are more likely to produce dormant seeds than those growing in warmer, wetter climates. Therefore, dormancy is more frequent among deciduous forest and montane tree species than among lowland evergreen forest tree species. In a survey of more than 2,000 climax tropical tree species in evergreen, semi-evergreen, deciduous, savanna and montane forests, Baskin and Baskin (2005) reported that 43%, 48%, 65%, 62% and 66% of species, respectively, exhibited dormancy periods of more than 4 weeks. Physiological dormancy (inhibited embryo development) is the most frequent mechanism of dormancy amongst evergreen, semi-evergreen and montane forest tree species, whereas physical dormancy (caused by impermeable seed coverings that restrict moisture absorption and gaseous exchange) is more prevalent among deciduous and savanna forest tree species.

Germination

The transition from seed to seedling is a dangerous time in a tree's life. Seed dormancy must end and appropriate levels of moisture and light must exist to trigger germination. Because of its small size, low energy reserves and low photosynthetic capability, a young seedling is very vulnerable to changes in environmental conditions, competition from other plants and attack by herbivores. A single caterpillar can completely destroy a young seedling in minutes, whereas larger plants are more resistant to attack.

Timing of seed germination

In the ever-wet tropics near the equator, where soil moisture is continuously high, conditions for seed germination remain favourable all year round. But in the seasonal tropics, the optimal time for tree seed germination is shortly after the start of the rainy season. Seedlings establishing during this period have the full length of the rainy season to build up energy reserves and grow their roots deep into the soil. An extensive root system enables seedlings to survive the desiccating heat of their first dry season by accessing moisture stored deep in the soil. Another reason for germination at the start of the rainy season is the release of soil nutrients at that time. Dry-season fires release nutrients as ash, which the first rains then wash into the soil. As soil moisture rises, the decomposition of organic matter accelerates, releasing yet more nutrients into the soil.

Although the number of germinating tree species peaks at the start of the rainy season, seed dispersal, at the tree community level, occurs throughout the year. This is because the optimal seed dispersal time for any individual tree species depends on a multitude of varying factors, such as the seasonal availability of pollinators, the time required to develop a mature fruit from a fertilised flower and the seasonal availability of dispersal agents. Variation among species in the length of seed dormancy allows each species to disperse its seeds at the optimal time and yet still germinate at the most favourable period, early in the rainy season. For example, seeds that are dispersed at the start of the rainy season tend to have very short dormancy or germinate immediately, whereas those dispersed six months earlier tend to have dormancy of around 6 months. This phenomenon has been well-documented for both Central America and Southeast Asia (Garwood, 1983; Forest Restoration Research Unit, 2006) and it is of crucial importance for the production of trees from seed in nurseries (see **Chapter 6**).

Conditions necessary for germination

Seed germination depends on many factors, the most important of which are sufficient soil moisture and adequate light conditions (not only total light levels, but also the quality of the light). Large, deforested sites, typically dominated by dense weeds, present a hostile environment to tree seeds. On these sites, temperatures fluctuate dramatically between night and day. Humidity is lower, wind speeds are higher and soil conditions are much harsher than those in a forest. Many seeds become trapped in the weed canopy, where they dry out and die, before even reaching the soil.

Even for seeds that penetrate through the weed canopy, weeds present another problem. A high ratio of red to far-red light stimulates the germination of many pioneer tree species, particularly those with small seeds (Pearson et al., 2003). By absorbing proportionately more red light than far-red light, a dense green canopy of weed foliage removes this vital stimulus. Therefore, the germination of most forest tree species depends on the presence of so-called 'germination micro-sites', where conditions are favourable. These are tiny sites with reduced weed cover and sufficient soil moisture to induce seed germination. They include decaying termite mounds, rocks covered in moss and especially rotting logs. The latter provide an excellent moist and nutrient-rich medium for seed germination and are usually weed free.

Rotting dead tree trunks provide excellent micro-sites in which tree seeds can germinate.

Animals can enhance seed germination

The passage of seed through an animal's digestive system can affect both total germination percentage and the pace of germination. For most tropical trees, passage through an animal's gut has no effect on germination, but for those species that show a response, germination is enhanced more often than it is inhibited. Travaset (1998) reported that ingestion by animals increased germination percentage of 36% of the tree species examined; it reduced germination percentage for only 7%. The seeds of 35% of the tree species included in the study germinated more rapidly after passage through an animal's gut; only 13% had delayed germination. Nevertheless, the responses are highly variable: the seeds of species within the same genus, or even from different individual plants of the same species, can have different responses. So, the consumption of seeds by animals can be essential for dispersal, but it is less important for enhancing germination.

Seedling establishment

After a seed has germinated, the greatest threats to seedling survival in deforested areas are competition with weeds, desiccation and fire.

Weeds can suppress regeneration

Deforested areas are usually dominated by a few species of light-demanding grasses, herbs and shrubs. These plants rapidly exploit the soil and develop a dense canopy, which absorbs most of the light available for photosynthesis. The abilities of a dense weed canopy to trap incoming seeds and to inhibit germination by altering light quality have already been mentioned. But even if seeds penetrate the weed canopy and germinate, the emerging seedlings are then overshadowed by the weeds and starved of light, moisture and nutrients.

Because trees evolved to grow tall, they must expend considerable energy and carbon to produce the woody substance, lignin, that supports their future massive size against gravity. Free of the need to make lignin, herbs can grow much faster than trees. Only when a tree's crown overshadows the surrounding weeds, and its root system penetrates below that of the weeds, does a tree gain an advantage. At this point, light-demanding weeds are quickly killed by the shade cast by the tree, but weed competition usually kills most tree seedlings long before they over-top the weeds.

Weeds also prevent forest regeneration by providing fuel for fires in the dry season. Most herbaceous weeds survive fire as seeds, corms or tubers buried in the soil, or they (e.g. grasses) possess well-protected growing points that re-sprout after fire. In trees, the growing points are unprotected, raised on the tips of branches. In a fire, therefore, small seedlings are often completely incinerated by the blazing dried weeds surrounding them. Re-sprouting of older saplings is possible, but only after they are about 1 year old.

The weeds that are most capable of suppressing forest regeneration are nearly always exotic species that have been deliberately introduced and now flourish outside the ranges of their natural enemies. Many weeds in Africa and Asia originate from Central or South America. Several are in the families Leguminosae and Asteraceae (Compositae) and they usually share the following characteristics: i) they are rapidly growing perennials that flower and fruit at a very young age; ii) they produce very large numbers of seeds (or spores) that can survive in a dormant state and thus build up in the soil seed bank; iii) they are resilient after burning (even though their above-ground parts may be totally destroyed, they can regenerate rapidly from rootstock); iv) they produce chemicals that inhibit the seed germination and/or seedling growth of other plant species (allelopathy); and v) they may also produce chemicals that are toxic to potential seed-dispersing animals. Reports of the toxicity of invasive exotic plants to cattle are common, and such plants are probably also toxic to wildlife. Some of the most widespread species are listed in **Table 2.2**.

Grasses, bracken fern (*Pteridium aquilinum*) and species of the family Asteraceae (Compositae) (e.g. *Chromolaena odorata* pictured here) are among the most ubiquitous of tropical weeds that are capable of suppressing natural forest regeneration.

Table 2.2. Dominant weeds capable of suppressing forest regeneration.

Species	Family	Habit	Origins	Invasive exotic	Allelopathic	Toxic to Ungulates	Notes
Dicranopteris linearis	Gleicheniaceae	Climbing fern	Asia, Africa Australasia, Pacific	—	Yes	Not known	Forms a 2-m-high thicket on bare degraded land. Not fire or shade tolerant.
Chromolaena spp.	Asteraceae (Compositae)	Herb or shrub	New World	West Africa, Asia, Australia	Yes	Yes	Syn. *Eupatorium* (Compositae). Wind-dispersed seeds.
Lantana camara	Verbenaceae	Prickly, scrambling shrub	New World	Central Africa, Australia, India, Southeast Asia, Pacific Islands	Yes	Yes	Introduced as ornamental. Bird-dispersed fruits, ornamental. poisonous to humans. Coppices well, resilient.
Leucaena leucocephala	Leguminosae	Small tree	Belize, Mexico	Pacific Islands, north Australia	Yes	Yes (in large doses)	Introduced for firewood, fodder and biomass production. Fire promotes seed germination.
Mikania micrantha	Asteraceae (Compositae)	Vine	New World	Nepal, India	Yes	No	Introduced for military camouflage. Wind-dispersed vine smothers trees. Threatens rhino and tiger habitat in Nepal.
Mimosa pigra	Leguminosae	Prickly shrub	New World	Africa, India, Southeast Asia, Australia, Pacific Islands	Yes	Not known	Introduced for riverbank stabilisation. Accumulates dense seed banks. Thrives in wet areas and on disturbed soils.
Pteridium aquilinum	Dennstaedt-iaceae	Fern	Pan-tropica	—	Yes	Yes	Fire promoting. Fire resilient. Carcinogenic.
Grasses (e.g. *Imperata, Pennisetum, Andropogon, Panicum, Phragmites, Saccharum* and many species of other genera)	Poaceae	Herbs	Many	Many	Some species	No	Fire promoting. Fire resilient.

Predators of seedlings

In terms of biomass and species, insects are by far the most abundant herbivores, but in tropical forests, most insect species eat only one or a few plant species. Therefore, herbivorous insects are only capable of causing high mortality among seedlings growing close to the parent tree. This is because insects that are attracted to the parent trees also find and eat the seedlings growing beneath them (Coley & Barone, 1996). In deforested sites, however, small scattered seedlings are much harder to find, so herbivorous insects rarely limit forest regeneration.

By contrast, large mammalian herbivores can have a serious impact on forest regeneration. Large wild herbivores, such as elephants, rhinos and wild cattle, are now so sparse that they rarely affect forest regeneration, except locally. Domestic cattle, on the other hand, are ubiquitous and they impede forest regeneration over large areas. In most tropical countries, it is common to find domestic livestock ranging freely across degraded forestlands. Their impact on forest regeneration depends on population density. A small herd of cattle might have no significant effect (or they might even be beneficial) but, where populations are dense, they can completely halt natural succession.

The most obvious impact of cattle is browsing on tree saplings. Cattle can be very selective, often eating the foliage of palatable tree species while ignoring that of unpalatable ones. Distasteful or thorny trees can thus become dominant, while edible ones are gradually eliminated. Cattle also trample young seedlings indiscriminately.

The potential beneficial effects of cattle include reducing competition for tree seedlings by grazing or browsing on weeds, although, as mentioned above, several of the typical weeds of deforested sites contain toxins that protect them against being eaten. Another potentially beneficial effect of cattle could be seed dispersal. Where large wild ungulates have been extirpated, domestic cattle can be the only animals present that are capable of dispersing large seeds from forest into gaps. Furthermore, their hoof prints can provide micro-sites for seed germination in which moisture and nutrients accumulate and weeds have been crushed.

The balance between these positive and negative effects and their relationship with herd density, site conditions and vegetation type are not fully understood. Therefore, further research is required to allow us to predict the overall effects of cattle on forest regeneration at any particular site.

Too many cattle can slowly devastate a forest by preventing regeneration, but they can also keep weeds in check and may act as seed dispersers.

Fire

Fires are a major constraint to forest regeneration. Infrequent, low-intensity fires may slow succession and alter the composition and structure of regenerating vegetation (Slik *et al.*, 2010; Barlow & Peres, 2007), but frequent burns can completely prevent it, leading to the persistence of grasslands where forests would otherwise grow.

Fires can occur naturally in all tropical forest types, even the wetter ones. In Amazonia, Borneo and Cameroon, layers of charcoal deposits deep in the soil profile show that rain forests have burnt at least periodically over the past few thousand years, at intervals of hundreds or thousands of years (Cochrane, 2003). Historically, such fires have been restricted to periods of severe droughts, but now, increased forest degradation, fragmentation and climate change are all contributing to increased fire frequency, even in the wet tropics (Slik *et al.*, 2010). The tree species of wet evergreen tropical forests usually have thin bark, making them highly vulnerable to fire damage. Even low-intensity fires in wet tropical forests result in high tree mortality and dramatic and rapid changes in tree species composition, especially where fires recur at short intervals (Barlow & Peres, 2007).

It is in the seasonally dry tropics where fires are the most prevalent threat to forest regeneration. By the end of the rainy season, weedy vegetation has often grown above head height and is practically impenetrable. In the hot season, this vegetation dies

Fire can burn in all types of tropical forest but is particularly frequent in seasonally dry forests.

back, dries out and becomes highly flammable. Each time it burns, most of the tree seedlings that may have gained a roothold amongst the weeds are killed, whereas the weeds and grasses survive, re-growing from rootstocks or seeds protected beneath the soil. Thus, the weedy vegetation creates conditions conducive to fire and in doing so prevents the establishment of trees that could shade out the weeds. Breaking this cycle is the key to restoring seasonally dry tropical forests.

Causes of fire

Fires can be started naturally by lightning strikes and volcanic eruptions. But, such natural fires are infrequent, allowing plenty of time in between each event for the trees to grow large enough to develop some resilience to burning. These days, however, most fires are started by humans. The most common reason for starting fires is to clear land for cultivation. The fires spread from cultivated land into surrounding areas, where they kill young trees, effectively halting forest regeneration. Fires are also used as a weapon in disputes over land tenure, to stimulate the growth of grasses for livestock and to attract wild animals for hunting. In addition to causing ecological damage, fires are a major health hazard. Smoke pollution causes respiratory, cardiovascular and eye problems in hundreds of thousands of people every year.

Human-caused fires are increasing throughout the tropics, both in frequency and intensity. The underlying cause is a growing human population that requires clearance of ever more agricultural land. This results in the fragmentation of forest areas, which exposes more forest edge into which fires can spread from surrounding areas. Within forests fragments, degradation creates more fire-prone conditions by opening up the forest canopy. This allows light-loving and highly flammable grasses and other weeds to invade and dead wood to accumulate. Furthermore, global climate change is resulting in hotter, drier conditions that favour fire in many tropical regions, particularly in the dry season.

Effects of fire on regeneration

Frequent fires reduce both the density and species richness of the tree seedling and sapling communities (Kodandapani et al., 2008). Burning reduces the seed rain (by killing seed-producing trees) and the accumulation of viable seeds in the soil seed bank. It favours the establishment of wind-borne, light-demanding pioneer tree species at the expense of shade-tolerant climax species (Cochrane, 2003; Meng, 1997; Kafle, 1997). Fire burns off soil organic matter, leading to a reduction in the soil's moisture-holding capacity (the drier the soil, the less favourable it is for tree seed germination). It also reduces soil nutrients. Calcium, potassium and magnesium are lost as fine particles in smoke, while nitrogen, phosphorus and sulphur are lost as gases. By destroying vegetation cover, fire increases soil erosion. It also kills beneficial soil micro-organisms, especially mycorrhizal fungi and microbes that break down dead organic matter and recycle nutrients. Studies that have compared frequently burnt areas with those protected from fire show that preventing fires accelerates forest regeneration.

Fire and germination

Direct exposure to fire either kills the seeds of the vast majority of tropical tree species or significantly reduces their germination. Seeds lying on the soil surface are nearly all killed by even low-intensity fires, but those buried even a few centimetres below the soil surface can usually survive (Fandey, 2009). The germination of a very small number of tree species can, however, be stimulated by fire. If burning disrupts the seed coat without killing the embryo, water entering the seed can trigger germination, and substances in smoke or from charred wood can sometimes stimulate germination chemically. Species whose germination can be stimulated by fire include teak (*Tectona grandis*) and some leguminous trees in dry tropical forests (Singh & Raizada, 2010).

Does fire kill trees?

Small seedlings and saplings are usually killed by fire, but larger trees can survive occasional low-intensity fires (i.e. burns restricted to the leaf litter or ground vegetation). So how large does a tree have to grow before it can survive fire? Bark thickness, rather than overall growth rate, appears to be the key survival factor (Hoffman *et al.*, 2009; Midgley *et al.*, 2010). Larger trees have thicker bark, which insulates their vital vascular system (the cambium layer) from the heat of fires, so they survive better than smaller trees. As a rough guide, trees with bark thicker than 5 mm have a greater than 50% chance of survival after a low-intensity fire (Van Nieuwstadt & Sheil, 2005). To develop bark of this thickness, trees must grow to at least 23 cm diameter at breast height (dbh), which takes a minimum of 8–10 years. Therefore, it is likely that forest regeneration will be severely impeded where fires burn more frequently than once every 8 years. In general, the trees of wet evergreen forests have relatively thin bark, and are therefore more susceptible to fire damage than those of seasonally dry or dry deciduous forests (Slik *et al.*, 2010).

Even if fire kills the above-ground parts of a tree, the roots may still survive, insulated from the heat by soil. Food reserves that are stored in the roots can then be mobilised to support the growth of re-sprouts (or coppices) from dormant buds near the root collar or stem (epicormic buds). Re-sprouting capability varies greatly among species and is more common among dry deciduous forest tree species than among evergreen tree species of wet forests. Usually, a tree must grow for at least a year before it can re-sprout. So, frequent fires also reduce the chances of forest regeneration from re-sprouting.

2.3 Climate change and restoration

Climate change severely threatens tropical forests, reducing the bioclimatically suitable area for certain species (Davis, 2012) and increasing the risk of large-scale forest 'dieback' in some areas (Nepstad, 2007). International negotiations to shift the global economy from carbon-dependence to carbon-neutrality have largely failed (but are continuing). The burning of fossil fuels and continued destruction of tropical forests both continue apace. So it seems inevitable that concentrations of carbon dioxide, methane and other greenhouse gases will continue to rise over the next few decades (IPCC, 2007).

The relationship between rising atmospheric concentrations of greenhouse gases and global warming is well established. Therefore, predictions of future warming depend on the future levels of greenhouse gas emissions, which in turn depend on human population size and economic activity. Computer models predict that, with moderate economic growth and rapid adoption of green technologies, surface air will warm by an average of 1.8°C (range 1.1–2.9°C) by the end of this century. But with rapid economic growth and continued dependence on fossil fuels this 'best estimate' climbs to 4.0°C (range 2.4–6.4°C) (IPCC, 2007). What is absolutely clear is that urgent and extreme action is needed now to deal with unprecedented changes in the environment.

Rainfall patterns will also change, but there is less agreement amongst meteorologists on how. Atmospheric warming will result in greater evaporation from water bodies and soil, causing some areas to become more arid. In those areas, forest fires will become more frequent, adding even more carbon dioxide to the atmosphere. On the other hand, increased water vapour in the atmosphere must result in more rainfall overall, but changes in global air currents are uncertain, so there is disagreement about when and where extra rain will fall. The latest computer models predict that rainfall will increase over tropical Africa and Asia and decrease slightly over tropical South America (by +42, +73 and –4 mm per year, respectively, with 2°C warming; double these values with 4°C warming) (Zelazowski *et al.*, 2011). In the seasonal tropics, dry seasons will

Climate change is predicted to result in reduced rainfall in South America where, in years of severe drought such as 2005 and 2010, areas of the Amazon rain forest switch from being a carbon sink to being a carbon source.

most likely become drier and rainy seasons wetter. Most computer models predict an increase in rainfall in the summer monsoon season of South and Southeast Asia and East Africa (IPCC, 2007). Droughts may also cause tropical forests to emit more carbon dioxide than they absorb (because of tree deaths and fire), thus exacerbating the problem of greenhouse gas emissions (Lewis *et al.*, 2011)

These changes in global climate may alter both the distribution of forest types and the mechanisms of forest regeneration described above. Since the climax forest type depends on the climate, changes in temperature and rainfall will alter the climax forest type suited to any particular site. Achievement of the climax forest type is the ultimate goal of forest restoration, and so climate change will have profound consequences for the planning and execution of forest restoration projects (see **Section 4.2**). The latest models predict that areas in South America that currently have a climate that is capable of supporting wet tropical forests will contract substantially, whereas such areas will expand in Africa and Southeast Asia (Zelazowski *et al.*, 2011). In South America, former ever-wet rain forests may become seasonally dry forests or even savannas. By contrast, in Africa, and Southeast Asia, it is not likely that wet tropical forests will spread naturally into new wetter areas because of limited seed dispersal and the existing occupation and use of the land. Climate change will also affect the distribution of forest types on mountains. In drier areas, higher temperatures might allow dry forest types to spread higher up mountains[1], displacing evergreen forests, but where rainfall increases, evergreen forest could spread to lower elevations.

The effects of global warming on the mechanisms of forest regeneration will also be significant. Changes in the climate, especially in the seasons, will result in changes in the flowering and fruiting times of plants, as well as to the life cycles of their pollinators and seed dispersers. This could result in a 'de-coupling' of reproductive mechanisms e.g. flowers opening when their insect pollinators are not flying. On the other hand, wind pollination and seed dispersal could benefit from global warming because wind-gust speeds and the frequency of storms capable of uplifting even large wind-dispersed seeds will both increase. Germination and early seedling development are both highly sensitive to temperature and moisture levels and could also be particularly vulnerable to the spread of weeds, pests and diseases that are favoured by climate change.

An increase in wildfires, with all the associated impacts described above, seems inevitable, particularly in the predicted drier areas of South America. There, more frequent wildfires are expected to bring about "substantial changes in forest structure and composition, with cascading shifts in forest composition following each additional fire event" (Barlow & Peres, 2007).

Does nature need help?

Some people take the view that deforested sites should be left to recover naturally and that forest restoration is "unnecessary interference with nature". This view fails to recognise that the situation today in most large deforested areas is far from 'natural'. Humans have not merely destroyed the forest, we have also destroyed the natural mechanisms of forest regeneration. All of the barriers to forest regeneration described in this chapter are caused by humans. Hunting threatens seed dispersal by animals,

[1] The upper limit of their preferred temperature will ascend, on average, about 100 m elevation for every 0.6°C increase in temperature.

most wildfires are anthropogenic in origin and humans introduced most of the invasive weeds that now prevent tree seedling establishment. Forest restoration is merely an attempt to remove or overcome these 'unnatural' barriers to forest regeneration.

Even under the most favourable circumstances, natural forest regeneration occurs slowly. In his definitive text, *The Tropical Rain Forest*, P. W. Richards (1996) comprehensively reviewed forest succession throughout the tropics. He concluded: "if the seral vegetation is left undisturbed, succession leads eventually to the restoration of forest similar to the climatic climax. This process … probably takes several centuries, even when the cleared area is only a short distance from intact forest."

Unprecedented rates of biodiversity loss and climate change require urgent action. Waiting centuries for forests to regenerate naturally is no longer an option if species that are on the verge of extinction are to be saved or if carbon storage by forests is to have any impact on climate change. Human-caused problems require human-made solutions … and forest restoration is one of them.

A degraded mountainous landscape. Watershed degradation, soil erosion and landslides threaten agriculture. Isolated remnant trees and forest fragments, clinging to the ridges, may yet provide seed for forest restoration, but without active restoration, this landscape, its wildlife and its communities have an impoverished future.

CHAPTER 3

RECOGNISING THE PROBLEM

Forest degradation reverses and impedes natural forest succession, whereas forest restoration releases succession and accelerates it towards achieving the climax forest condition. The overall management approach that is required to restore a climax forest ecosystem depends on how far back in the successional sequence the vegetation has been 'pushed' by degradation and the factors that are limiting succession. The complexity, intensity and cost of restoration all increase as the level of degradation increases.

The diagrams and notes in this chapter will help you to recognise the general level of degradation in your restoration area and to decide which broad restoration strategy to adopt (i.e. protection, accelerated natural regeneration, framework forestry, maximum diversity techniques, foster ecosystems etc.). Once you have chosen a restoration strategy, the next step is to carry out a site assessment that will allow you to determine the detailed management operations necessary to implement that strategy (see Section 3.2). You will then be ready to plan your restoration project (see Chapter 4). The implementation of each restoration strategy is then explained in detail in Chapter 5, whereas the growing and planting of trees (which may be necessary for the restoration of sites where the degradation has reached stages 3–5) are described in Chapters 6 and 7.

3.1 Recognising levels of degradation

There are five broad levels of degradation, each of which requires a different restoration strategy. They can be distinguished by recognising six critical 'thresholds' of degradation; three pertain to the site being restored and three to the surrounding landscape.

Site-critical thresholds:

1) The density of trees is reduced such that herbaceous weeds dominate the site and suppress tree seedling establishment (see **Section 2.2**).

2) On-site sources of forest regeneration (i.e. the seed or seedling bank, live stumps, seed trees etc.) decline below the levels needed to maintain viable populations of climax forest tree species (see **Section 2.2**).

3) Soil degradation has proceeded to such an extent that poor soil conditions limit the establishment of tree seedlings.

Landscape-critical thresholds:

4) There are only small and sparse remnants of climax forest in the landscape, such that the diversity of tree species within dispersal distance of the forest restoration site is not sufficient to represent the climax forest.

5) Populations of seed-dispersing animals are reduced to the point that seeds are no longer transported to the forest restoration site in sufficiently high densities to re-establish all of the required tree species (see **Section 2.2**).

6) Fire risk is increased such that naturally established trees are unlikely to survive because of the increased cover of combustible herbaceous weeds in the landscape immediately surrounding the restoration site.

Dominance of a site by herbaceous weeds marks a critical point, at which protection alone ceases to be sufficient to restore forest. Tree seedlings beneath the weed canopy must be 'assisted' by weeding or supplemented by planting framework tree species.

Stage-1 degradation

SITE-CRITICAL THRESHOLDS		LANDSCAPE-CRITICAL THRESHOLDS	
Vegetation	Trees dominate over herbaceous weeds	**Forest**	Large remnants remain as seed sources
Sources of regeneration	Plentiful: soil seed bank viable; dense seedling bank; dense seed rain; live tree stumps	**Seed dispersers**	Common; both large and small species
Soil	Little localised disturbance; remains mostly fertile	**Fire risk**	Low to medium

RECOMMENDED RESTORATION STRATEGY:

- Protection from encroachment, cattle, fire and any other further disturbances and prevention of the hunting of seed-dispersing animals
- Re-introduction of locally extirpated species

OPTIONS TO INCREASE ECONOMIC BENEFITS:

- Extractive reserves for sustainable use of forest products
- Ecotourism

Stage-2 degradation

SITE-CRITICAL THRESHOLDS		LANDSCAPE-CRITICAL THRESHOLDS	
Vegetation	Mixed trees and herbaceous weeds	**Forest**	Remnants remain as seed sources
Sources of regeneration	Seeds and seedling banks depleted live tree stumps common	**Seed dispersers**	Large species becoming rare, but small species still common
Soil	Remains mostly fertile: erosion low	**Fire risk**	Medium to high

RECOMMENDED RESTORATION STRATEGY:

- Protection + ANR
- Re-introduction of locally extirpated species

OPTIONS TO INCREASE ECONOMIC BENEFITS:

- Enrichment planting with economic species that have been lost through unsustainable use
- Establishment of extractive reserves to ensure the sustainable use of forest products
- Ecotourism

Stage-3 degradation

SITE-CRITICAL THRESHOLDS		LANDSCAPE-CRITICAL THRESHOLDS	
Vegetation	Herbaceous weeds dominate	**Forest**	Remnants remain as seed sources
Sources of regeneration	Mostly from incoming seed rain; a few saplings and live tree stumps might remain	**Seed dispersers**	Mostly small species dispersing small seeds
Soil	Remains mostly fertile; erosion low	**Fire risk**	High

RECOMMENDED RESTORATION STRATEGY:

- Site protection + ANR + planting framework species

OPTIONS TO INCREASE ECONOMIC BENEFITS:

- Planting framework species that have economic benefits
- Ensuring local people benefit from funding of tree planting and site maintenance
- Analogue Forestry[1] or 'Rainforestation' farming[2]

[1] en.wikipedia.org/wiki/Analog_forestry
[2] www.rainforestation.ph/index.html

Stage-4 degradation

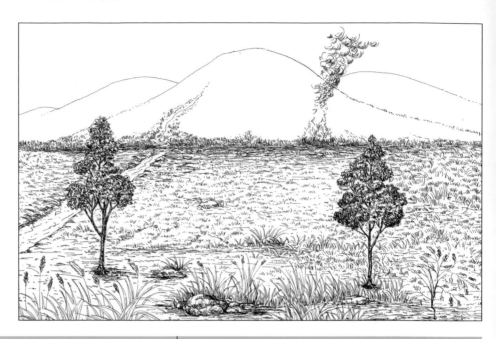

SITE-CRITICAL THRESHOLDS		LANDSCAPE-CRITICAL THRESHOLDS	
Vegetation	Herbaceous weeds dominate	**Forest**	Remnants too few or too distant to disperse tree seeds to site
Sources of regeneration	Low	**Seed dispersers**	Mostly gone
Soil	Erosion risk increasing	**Fire risk**	High

RECOMMENDED RESTORATION STRATEGY:

- Site protection + ANR + planting framework species + enrichment planting with climax species
- Maximum diversity methods (Goosem & Tucker, 1995) such as the Miyawaki method[3]

OPTIONS TO INCREASE ECONOMIC BENEFITS:

- Enrichment planting with economic species + sustainable harvesting of non-timber forest products
- Employment of local people in the restoration program
- Analogue forestry or 'Rainforestation' farming[4]

[3] www.rainforestation.ph/news/pdfs/Fujiwara.pdf
[4] www.rainforestation.ph/index.html

Stage-5 degradation

SITE-CRITICAL THRESHOLDS		LANDSCAPE-CRITICAL THRESHOLDS	
Vegetation	No tree cover. Poor soil might limit growth of herbaceous weeds	**Forest**	Usually absent within seed dispersal distances of the site
Sources of regeneration	Very few or none	**Seed dispersers**	Mostly gone
Soil	Poor soil conditions limit tree establishment	**Fire risk**	Initially low (soil conditions limit plant growth); higher as the vegetation recovers

RECOMMENDED RESTORATION STRATEGY:

- Soil improvement by planting green mulches and the addition of compost, fertilisers or soil micro-organisms
- ... followed by planting 'nurse trees' — i.e. hardy nitrogen-fixing trees that will further improve the soil (also known as the "plantations as catalysts" method (Parrotta, 2000))
- ... and then thinning of nurse trees and their gradual replacement by planting a wide range of native forest tree species

OPTIONS TO INCREASE ECONOMIC BENEFITS:

- There will be few economic benefits until the soil ecosystem has recovered
- Plantations of commercial tree species can be used as nurse trees to generate revenue from thinning
- Mechanisms to ensure that local people benefit from the harvesting of commercial tree species
- Once the nurse-tree crop is ready for thinning and modification, the option for achieving economic benefits are the same as those for stage-4 degradation

3.2 Rapid site assessment

Recognising which of the five stages of degradation has been reached at a site will determine which broad restoration strategy will be most suitable (**Table 3.1**). A more detailed site assessment is then needed to determine the existing potential for natural forest regeneration and to identify factors that could be limiting it. These factors will determine which activities should be implemented and the intensity of the work required for each site (and hence the labour requirements and costs). The project plan can then begin to take shape.

To carry out a simple site assessment, you will need the following equipment: a compass, a topographic map, a global positioning system (GPS), a camera, plastic bags, a 2m bamboo cane or similar, a piece of string with a mark exactly 5 m from the end and datasheets (see opposite and **Appendix A1.1**) on a clip board with a pencil.

Invite all stakeholders (particularly local people) to participate in the site survey and begin by marking the boundaries of the site on a map and recording the GPS co-ordinates. Next, survey natural regeneration along a transect across the site at its widest point. Select the starting point and decide on a compass bearing to follow for the line.

At the starting point, position the bamboo pole and use the string attached to it to mark out a circular sample plot of radius 5 m. If there are any signs that livestock have been present in the sample plot (e.g. dung, hoof prints, bite marks on vegetation etc.) place a tick in the 'livestock' column on the datasheet; likewise for signs of fire (ash, or black marks at the base of woody vegetation). Record any information provided by local participants when asked about the land-use history of the site. Estimate the extent of exposed soil in the circle (as a percentage of the area), ask local participants to rank the soil condition (good, medium, poor etc.) and make note of any signs of soil erosion. Estimate the percentage cover and average height of grasses and herbaceous weeds across the plot and note whether tree seedlings are strongly represented in the ground flora.

Record the number of a) trees larger than 30 cm girth at breast height (gbh) (i.e. 1.3 m from the ground), b) saplings taller than 50 cm (but smaller than 30 cm gbh) and c) live tree stumps (with green shoots) within the circle. The number of 'regenerants' per circle is the total number of trees in all three of these categories. Place leaf samples from each of the tree species you find into plastic bags. Finally take photos facing due north, south, east and west from the centre pole.

Pace out the required distance along the pre-determined compass bearing to the next sample point. Collect data at a minimum of 10 sample points across the site, at least 20 paces apart. If the site is large, position the sample points further apart and use more points (at least 5 per hectare). If the site is small and the required number of points cannot be fitted into a single transect, then use two or more parallel lines placed at representative locations across the site. Once you have decided on a compass bearing for the transect heading and a distance between sample points for each line, stick strictly to these parameters during the survey.

At the end of the survey, find a clear space and sort through the leaf samples. Group leaves of the same species together and count the number of common tree species on the site (i.e. those represented in more than 20% of the circles). Ask local people to

Example of rapid site assessment

Circle	Livestock signs	Fire signs	Soil – % exposed/ condition/ erosion	Weeds – % cover/mean height/± tree seedlings	No. trees >50 cm tall (<30 cm gbh)	No. live tree stumps	No. trees >30 cm gbh	Total no. regenerants
1	✓	✓	5%/poor/no	95%/1.0 m/none	6	14	0	20
2	✓	✗	15%/poor/no	85%/0.5 m/few	9	15	0	24
3	✓	✗	5%/poor/no	95%/1.5 m/none	12	12	1	25
4	✓	✓	30%/poor/no	70%/0.3 m/none	4	3	0	7
5	✓	✓	5%/poor/no	95%/1.5 m/many	14	15	2	31
6	✗	✓	0%/poor/no	100%/1.5 m/few	7	13	1	21
7	✓	✓	5%/poor/no	95%/0.8 m/many	10	15	1	26
8	✓	✓	10%/poor/no	90%/1.2 m/many	9	12	2	23
9	✓	✓	20%/poor/yes	80%/0.5 cm/none	9	5	1	15
10	✗	✓	20%/poor/no	80%/1.2 m/none	6	10	0	16

Location, GPS	Siem Reap, Cambodia, 13°34'3.24"N, 104° 2'59.80"E		
Recorder	Kim Sobon		
Date	1st June 2010		
Total no. of regenerant species 18	Pioneers 16	Climax 2	

Total	208	
Mean	20.8	(= total/10)
Average/ha	2,667	(= mean × 10,000/78)
No. of trees to plant per ha	433	(= 3,100 – Average/ha)

Other Comments: Villagers said that large mammal seed dispersers are absent, but fruit-eating birds and small mammals are commonly seen. Hunting is common in the area. Villagers want to use the forest to make charcoal.

Rapid site surveys to record existing natural forest regeneration and the most prominent factors preventing it are carried out using circular sample plots of 5 m radius.

provide local names for the species and try to determine if they are pioneer or climax species. If possible, make dried specimens, including flowers and fruits, and ask a botanist to provide scientific names. Then calculate the average number of regenerants per circle and per hectare.

At the end of the survey, hold a short discussion session with the participants to identify any other factors that could impede forest regeneration that have not already been recorded on the data sheets, especially the activities of local people such as the collection of fuel-wood. The abundance of seed-dispersing animals in the area cannot be assessed in the rapid site survey, but local people will probably know which seed dispersers remain common in the area. Try to determine if the seed dispersers are threatened by hunting.

3.3 Interpreting data from a rapid site assessment

Initial restoration activities should aim to:

i)   counteract the factors that impede forest regeneration (e.g. fire, cattle, hunting of seed dispersers etc.);

ii)   maintain or increase the number of regenerants to 3,100/ha;

iii) maintain the density of common tree species (if already high) or increase tree-species richness until at least 10% of the tree species characteristic of the target climax forest are represented.

Table 3.1. Simplified guide to choosing a restoration strategy

LANDSCAPE-CRITICAL THRESHOLDS			SUGGESTED RESTORATION STRATEGY	SITE-CRITICAL THRESHOLDS		
Forest in landscape	Seed-dispersal mechanisms	Fire risk		Vegetation cover	Natural regenerants	Soil
Remnant forest remains within a few km of the restoration site	Mostly intact, limiting the recovery of tree species richness	Low to medium	PROTECTION	Tree canopy cover exceeds herbaceous weed cover	Natural regenerants exceeds 3,100/ha with more than 30[5] common tree species represented	Soil does not limit tree seedling establishment
		Medium to high	PROTECTION + ANR	Tree crown cover insufficient to shade out herbaceous weeds		
		High	PROTECTION + ANR + PLANTING FRAMEWORK TREE SPECIES	Herbaceous weed cover greatly exceeds tree crown cover	Natural regenerants sparser than 3,100/ha with fewer than 30[5] common tree species represented	
			PROTECTION + ANR + MAXIMUM DIVERSITY TREE PLANTING			
Remnant forest patches very sparse or absent from the surrounding landscape	Seed-dispersing animals rare or absent such that the recruitment of tree species to the restoration site will be limited	Initially low (soil conditions limit plant growth); higher as the vegetation recovers	SOIL AMELIORATION + NURSE TREE PLANTATION, FOLLOWED BY THINNING AND GRADUAL REPLACEMENT OF MAXIMUM DIVERSITY TREE PLANTING	Herbaceous weed cover limited by poor soil conditions		Soil degradation limits tree seedling establishment

[5] Or roughly 10% of the estimated number of tree species in the target forest, if known.

The site survey will determine which factors are preventing natural forest regeneration. Achieving a density of 3,100 regenerants per hectare results in an average spacing of 1.8 m between them. For most tropical forest ecosystems, this is close enough to ensure that weeds are shaded out and canopy closure is achieved within 2–3 years after the restoration work commences. The guideline of "approximately 10% of the species richness of the target forest ecosystem" is adjustable, depending on the diversity of the target ecosystem. If you do not know the species richness of the target ecosystem, aim to re-establish roughly 30 tree species (by planting and/or encouraging natural regeneration). This is usually sufficient to 'kick start' biodiversity recovery in most tropical forest ecosystems, and 30 tree species is about the maximum that can be produced in a small-scale tree nursery. The overall rate of biodiversity recovery will increase with the number of tree species that can be re-instated at the start of restoration, but some low-diversity tropical forest ecosystems (e.g. high-altitude montane forests and mangrove forests) can be restored by initially establishing fewer than 30 tree species.

Coppicing tree stumps must be counted in the site survey along with saplings and larger trees.

Compare the site assessment results with the guidelines below to confirm the degradation level of your restoration site. Select the broad restoration strategy to match the recorded conditions and start to plan management tasks, including protective measures (e.g. livestock exclusion and/or fire prevention), the balance between tree planting and nurturing natural regeneration, the types of tree species to plant, the need for soil improvement and so on.

Stage-1 degradation

Survey results: The total number of regenerants averages more than 25 per circle, with more than 30 tree species (or roughly 10% of the estimated number of tree species in the target forest, if known) commonly represented across 10 circles, including several climax species. Saplings taller than 50 cm are common in all circles, with larger trees found in most. Small tree seedlings are common amongst the ground flora. Herbs and grasses cover less than 50% of circles and their average height is usually lower than that of the regenerants.

Strategy: Neither tree planting nor ANR are needed. Protection, i.e. preventing encroachment and any further disturbance to the site, should be sufficient to restore climax forest conditions fairly rapidly. The site survey and discussion with local people will determine if fire prevention, removal of livestock, and/or measures to prevent the hunting of seed-dispersing animals are necessary. If crucial seed-dispersing animals have been extirpated from the area, consider re-introducing them.

Stage-2 degradation

Survey results: The average number of regenerants remains higher than 25 per circle, with more than 30 tree species (or roughly 10% of the estimated number of tree species in the target forest, if known) represented across 10 circles, but pioneer tree species are more common than climax species. Saplings taller than 50 cm remain common in all circles, but larger trees are rare and the crown cover is insufficient to shade out weeds. Herbs and grasses therefore dominate, covering more than 50% of the circle areas on average, although small tree seedlings might still be represented amongst the ground flora. Herbs and grasses overtop the tree seedlings and often also the saplings and sprouts from tree stumps.

Strategy: Under these conditions, the protective measures described for stage-1 degradation must be complemented with additional measures to 'assist' natural regeneration in order to accelerate canopy closure. ANR is necessary to break the feedback loop whereby the high light levels, created by the open canopy, promote the growth of grasses and herbs, which discourages tree seed dispersers and makes the site vulnerable to burning. This in turn inhibits further tree establishment. ANR measures can include weeding, fertiliser application and/or mulching around natural regenerants. If several climax forest species do not naturally colonise the site after canopy closure has been achieved (because the nearest intact forest remnants are too far away, and/or seed dispersers have been extirpated), then enrichment planting may be necessary.

Stage-3 degradation

Survey results: The total number of regenerants falls below 25 per circle, with fewer than 30 tree species represented across 10 circles (or roughly 10% of the estimated number of trees in the target forest, if known). Climax tree species are absent or very rare. Tree seedlings are rarely found amongst the ground flora. Herbs and grasses dominate, covering more than 70% of the circle areas on average, and they usually grow taller than the few natural regenerants that may survive. Remnants of intact climax forest remain in the landscape within a few kilometres of the site and viable populations of seed-dispersing animals remain.

Strategy: Under these conditions, protection and ANR must be complemented with the planting of framework tree species. Prevention of encroachment and the exclusion of livestock (if present) remain necessary and fire prevention is important because of the abundance of highly flammable grasses. The ANR methods needed to repair stage-2 degradation must be applied to the few natural regenerants that remain, but in addition, the density of regenerants must be increased by planting framework tree species to shade out weeds and attract seed-dispersing animals.

The number of trees planted should be 3,100 per hectare minus the estimated number of natural regenerants per hectare (not counting small seedlings in the ground flora). The number of species planted across the whole site should be 30 (or roughly 10% of the estimated number of tree species in the target forest, if known), minus the total number of species recorded during the site assessment. For example, the site assessment data presented on p. 73 suggest that 433 trees per hectare of 12 species should be planted at this site. These trees should mostly be of climax species because the assessment shows that 18 pioneer species are already represented by surviving regenerants.

Framework tree species should be selected for planting using the criteria defined in **Section 5.3**. They might include both pioneer and climax species, but should be different species to those recorded during the site assessment. The planting of framework species recaptures the site from invasive grasses and herbs and re-establishes seed-dispersal mechanisms, thereby enhancing the re-colonisation of the restoration site by most of the other trees species that comprise the target climax forest ecosystem. If any important tree species fail to re-colonise the site, they can be re-introduced in subsequent enrichment plantings.

Stage-4 degradation

Survey results: The conditions recorded during the site assessment are similar to those of stage-3 degradation, but at the landscape level, intact forest no longer remains within 10 km of the site and/or seed-dispersing animals have become so rare that they are no longer able to bring the seeds of climax tree species into the site in sufficient quantities. Re-colonisation of the site by the vast majority of tree species is therefore impossible by natural means.

Strategy: Protective measures, ANR actions and the planting of framework tree species should all be carried out as for stage-3 degradation. These measures should be sufficient to re-establish basic forest structure and functioning, but with insufficient seed sources and seed dispersers in the landscape, the tree species composition can fully recover only when all of the absent tree species that characterise the target climax forest are manually established, either by planting and/or by direct seeding. This 'maximum diversity approach' (Goosem & Tucker, 1995; Lamb, 2011) is technically challenging and costly.

Stage-5 degradation

Survey results: The total number of regenerants falls below 2 per circle (average spacing between regenerants >6–7 m), with fewer than 3 tree species (or roughly 1% of the estimated number of tree species in the target forest, if known) represented across 10 circles. Climax tree species are absent. Bare earth is exposed over more than 30% of the circle areas on average and the soil is often compacted. Local people regard the soil conditions as exceedingly poor, and signs of erosion are recorded during the site assessment. Erosion gullies can be present, along with siltation of watercourses. The ground flora is limited by the poor soil conditions to less than 70% cover on average and is devoid of tree seedlings.

Strategy: Under such conditions, soil improvement is usually necessary before tree planting can commence. Soil conditions may be improved by ploughing, adding fertiliser and/or by green mulching (e.g. establishing a crop of leguminous herbs to add organic matter and nutrients to the soil). Additional soil improvement techniques can be applied during tree planting, such as the addition of compost, water absorbent polymers and/or mycorrhizal inocula to planting holes and mulching around planted trees (see **Section 5.5**).

Further improvements to site conditions might be achieved by first planting 'nurse' trees (Lamb, 2011): tree species that are tolerant of the harsh soil conditions, but which are also capable of improving the soil. These should gradually be thinned out as the site conditions improve; a wider range of native forest tree species should be planted in their place. To achieve full biodiversity recovery, the maximum diversity approach must be used in most cases, but where forest and seed dispersers remain in the landscape, planting a smaller range of framework tree species might suffice. This is known as the "plantations as catalysts" or "foster ecosystem" approach (Parrotta, 2000).

Nurse trees can be specialist framework species that are capable of growing in very poor site conditions, particularly nitrogen-fixing trees of the family Leguminosae. Plantations of commercial tree species have sometimes been used as nurse crops because their thinning generates early revenue that can help to pay for this expensive process. Protective measures, such as fire and encroachment prevention and the exclusion of cattle, all remain essential throughout the lengthy process to protect the substantial investment required to repair stage-5 degradation.

Due to the very high costs involved, forest restoration is rarely carried out on sites with stage-5 degradation. An exception is where wealthy companies are legally required to rehabilitate open caste mines.

Rehabilitation of an open-cast lignite mine in northern Thailand. Usually, only wealthy companies can afford the high costs of forest restoration on stage-5 sites.

Box 3.1. Origins of the framework-species method.

The framework species method of forest restoration originated in the Wet Tropics of Queensland, in Australia's tropical zone. Nearly 1 million hectares of tropical forest remain (some in fragments) in this region and the restoration of rain forest ecosystems to degraded areas began in the early 1980s, shortly before the region was collectively declared a UNESCO World Heritage Area in 1988. The challenging task of restoration was the responsibility of the Queensland Parks and Wildlife Service (QPWS) and much of the work was delegated to QPWS officer Nigel Tucker and his small team, based at Lake Eacham National Park. There, the team set up a tree nursery to grow many of the area's native rain forest tree species.

Nigel Tucker points to dense undergrowth, 27 years after restoration work at Eubenangee Swamp.

One of the early restoration trials began in 1983 at Eubenangee Swamp National Park on the coastal plain. This swamp forest area had been degraded by logging, clearing and agriculture, which had disrupted the water flow needed to maintain the swamp. The project aimed to restore the riparian vegetation along the stream that feeds into the swamp. A mix of native rain forest tree species was planted, including *Homalanthus novoguineensis*, *Nauclea orientalis*, *Terminalia sericocarpa* and *Cardwellia sublimis*. The seedlings were planted among grasses and herbaceous weeds (without weeding for site preparation) and fertiliser was applied. After 3 years, the initial results were disappointing. Canopy closure had not been achieved and the density of naturally established

Restored forest, fringing Eubenangee Swamp, now blends imperceptibly with natural forest.

Box 3.1. Origins of the framework-species method

Box 3.1. continued.

Homalanthus novoguineensis, one of the first recognised framework species.

seedlings was lower than hoped for. However, the experiment resulted in the crucial observation that natural regeneration occurred under certain tree species far more than under others. Species that fostered most natural regeneration were often fast-growing pioneers with fleshy fruits, and top of the list was the bleeding heart tree (*Homalanthus novoguineensis*).

From those early observations at Eubenangee Swamp, the idea of selecting tree species to attract seed-dispersing wildlife became established. This, along with recognising the need for more intensive site preparation and weed control, developed into the framework species method of forest restoration. Today, more than 160 of Queensland's rain forest tree species are recognised as framework tree species. The term first appeared in a booklet, '*Repairing the Rainforest*'[6] published by the Wet Tropics Management Authority in 1995, which Nigel Tucker co-authored with QPWS colleague Steve Goosem. The concept recognises that where remnant trees and seed-dispersing wildlife remain (i.e. stages of degradation 1–3), planting relatively few tree species, which are selected to enhance natural seed dispersal mechanisms and re-establish basic forest structure, is enough to 'kick start' forest succession towards the climax forest ecosystem, with a minimum of subsequent management inputs. Now, more than 20 years after its inception, the framework species approach is widely accepted as one of the standard methods of tropical forest restoration. It has been adapted for restoring other forest types, well beyond the borders of Queensland.

Forest restoration at Eubenangee Swamp created habitat for thousands of wildlife species, including this 4 o'clock moth caterpillar.

By Sutthathorn Chairuangsri

[6] www.wettropics.gov.au/media/med_landholders.html

CASE STUDY 2   Littoral forest restoration in south-eastern Madagascar

Country: Madagascar.

Forest type: Humid littoral forest, nutrient-poor sandy soil.

Nature of ownership: State-owned land with a long-term lease for ilmenite mining.

Management and community use: The landscape of open bush land, fragmented degraded forests, wetlands and protected forest was co-managed by the community, the government and QIT Madagascar Minerals (QMM). Years of harvesting and non-sustainable management for construction, firewood, and charcoal have led to the current landscape. Uses are now regulated by a Dina, a credible social contract for natural resources management.

Level of degradation: Open degraded bush land with residual fragments of highly degraded forests.

Background

The study area, close to the Mandena QMM mine site, lies in the south-eastern region of Madagascar near Tolagnaro (Fort Dauphin). QMM is 80% owned by the international mining group Rio Tinto and 20% owned by the Government of Madagascar, and will exploit the mineral sands in the Anosy region over the next 40 years. One of the world's most important biodiversity hotspots, Madagascar continues to experience environmental trauma. The restoration of natural forests has become an important issue in Madagascar's forestry and conservation activities. There are a few initiatives in which native trees have been being planted as buffer zones around natural forests or as corridors to provide continuity of forest habitats. There have also been a few attempts to restore natural forests after exploitation or complete destruction, but the work and knowledge in this field is still very preliminary. One of the commitments within QMM's Environmental Management Plan, which must be carried out under the terms of its

Location of the study area.

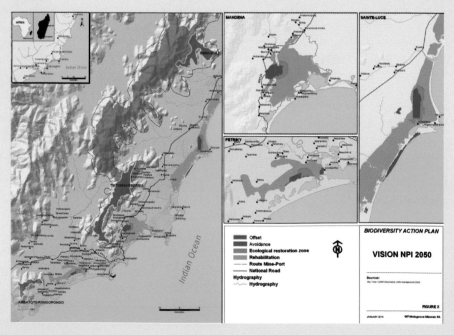

Tree species included in the study, according to their assigned category.

Characteristics	Sun-loving	Pioneer	Intermediate	Climax and shade-loving
Characteristics	• True forest species • Require sunlight	• Full sunlight needed for optimal growth	• Neither sun-loving nor shade-loving • Mediocre germination rate under unshaded tree-nursery conditions	• Shade for optimal growth
Species	*Canarium bullatum* inedit (Burseraceae), *Eugenia cloiselii* (Myrtaceae), and *Rhopalocarpus coriaceus* (Sphaerosepalaceae)	*Vernoniopsis caudata* (Asteraceae), *Gomphia obtusifolium* (Ochnaceae), *Dodonaea viscosa* (Sapindaceae), *Aphloia theiformis* (Aphloiaceae), *Scutia myrtina* (Rhamnaceae), and *Cerbera manghas* (Apocynaceae)	*Tambourissa castri-delphini* (Monimiaceae), *Vepris elliottii* (Rutaceae), *Dracaena reflexa* var. *bakeri* (Asparagaceae), *Psorospermum revolutum* (Hypericaceae), *Eugenia* sp. (Myrtaceae), and *Ophiocolea delphinensis* (Bignoniaceae)	*Dypsis prestoniana* and *D. lutescens* (Arecaceae), *Pandanus dauphinensis* (Pandanaceae), *Podocarpus madagascariensis* (Podocarpaceae), *Diospyros gracilipes* (Ebenaceae), *Apodytes bebile* (Icacinaceae), and *Dombeya mandenensis* (Malvaceae)

mining permit, is the restoration of natural forests and wetlands after mining. The plan is to double the surface area of the existing conservation zone at Mandena by restoring approximately 200 ha of natural forests and 350 ha of wetlands after mining. Trials have been underway for the past 15 years.

Investigating species selection

This case study summarises 10 years of restoration experiments. During the first round of qualitative data collection, the growth characteristics of the saplings of several species of littoral forest trees growing in a tree nursery were observed and described qualitatively. The objective of the first stage of the plantation program was to install vegetation that could serve as a starting point for a natural or facilitated succession towards restoration of the desired forest components. The tree species were categorised according to their tolerance of sun exposure, high evaporation and poor soil conditions, and their capacity to develop an extensive and dense root system rapidly. Ninety-two species of native trees were examined and assigned as sun-loving, pioneer, intermediate or late-successional (climax or shade-loving) species.

Exploring impact factors for restoration

Trials were conducted to test the effects of various factors on tree growth and survival rates:

1. The effects of the extent of demineralisation on restoration and succession were examined in an experiment in which post-mining soil conditions were simulated. Plants were grown in soils that were demineralised to one of three levels: a) large-scale demineralisation to a depth of 2 m (mimicking the mining process), b) simulated demineralisation (mimicking the removal of humus after exploitation) or c) no demineralisation.

2. The effects of adding topsoil were tested in an experiment in which saplings were planted in topsoil that had been either a) added to continuously cover the plantation area to a depth of 20 cm or b) added only to the hole in which the sapling was planted.

3. A further study looked at the effects of distance to natural forests as a source for regeneration.

4. Native species with or without exotic species (including *Eucalyptus robusta* and *Acacia mangium*) were planted as shade trees in an attempt to promote succession.

5. In accordance to the results of studies of forest succession, forest tree species were assigned into one of three classes: pioneer (sun-loving), intermediate and late successional (climax or shade-loving).

6. The influences of ectomycorrhiza, nitrogen fixation and unknown microbial associations upon succession were also considered.

Lessons learned

The demineralisation of sandy soils as during mining (i.e. the removal of the heavy minerals, such as ilmenite ($FeTiO_2$) and zircon) did not have any measurable effect on the survival rates of trees. These minerals are stable and do not seem to be taken up by plants, which need the ions to be in water solution for assimilation. Several trees that were planted on demineralised soils produced flowers and fruits; hence, demineralisation did not seem to affect the reproductive state of the plants.

Native trees that were planted in combination with the exotic species *Eucalyptus robusta* and *Acacia mangium* had a very low survival rate or were totally overtaken by the exotic species. Within five years, the exotic species reached heights of at least 5 m. Only a few shade-tolerant species such as *Apodytes bebile*, *Astrotrichilia elliotii* and *Poupartia chapelieri* survived under these conditions. It is unclear, however, whether the low survival rate of the native species is due to competition for light or to allelochemical interactions with products from the exotic trees. In the experimental plots without exotic tree species, native species from the sun-loving/pioneer and intermediate classes survived well. These plants will probably be important for the first stage of the restoration of native littoral forest after mining.

Saplings that were close to the edge of the natural forest grew more rapidly than those further away from the forest edge. In addition, trees growing in small isolated patches of forest (i.e. groups of trees growing in an open landscape) were generally much smaller than those in the larger blocks of forest. These observations support the idea that restoration activities should commence by enlarging existing forest blocks rather than by starting with isolated plantations.

The addition of topsoil has a major impact on the growth of the saplings. Saplings planted with a 20 cm layer of topsoil concentrated around them grew at the same rate as those planted in an area covered with a continuous 10 cm layer of topsoil. In Mandena, the supply of topsoil has become a significant management issue because most of the natural forests outside the conservation zone were destroyed. It is therefore important that the remaining topsoil is used as efficiently as possible.

The idea of using exotic trees to provide shade and a suitable microclimate for the saplings of native trees needs to be abandoned. Competition for light and growth-enabling components make exotic species unsuitable pioneer plants for the restoration of native littoral forests.

One further consideration that is of paramount importance for the growth and survival of trees is the ubiquitous association of trees with nitrogen-fixing bacteria and mycorrhiza. Specific fungi can penetrate the root cells of their symbiotic partners forming endomycorrhiza, or remain in close association with the roots without penetrating the cell, forming ectomycorrhiza. Ectomycorrhiza form mixed mycelium-root structures that effectively increase the resorptive surface of the tree and facilitate the uptake of nutrients. Furthermore, the fungi also seem to be able to mobilise essential plant nutrients directly from minerals. This could be important for forest restoration as it might enable ectomycorrhizal plants to extract essential nutrients from insoluble mineral sources through the excretion of organic acids.

Ectomycorrhizal symbioses are known for less than 5% of terrestrial plant species and are more common in temperate zones than in the tropics. Follow-up research on the ectomycorrhizal associations of the Sarcolaenaceae, a tree family that is endemic to Madagascar and has eight species that occur in the littoral forest, is recommended. Whether ectomycorrhizal associations provide an advantage over the formation of endomycorrhiza for plants growing on nutrient-poor sand remains to be studied in more detail. The importance of either form of mycorrhiza to the tree species of the littoral forests of south-eastern Madagascar is unknown. It has been observed, however, that saplings planted on demineralised soil hardly grew for several years and then suddenly increased in height. This might indicate that the plants first had to acquire their mycorrhizal fungi or nitrogen-fixing bacteria before they could start growing. Species with ectomycorrhiza or nitrogen-fixing bacteria seem to have enhanced growth on demineralised soil, growing about three times faster than other species. Under normal conditions, this advantage was not as obvious. Thus, knowledge of microbial symbioses and their species specificities could facilitate forest restoration programmes.

By Johny Rabenantoandro

Chapter 4

Planning forest restoration

The planning of forest restoration is a lengthy and complex process involving many stakeholders, who often have contradictory opinions about where, when and how the restoration project should be implemented. The project must be supported by local people and the relevant authorities, and issues of land tenure and benefit-sharing must be sorted out. Where tree planting is needed, seeds of the required tree species must be found, nurseries constructed and trees grown to a suitable size in time for the optimum planting season. If starting from scratch, these preparations will take 1–2 years, so it is important to begin the planning process well in advance.

As the need to solve environmental problems becomes ever more urgent, funders often demand to see results on the ground within one to three years. This pressure can lead to hurried and largely unplanned restoration projects, which often result in the wrong tree species being planted in the wrong places at the wrong time of year. Project failure then discourages both stakeholders and funders from becoming involved in further restoration projects. Advanced planning is therefore essential for success.

The technical challenges that must be overcome by the project plan are decided by undertaking a site assessment and by recognising the degradation level (see Chapter 3). In this chapter, we discuss the 'who', 'what', 'where' and 'how' of project planning. Specifically, we discuss how to involve stakeholders, how to clarify the project's objectives, how to fit forest restoration into human-dominated landscapes, the timing of management activities, and finally how to combine all of these considerations into a coherent project plan.

4.1 Who are the stakeholders?

Stakeholders are individuals or groups of people who have any kind of interest in the landscape in which the proposed restoration will take place, as well as those who may be affected by the wider consequences of restoration, such as water-users downstream. They may also include those who could influence the long-term success of the restoration project, such as technical advisors, local and international conservation organisations, funders and government officials. Stakeholders should represent all those who may benefit from the full range of benefits offered by the forest (see **Section 1.3**), as well as those likely to be disadvantaged by continued degradation (see **Section 1.1**).

It is essential that all stakeholders have the opportunity and are encouraged to participate fully in negotiations at all stages of project planning, implementation and monitoring (see **Section 4.3**). Different opinions about the eventual use of the restored forest, and whose interests will be served by it, will inevitably arise. Stakeholders might also disagree about which restoration methods will be most successful. When the benefits of forest restoration are poorly understood, some stakeholders might favour traditional plantation forestry (i.e. the planting of monocultures, often of exotic species) but, by allowing all views to be heard, the case for conservation can be clearly communicated from the outset and common goals can usually be found. Successful forest restoration often depends on resolving conflicts early in the planning process by holding regular stakeholder meetings, at which records are kept for future reference. The purpose of such meetings should be to reach a consensus on a project plan that clearly defines the responsibilities of each stakeholder group, thereby preventing confusion and replication of effort.

The strengths and weaknesses of each of the stakeholders must be recognised, so that a joint strategy can be devised while each stakeholder group is allowed to maintain its own identity. Once the capabilities of each stakeholder group have been identified, their roles can be defined and the allocation of tasks agreed upon.

This is often a tricky process, which may best be led by a facilitator. This is a neutral person or organisation who is familiar with the stakeholders but is not seen as authoritarian or gaining any benefit from involvement in the project. Their role is to ensure that all opinions are discussed, that everyone agrees with the aim of the project and that responsibility for the various tasks is accepted by those most able and willing to carry them out.

Success is most likely when all of the stakeholders are content with the benefits they might receive from the project and believe that their contribution is beneficial to the project's success. When everyone is satisfied that they have had input into project planning, a sense of 'community stewardship' is generated (even though this does not necessarily mean actual legal ownership of the land or trees). This helps to establish essential working relationships amongst the stakeholders that must be maintained throughout the project.

4.2 Defining the objectives

What is the aim?

Forest restoration directs and accelerates natural forest succession with the eventual aim of creating a self-sustained, climax forest ecosystem, i.e. the target forest ecosystem (see **Section 1.3**). So, a survey of an example of the target forest ecosystem is an important part of setting a project's objectives.

Locate remnants of the target forest ecosystem using topographic maps, Google Earth or by visiting viewpoints. Select one or more remnants as reference site(s). The reference site(s) should:

- have the same climax forest type as that to be restored;
- be one of the least-disturbed forest remnants in the vicinity;
- be located as close as possible to the restoration site(s);
- have similar conditions (e.g. elevation, slope, aspect etc.) as those of the proposed restoration site(s);
- be accessible for survey and/or seed collection etc.

Invite all stakeholders to join in a survey of the reference site(s). Before the survey, prepare metal labels and 5 cm galvanised zinc nails with which to tag the trees. To make labels, cut the top and bottom off drinks cans, slice open the cans and cut 6–8 square labels from the soft aluminium of each can. Lay the labels on a soft surface and use a metal stylus to indent sequential numbers into the metal (inside surface), then trace over the engraved numbers with an indelible pen.

Walk slowly along trails through the remnant forest and label mature trees growing within 5 m left or right of the trail. Tag the trees with the numbered metal labels in the order in which they are encountered, 1, 2, 3, 4, ... etc. Place the top edge of the labels exactly 1.3 m above ground level and nail them into place. Hammer in the nails only half way, because as the trees grow they will expand along the exposed half of the nails. Measure the girth of each tree 1.3 m above the ground and record the local names of the tree species. Collect leaf, flower and fruit specimens (where available) for formal identification. Continue until about five individuals of each tree species have been recorded. Take plenty of photos to illustrate the structure and composition of the target forest ecosystem and record any observations or signs of wildlife.

Use the opportunity to discuss with stakeholders:

- the history of the forest remnant and why it has survived;
- any uses of the tree species recorded;
- the value of the forest for non-timber products, watershed protection etc.;
- wildlife they have seen in the area.

After the survey, take the tree specimens to a botanist to obtain scientific names. Then use a flora or web search to determine the successional status of the species identified (pioneer or climax trees), the typical flowering and fruiting times of the species and their seed dispersal mechanisms. This information will be useful for planning species selection and seed collection later.

Select nearby remnants of the target forest ecosystem as reference sites and survey the plants and wildlife within them to help set project objectives.

The reference site can then be used for seed collection (see **Section 6.2**) and, if included in the project (see **Section 6.6**), for studies of tree phenology. Most importantly, it becomes a bench mark against which progress and the ultimate success of forest restoration can be measured.

Aiming at a moving target?

We have already stated that the target of forest restoration should be the eventual re-establishment of the climax forest ecosystem, i.e. forest with the maximum biomass, structural complexity and species diversity that can be supported by the soil conditions and prevailing climate. Since the climax forest type depends on the climate, global climate change might mean that the climax forest type for a particular site at some time in the future might differ from that best-suited to the site in the present climate (see **Section 2.3**). The problem is that we don't know how far global climate change could proceed before measures to halt it become effective, especially as (at the time of writing) international negotiations to implement such measures are stalled. With such uncertainty, it becomes impossible to know exactly what the future climate will be at any particular site and consequently which climax forest type to aim for. It is therefore possible that at least some of the tree species selected from today's remnants of climax forest might not be suitable for tomorrow's climate. Some species might be tolerant of climate change, but some may not be. So in addition to aiming for ecological richness, forest restoration should also seek to establish forest ecosystems that are capable of adapting to future climate changes.

Increasing ecological adaptability

The keys to securing the adaptability of tropical forest ecosystems to a changing global climate are i) diversity (both species and genetic diversity) and ii) mobility.

Tree species vary considerably in their responses to temperature and soil moisture. Some can tolerate large fluctuations in conditions (and are said to have a 'wide niche'), whereas others die when conditions waver even slightly from optimal ('narrow niche'). The more tree species are present at the start of restoration, the more likely it is that at least some of them will be suited to the future climate, whatever it turns out to be. So, in any restoration project, try to increase tree species diversity early in the succession as much as possible.

Genetic diversity within tree species is also important. Responses to climate change among individual trees within a species can also vary. So maintaining high genetic diversity within species can increase the probability that at least some individuals will survive to represent the species in the future forest. These genetic variants will then be able to pass on the genes that enable survival in a warmer world to their offspring. Until recently, it has been recommended that seeds should be collected from trees growing as close as possible to the restoration site (because they are genetically adapted to local conditions and they maintain genetic integrity). Now, the idea of including at least some seeds from the warmer limits of a species' distribution is being considered in order to broaden the genetic base from which genetic variants that are suited to a future unknown climate might emerge through natural selection (see **Box 6.1**, p. 159). The warmer limits of a species distribution would typically include the southern-most populations of species in the northern hemisphere, the northern-most populations of species in the southern hemisphere and the lower elevation limit of montane species.

Trees cannot 'run away' from climate change, but their seeds can (see **Section 2.2**). So, any actions that facilitate seed dispersal across landscapes will increase the probability that more tree species will survive. The mobility of seeds across landscapes can be maximised by planting framework tree species, as they are specially selected for their attractiveness to seed-dispersing wildlife. Tree species that have large seeds, particularly those that would have depended on extirpated large animals (e.g. elephants or rhinos) for their dispersal, should also be targeted for planting. Without their seed dispersers, human intervention to move their seeds (or seedlings) might be their only remaining chance of dispersal. Campaigns to prevent the hunting of seed-dispersing animals are obviously important in this regard (see **Section 5.1**). Increasing forest connectivity at the landscape level also facilitates seed dispersal because many seed-dispersing animal species are reluctant to cross over large open areas. This can be achieved by restoring forest in the form of corridors and 'stepping stones' (see **Section 4.4**).

It is fanciful to suppose that something as dynamic and variable as a tropical forest can be 'climate proofed', but some of the measures suggested above might at least help to secure the long-term future of some form of tropical forest ecosystem at today's restoration sites.

4.3 Fitting forests into landscapes

Today, no forest restoration project is carried out in isolation. Forest destruction is a feature of human-dominated landscapes, and consequently, restoration is always implemented within a matrix of other land uses. Therefore, considering the effects of restoration projects on the character of the landscape, and *vice versa*, is often one of the first considerations when putting together a restoration project plan (see Chapter 11 of Lamb, 2011). Consideration of the whole landscape in restoration planning has now been formalised within the framework of forest landscape restoration (FLR).

Forest landscape restoration

Forest landscape restoration is "a planned process, which aims to regain ecological integrity and enhance human well-being in deforested or degraded landscapes"[1] (Rietbergen-McCracken *et al.*, 2007). It provides procedures whereby site-level restoration decisions conform to landscape-level objectives.

The goal of FLR is a compromise between meeting the needs of humans and wildlife, by restoring a range of forest functions at the landscape level. It aims to strengthen the resilience and ecological integrity of landscapes and hence to keep future management options open. Local communities play a crucial role in shaping the landscape, and they gain significant benefits from restored forest resources, so their participation is central to the process. Therefore, FLR is an inclusive, participatory process.

FLR combines several of the existing principles and techniques of development, conservation and natural resource management, such as landscape character assessment, participatory rural appraisal and adaptive management, within a clear and consistent evaluation and learning framework. ANR and tree planting are just two of many forestry practices that might be implemented as part of an FLR program. Others include the protection and management of secondary and degraded primary forests, agro-forestry and even conventional tree plantations.

The achievements of FLR can include:
- identification of the root causes of forest degradation and prevention of further deforestation;
- positive engagement of stakeholders in the planning of forest restoration, resolution of land-use conflicts and agreement on benefit-sharing systems;
- compromises and land-use trade-offs that are acceptable to all stakeholders;
- a repository of biological diversity of both local and global value;
- delivery of a range of utilitarian benefits to local communities including —
    - a reliable supply of clean water;
    - a sustainable supply of a diverse range of foods, medicines and other forest products;
    - income from ecotourism, carbon trading and from payments for other environmental services;
    - environmental protection (e.g. flood or drought mitigation and the control of soil erosion).

[1] A forested landscape is considered to be degraded when it is no longer able to maintain an adequate supply of forest products or ecological services for human well-being, ecosystem functioning and biodiversity conservation. Degradation can include declining biodiversity, water quality, soil fertility and supplies of forest products as well as increased carbon dioxide emissions.

The concept of FLR is the result of collaboration among the world's leading conservation organisations including The World Conservation Union (IUCN), the World Wide Fund for Nature (WWF) and the International Tropical Timber Organization (ITTO); several comprehensive text books about the concept have recently been published (e.g. Rietbergen-McCracken *et al.*, 2007; Mansourian *et al.*, 2005; Lamb, 2011).

Landscape character

A landscape character assessment is often the first step in an FLR initiative. Landscape character is the combination of landscape elements (e.g. geology, land form, land cover, human influence, climate and history) that defines the unique local identity of a landscape. It results from interactions between physical and natural factors, such as geology, landform, soils and ecosystems, and social and cultural factors, such as land use and settlement. It identifies the distinctive features of the landscape and guides decisions about where forest can be restored in a positive and sustainable way that is relevant to all stakeholders.

Assessment of landscape character

Landscape character assessment is essentially a participatory mapping exercise carried out with the aim of reaching consensus on where forest can be restored while the landscape characteristics that stakeholders consider desirable are conserved or enhanced.

It begins with a review of existing information about the area, including its geology, topography, climate, distribution of forest types, plant and animal diversity, previous conservation or development projects, human population and socio-economic conditions. This information may be gleaned from maps (especially those showing forest cover), published research papers and/or unpublished reports. Such documents might be obtained from government offices (particularly the local or national forest or conservation authority, the meteorological office and the social welfare department), any NGO's that have worked in the area, and any universities that have done research. A considerable amount of information is also available on-line. Google Earth is a useful source of information on areas with limited map accessibility.

The next step is to hold a series of stakeholders' meetings to combine information from the review with local knowledge and field observations. Local people, particularly the older generations, can offer invaluable information on landscape character, particularly if they have memories of the area prior to disturbance. They may be able to identify changes in forest products and ecological processes that have occurred as a result of degradation, such as reduced dry-season stream flow, and might have other knowledge that can help to prioritise certain land uses. The stakeholders should work together to build a map that identifies potential forest restoration sites within a matrix of other desirable land uses. The processes and skills required to run effective participatory appraisals is beyond the scope of this book, but decision-support tools, such as participatory mapping, scenario analysis, role-playing games and market-based instruments have all been well reviewed by Lamb (2011), and a comprehensive body of literature has emerged from practitioners of community forestry (e.g. Asia Forest Network, 2002; www.forestlandscaperestoration.org and www.cbd.int/ecosystem/sourcebook/tools/).

The landscape character assessment should identify i) desirable landscape characters that should be conserved, ii) problems with the current landscape management and iii) the potential benefits of restoration. Field trips should include participatory assessments of i) remnants of the target forest ecosystem if present (see **Section 4.2** above) and ii) potential restoration sites (see **Section 3.2**).

The main output of landscape character assessment is a map, showing current land-uses, desired landscape features that should be conserved and degraded sites that require restoration. The map may show several sites that are potentially suitable for restoration, so the next step is prioritisation. It may be tempting to restore the less-degraded areas first, because their restoration will cost less and is perceived as having a better chance of success, but this may not be the best option. Consider each of the following issues:

- the condition of each degraded site and the time and effort required to restore each of them;
- whether forest restoration could adversely impact an existing habitat of high conservation value (e.g. wetlands or natural grassland) on the site or in the vicinity;
- whether a restored site will contribute to the conservation of biodiversity in the wider landscape, by expanding the area of natural forest, by serving as a buffer, or by reducing forest fragmentation.

Forest fragmentation

Fragmentation is the sub-division of large forest areas into ever-shrinking pieces. It occurs when large, continuous areas of forest become dissected by roads, cultivated land and so on. Small, disconnected forest patches can shrink even further because of edge effects: damaging factors that penetrate a forest fragment from the outside. These might include light that promotes weed growth, hot air that desiccates young tree seedlings, or domestic cats that prey on nesting birds. Small fragments are more vulnerable to edge effects than large ones, because the smaller the fragment, the greater is the edge to total area ratio.

A well-known example of fragmentation is the result of road construction in Brazilian Amazonia. The roads, often constructed to facilitate oil and gas exploration, allowed loggers, illegal hunters and cattle ranchers to follow. The resulting forest fragments are prone to edge effects, which can impact ecological processes over a perimeter area of at least 200 m in depth (Bennett, 2003). If such fragmentation continues, much of the Amazon could be converted to fire-prone scrub vegetation (Nepstad et al., 2001).

Fragmentation has important implications for wildlife conservation because many species require a certain minimum area of continuous habitat in order to maintain viable populations. Often, these species cannot disperse across inhospitable farmland, roads or other barriers of 'non-habitat'. Few forest animal species can traverse large non-forested areas (the exceptions being some birds, bats and other small mammals). Up to 20% of the bird species found within tropical forests are unable to cross gaps of more than a few hundred metres (Newmark, 1993; Stouffer & Bierregaard, 1995). This means that large animal-dispersed seeds are rarely transported between forest fragments.

DISSECTION

Roads, railways, power lines etc. cut into a large expanse of forest..

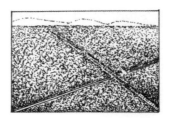

PERFORATION

Holes develop in the forest as settlers exploit the land along the lines of communication.

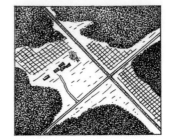

FRAGMENTATION

The gaps become larger than remaining forest.

ATTRITION

Isolated forest remnants are gradually eroded by edge effects.

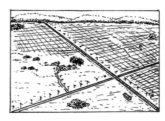

Tiny forest fragments can support only very small populations of animals, which are highly vulnerable to extirpation. Once gone, species cannot return because migration between forest patches is hindered by vast areas of agricultural land or dangerous barriers such as roads. Restoring wildlife corridors to re-link forest fragments can overcome some of these problems and help to create viable wildlife populations in a fragmented landscape.

The resulting small, isolated plant and animal populations are easily wiped out by hunting, diseases, droughts and fires, which would not normally eliminate larger, more resilient, populations in larger forest areas. Genetic isolation and inbreeding further increase the risk of extirpation. In one fragment after another, small species populations disappear and cannot be re-established by migration, so that eventually the species becomes extirpated across the whole landscape (see **Section 1.1**). Re-colonisation is made impossible because inhospitable terrain (such as agricultural or urban land) between forest fragments blocks the dispersal of potential new founder individuals of extirpated species.

4.4 Choosing sites for restoration

Forest restoration can be relatively costly in the short term (although it is more cost-effective than allowing degradation to be continued), so it makes sense to implement it first where it will generate maximum ecological benefits, such as protecting watercourses, preventing soil erosion and reversing fragmentation.

How can fragmentation be reversed?

Small fragments of forest that are re-linked have a greater conservation value than those that are left isolated (Diamond, 1975). Forest restoration can be used to establish 'wildlife corridors' that reconnect forest fragments. They provide wildlife with the security needed to move from one forest patch to another. Genetic mixing recommences and, if a species population is extirpated from one forest patch, it can be re-founded by immigration of individuals along the corridor from another forest patch. Wildlife corridors can also help to re-establish natural migration routes, particularly for species that migrate up and down mountains.

The concept of wildlife corridors is not without controversy. For example, the corridors could be become 'shooting galleries', encouraging wildlife out from the safety of conservation areas and making them easy targets for hunters. Corridors might also facilitate the spread of diseases or fire. Early corridors were created with little guidance as to their location, design and management (Bennett, 2003), but there is growing evidence to suggest that the benefits of corridors outweigh the potential disadvantages. In Costa Rica, for example, riparian corridors have successfully connected fragmented bird populations (Sekercioglu, 2009), and in Australia, it was recently confirmed that genetic mixing among small mammals can be re-established by linking forest patches by even narrow corridors (Tucker & Simmons, 2009; Paetkau *et al.*, 2009) (see **Box 4.1**, p. 96). Also in Australia, linear forest remnants of 30–40 m wide have been found to support the movement of most arboreal mammals, although the quality of the forest is very important (Laurance & Laurance, 1999).

How wide should a corridor be?

The wider the corridor, the more species will use it. Bennett (2003) recommended that corridors should be 400–600 m wide so that the core vegetation is buffered against edge effects and thus animals and plants of the forest interior are attracted. Nevertheless, the Australian example (see **Box 4.1**) shows that corridors as narrow as 100 m can effectively reverse genetic isolation, provided they are well-designed to minimise edge effects. Corridors of this width can be used by small to medium-sized mammals and forest floor birds, which cannot cross open land (Newmark, 1991). Large vertebrate herbivores are more likely to use corridors that are wider than 1 km, whereas large mammalian predators prefer even wider corridors (of 5–10 km in width). A reasonable strategy is to start by restoring a narrow forest corridor and then gradually widen the corridor each year by planting more trees while keeping records of the species observed travelling along it.

Box 4.1. Framework species for creating corridors.

The Atherton Tablelands in Queensland, Australia, was once covered in upland rain forest, providing habitat for a huge diversity of plant and animal species. Among these, the spectacular southern cassowary (*Casuarius casuarius johnsonii*), a large flightless bird, is a major seed disperser within these forests and now a critically endangered species. European settlers were first attracted to the area in the 1880s by the opportunities for logging and subsequently land was cleared for livestock and crop production. By the 1980s only a few fragments of the original rain forest remained in parts of the Atherton Tablelands, and these contained small genetically isolated wildlife populations each heading towards an uncertain future.

Wildlife corridors were planned to reconnect isolated forest fragments and to monitor wildlife migration through these new linkages. Donaghy's Corridor was the first such linkage, intended to link the isolated Lake Barrine National Park (491 ha) to the much larger Gadgarra State Forest block (80,000 ha). The corridor was established by planting framework trees species in a belt 100 m wide along the banks of Toohey Creek, which meandered for 1.2 km through grazing lands. With its emphasis on enhancing seed dispersal from nearby forest, the framework species method was the obvious choice for creating such a corridor.

Agreement was reached with the farm owners, by incorporating their needs into the project; for example, by providing watering points and shade trees for cattle. The Queensland Parks and Wildlife team at the Lake Eacham National Park tree nursery formed a partnership with a community group, TREAT (Trees for the Evelyn and Atherton Tablelands), that would grow and plant over 20,000 trees between 1995 and 1998. In addition to cattle management, other key design points included planting windbreaks to minimise edge effects, a rigorous maintenance program (including weeding and fertiliser application) and long-term monitoring of plant and animal colonisation.

Trees planted to establish Donaghy's corridor, February 1997.

BOX 4.1. FRAMEWORK SPECIES FOR CREATING CORRIDORS

Box 4.1. continued.

The same area in February 2010.

Recovery of the vegetation along the habitat linkage was rapid, with 119 plant species colonising transects within the corridor after 3 years. Several planted tree species fruited very quickly after planting; for example, *Ficus congesta* produced figs after 6–12 months. Several studies, using mark-recapture and genetic analysis, showed that the corridor did indeed promote the migration of wildlife and re-established genetic mixing (Tucker & Simmons, 2009; Paetkau *et al.*, 2009), providing a more secure basis for longer-term population viability.

The involvement of the community group, right from the start, has resulted in widespread interest in both the framework species method and in habitat linkages. Several other linkages are now being restored throughout the region and beyond, some of them many kilometres long.

One of the most difficult aspects of creating long corridors across private land is securing the collaboration of all of the landowners along the route. But according to Nigel Tucker (see **Box 3.1**, p. 80), it may not be necessary to get everyone on board before the project starts. "We work first with the landowners who agree. The other landowners are won over later, when they see their neighbours benefiting from the corridor. It's all about building relationships and securing collaboration with a handshake — more important than formal contracts".

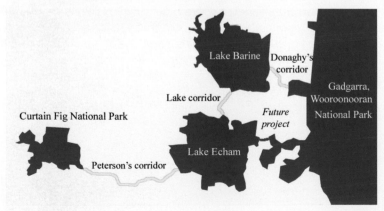

This well-studied demonstration site proved that corridors support the conservation of biodiversity. Now, several corridors link forest fragments across the Atherton Tablelands.

By Kwankhao Sinhaseni

Where should corridors be created?

Not all forest fragments have equal ecological value. Large fragments and those that have most recently become isolated from larger forest areas retain more biodiversity than smaller and older fragments. So, forest corridors that reconnect large and recently formed forest fragments have greater ecological value than those that reconnect smaller, older fragments. If fragments are known to retain populations of endangered species, their reconnection with large forest patches should also receive high priority (Lamb, 2011).

What about stepping stones?

There may be insufficient funds to link all forest fragments with continuous corridors, and in this situation 'stepping stones' might be more achievable. Stepping stones are islands of restored forest, created primarily to facilitate the movement of wildlife through hostile landscapes such as farmland. Stepping-stone habitats might also enhance natural regeneration in surrounding degraded areas by encouraging visits from seed-dispersers, which might deposit seeds from remnant forest areas in which they had previously fed. Once the planted and naturally regenerating trees reach maturity, they will also become sources of seed in their own right, leading to continued forest regeneration both within and outside the boundaries of the 'stepping stone'.

Size and shape of 'stepping stones'

Any small-scale restored site can suffer the disadvantages of small forest fragments, so the design of 'stepping stones' is important. The shape of the restoration plot should have a minimal edge to area ratio. As a rough guide, try to make the length and width of 'stepping stones' approximately equal and do not plant trees in long, narrow plots, unless your objective is to establish a wildlife corridor. A buffer zone of dense, fruiting shrubs and small trees should be planted around the edge of the restoration site to act as a wind break and to reduce edge effects still further. The rest of the 'stepping stone' can be planted with framework tree species to re-establish forest structure and attract seed dispersers.

Generally speaking, large forest plots support more biodiversity recovery than small ones. Soule and Terborgh (1999) suggest that, ideally, rapidly increasing forest cover to 50% of the landscape minimises further loss of species. Nevertheless, small restoration plots can have significant positive benefits for biodiversity conservation, especially if they are well designed in terms of tree species composition, minimisation of edge effects (buffer zones) and increasing forest connectivity. Thus, the quality and positioning of restoration plots can help to compensate for their small size (p. 448 of Lamb, 2011).

Restoring large sites

The size of the plots that are restored each year will depend on the availability of land, funding, and labour for weeding and caring for the planted trees during the first two years after restoration work commences (see **Section 4.5**). Large sites will require large quantities of seed. Seed of the relatively low number of framework species can be acquired by carefully planned advance collection and storage. But where the maximum diversity approach is to be used on heavily degraded land (see **Section 3.1**), it may be

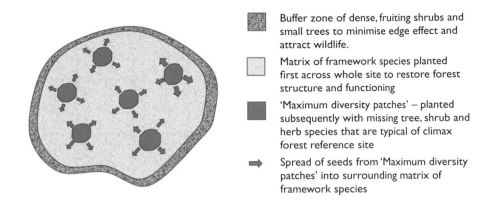

Buffer zone of dense, fruiting shrubs and small trees to minimise edge effect and attract wildlife.

Matrix of framework species planted first across whole site to restore forest structure and functioning

'Maximum diversity patches' – planted subsequently with missing tree, shrub and herb species that are typical of climax forest reference site

Spread of seeds from 'Maximum diversity patches' into surrounding matrix of framework species

Suggested plan for a large forest restoration site that is far away from nearest area of remnant forest. NB: Planted area is roughly circular in shape to minimise edge effects.

impossible to acquire sufficient seed to plant all the required species across the whole site. In such cases, an alternative approach is to plant the entire site with framework tree species so as to re-establish forest structure and attract seed dispersers, and then to create smaller 'maximum diversity patches' within the framework tree matrix using the 'maximum diversity' technique (see **Section 5.4**).

Restoration for water and soil conservation

The effects of deforestation and forest restoration on water and soil are explained in **Sections 1.2** and **1.4**. Both the regularity of water supplies and water quality can be improved by targeting upper watershed sites, particularly those around springs, for restoration. Although trees do remove water from the soil by transpiration, they more than make up for this by increasing the soil's water-holding capacity through the addition of organic matter, so that it can absorb more water during the rainy season and release it during dry periods. In this way, forest restoration can convert seasonally dry streams into permanently flowing ones, and can also help to reduce the amount of sediment in water supplies.

Planting along stream banks creates riparian habitats, which are essential for specialised species (from dragonflies to otters) that live in or beside sheltered streams. Such habitats also serve as essential refuges for many other, less specialised animal species during the dry season, when neighbouring habitats dry out or burn. Riparian tree-planting also prevents stream bank erosion and the clogging of stream channels with silt. This reduces the risk of streams bursting their banks leading to flash floods in the rainy season.

Soil erosion reduces the capacity of a water catchment to store water, which contributes to both floods in the rainy season and droughts in the dry season. Landslides may be considered to be the most extreme form of soil erosion. They can occur with such suddenness and force that they can completely destroy villages, infrastructure and agricultural land and can lead to loss of human life. Forest restoration can help reduce both soil erosion and the frequency and severity of landslides because tree roots bind the soil, preventing the movement of soil particles. Leaf litter also helps to improve soil structure and drainage. It increases the penetration of rainwater into the soil (infiltration) and reduces surface run-off.

For maximum conservation value, restore forest wildlife corridors to link forest patches and create permanent forest to reduce the risk of soil erosion or landslides and to protect water courses and their associated riparian wildlife.

To prevent soil erosion and landslides, restoration should be targeted at mountainous sites with long, steep, uninterrupted slopes. Erosion gullies and cleared sites with slopes exceeding 60% should be completely restored with dense vegetation (Turkelboom, 1999). Sites with more moderate slopes may be stabilised with less than 100% cover if the restoration plots are strategically placed to follow the slope's contours. Most countries have a national watershed classification system, with maps showing the relative risk of soil erosion in any particular area. Ask your local agricultural extension service to consult such maps to determine the extent to which forest restoration might help to reduce erosion in your locality.

Who owns the land?

When undertaking conservation activities, the last thing you want is a land dispute.

When restoring forest on public land, obtain written permission that includes a map to confirm the location of the site from the relevant authorities. Most authorities welcome help with forest restoration from community groups and NGOs, but obtaining written permission can take a long time, so start discussions at least a year before the intended planting date. Ensure that all relevant officials are fully involved in project planning. Everyone involved should understand that planting trees does not necessarily constitute a legal claim to the land, and local people will require assurance that they can access the site to implement restoration activities and/or to harvest forest products.

If planting on private land, make sure that the landowner (and his/her heirs) are fully committed to maintaining the area as forest by obtaining a memorandum of understanding or conservation agreement. Tree planting considerably increases the value of private property, so private landowners should fully cover the costs.

With the potential looming on the horizon for huge sums of money to be made by selling carbon credits under REDD+, part of the UN's Reducing Emissions from Deforestation and Forest Degradation (REDD) programme, the issue of 'who will own the carbon' has become almost as important as 'who owns the land'. Arguments over how the benefits from carbon trading will be shared amongst the various stakeholders can lead to project failure. If any of the stakeholders who contribute to the project are subsequently excluded from sharing carbon revenue, they may decide to burn the restored forest. It is therefore essential to resolve issues of ownership and/or access to land, carbon and other forest products with all stakeholders during the project-planning process.

4.5 Drafting a project plan

Once all of the stakeholders have contributed to the pre-planning activities, it is time for formal meetings to draft the project plan.

A project plan should include:
- the aim and objectives of the project;
- a clear statement of the expected benefits from the project and an agreement as to how these benefits will be shared amongst all stakeholders;
- a description of the site to be restored;
- the methods that will be used to restore forest to the site, including provisions for monitoring (and research);
- a task schedule, detailing who will be responsible for each task and calculation of the labour required to complete each task;
- a budget.

Aim and objectives

All activities depend on the project's aim and objectives. Outline the overall aim of the project (e.g. 'to secure water supplies', 'to conserve biodiversity' or 'to reduce poverty'), followed by more specific statements of the immediate project objectives (e.g. 'to restore 10 hectares of evergreen forest in location X to create a wildlife corridor between Y and Z'). The 'target forest' survey (see **Section 4.2**) will provide the detailed technical objectives, such as the forest type, structure and species composition, that the project aims to achieve.

Benefit-sharing agreement

List the full range of benefits from the project and how each benefit will be shared among the stakeholders. Once consensus is reached, all stakeholders should sign the agreement.

Table 4.1. Example of a benefit-sharing matrix.

Benefit	Protected area authority	Local villagers	Funder	NGO	University
Payments for project labour	30%	60%	0%	10%	0%
Non-timber forest products	0%	100%	0%	0%	0%
Water	50%	50%	0%	0%	0%
Ecotourism income	40%	50%	0%	10%	0%
Sale of carbon credits	30%	40%	10%	20%	0%
Research data	30%	0%	0%	10%	60%
Good publicity	20%	20%	20%	20%	20%

Where the benefits are monetary (e.g. income from carbon trading, income from ecotourism) the shares agreed in the project plan may serve as the basis for more formal legal contracts when that income is realised. A table like this serves to emphasise the range of different non-monetary benefits and their various values to the various stakeholder groups. For example 'good publicity' might result in an unquantified increase in revenue for a corporate sponsor, whereas to local villagers, it may serve to strengthen their right to remain living in a protected area or it could attract ecotourists.

When drafting the benefit-sharing agreement, it is also necessary to ensure that potential beneficiaries are aware of any legal restrictions to realising any of the benefits (e.g. laws that prohibit the collection of certain forest products), as well as any further investment that might be required before a benefit can be realised (e.g. investment in ecotourist infrastructure). Each stakeholder group can then decide for themselves how the project benefits will be shared amongst their members (e.g. how water is shared amongst downstream landowners).

Intangible benefits may be valued differently by different stakeholder groups. Good publicity might strengthen the right of ethnic minorities to live within a protected area, whereas for a corporate sponsor, it may attract new customers.

Site description

The restoration site survey report (see **Section 3.3**) provides all the details needed for the site description. It should be supplemented with annotated maps and/or satellite images and photographs. A sketch of how the landscape might appear after restoration is also useful.

Methods

The restoration site survey report also provides most of the information needed to determine the methods required to implement the restoration project. For example, it will help to determine which protective measures are required, the balance between tree planting and ANR, which ANR actions to implement, how many trees and which species should be planted and so on. Formally listing the methods that will be used in the project plan makes it easy to identify the actions required to implement them and thus to develop a task schedule. More details about the methods needed to implement the major forest restoration strategies are provided in **Chapter 5**.

Table 4.2. Example task schedule for the restoration of a seasonally dry tropical forest in a protected area by planting framework tree species in combination with ANR.

Task	When	Stakeholder with responsibility for organisation
Time before planting event		
Stakeholder consensus reached, survey target forest and potential restoration sites, start nursery establishment	18–24 months	Protected area authority
Draft project plan, final decision on restoration sites	18–24 months	Protected area authority
Start seed collection and germination	18 months	NGO and local community
Monitor sapling production, supplement with trees from other nurseries if necessary	6 months	NGO
Harden-off saplings, arrange planting teams	2 months	Local community
Label saplings to be monitored	1 month	Local community
Site preparation: identify and protect natural regenerants, clear site of weeds	1 month	Local community
Transport saplings and planting equipment to site, brief planting team leaders	1–7 days	Protected area authority
Planting event	0 days (early wet season)	Protected area authority
Time after planting event		
Check planting quality, adjust any badly planted saplings, remove rubbish from site	1–2 days	Local community
Collect baseline data on trees to be monitored	1–2 weeks	University researchers
Weeding and fertiliser application as required	During first wet season	Local community
Monitor growth and survival of planted trees	End of first wet season	University researchers
Cut fire breaks if necessary, organise fire patrols	Start of first dry season	Local community
Monitor the growth and survival of planted trees, weeding and fertiliser application as required, assess the need to replant any dead trees	End of dry season	University researchers
Maintenance planting as required	Start of second wet season	NGO
Continue weeding and fertiliser application as required	Second wet season	Local community
Monitor the growth and survival of planted trees	End of second wet season	University researchers
Continue weeding in wet season until canopy closure, monitor tree growth as necessary, monitor biodiversity recovery	Subsequent years	Local community

Task schedule

List the tasks needed to implement the methods chronologically and assign responsibility for organising each task to the stakeholder group that has the most suitable skills and resources (see **Table 4.2** for an example).

Note that a monitoring program is included in the schedule. Monitoring is an essential component of the project plan, important both to demonstrate project success (hopefully) and to identify mistakes and ways to avoid them in the future. It should involve assessments of both tree performance (both of planted trees and natural trees subjected to ANR treatments) and biodiversity recovery (see **Section 7.4**).

Underestimation of the total time required to implement forest restoration projects is a common mistake. If trees are grown locally from seed, nursery construction and seed collection must begin 18 months to 2 years before the first planned planting date.

Budget

Calculating labour requirements

The availability of labour is the crucial factor that determines the maximum area that can be restored each year. It is also likely to be the most costly item in the project budget, so calculation of labour requirements determines overall project viability.

Grand schemes, with ambitious aims to replant vast areas, often fail because they do not take into account the limited capacity of local stakeholders to carry out weeding and fire prevention. The effort required to produce very large numbers of saplings of the correct species is also commonly underestimated. It is therefore better to restore smaller areas (which can be adequately cared for by the locally available labour force) annually, over many years, than to plant trees over a large area in one high-profile event, only to see the planted trees subsequently die of neglect.

Where local villagers provide most of the labour for a forest restoration project, tasks can be organised as community activities. For example, a village committee might request that each family in the village provides one adult to work on each day that a scheduled task is to be carried out. The maximum area that can be restored each year, therefore, depends on the number of participating households. As community size increases, an 'economy of scale' comes into effect, meaning that a larger area can be planted with fewer days labour input per household.

At the outset of any forest restoration project, all stakeholders must be aware of the labour commitments. Project planners must also address the crucial issue of whether labour will be donated voluntarily or whether daily rates for casual labour must be paid. If the latter, then labour costs will dominate the budget. If local villagers appreciate the benefits of forest restoration and an equitable benefit-sharing scheme is included in the project plan, they are often willing to work on a voluntary basis to secure those benefits.

Table 4.3 outlines the labour requirements for some of the most common forest restoration tasks. Note that some tasks are required only during the first year of the project, whereas others must be repeated for up to 4 years after the first planting, depending on conditions.

Table 4.3. Checklist of the main labour requirements for the most common forest restoration tasks (for sites with stage 1–3 degradation (see Section 3.1)).

	Labour required (person days) per hectare per year	Explanation	Annual requirement (years 1 to 4)			
			Y1	Y2	Y3	Y4
PROTECTION						
Fires breaks	Fire break length (m) divided by 30 to 40	Assumes 1 person can cut 30–40 m of fire break (8 m wide) per day (depending on vegetation density). Calculate from the length of the perimeter of the restoration site.	+	+	+	?
Fire 'look outs' and suppression team	16 × no. of days in the fire season	Teams of 8 people working in two 12-hour shifts (day and night), throughout the hot dry season can take care of sites from 1–50 ha.	+	+	+	?
ANR						
Locating and marking regenerants	12	3,100 regenerants/ha ÷ 250 (average/person/day)	+	–	–	–
Weed pressing	30	1,000 m² (average/person/day) × 3 times/year (for 3 years).	+	+	+	?
Ring-weeding	50	3,100 regenerants/ha ÷ c. 180 (average/person/day) × 3 times per year (for 3 years).	+	+	+	?
TREE PLANTING						
Site preparation	25	Slashing weeds followed by glyphosate application (see **Section 7.1**).	+	–	–	–
Planting	No. of trees to plant/ha divided by 80	No. of trees to plant = 3,100 – the no. of regenerants/ha (see **Section 3.3**). One person can plant about 80 trees/day (following the methods described in **Section 7.2**).	+	–	–	–
Weeding and fertiliser application	50	3,100 trees/ha (including natural regenerants + planted trees) ÷ c. 180 (average/person/day) × 3 times per year (for 2 years).	+	+	–	–
Monitoring	32	16 people can monitor 1 ha/day. Monitor twice per year (at the beginning and end of main growing season). For large sites, randomly select a few sample hectares for monitoring.	+	+	+	+

Calculating costs

The costs of restoration vary considerably with local conditions (both ecological and economic) and increase markedly with degradation stage. Therefore, we can only present guidelines for cost calculations as any estimate of actual costs would quickly become out of date. Make sure that all expenditure is carefully recorded, to enable a cost–benefit evaluation of the project in the future and to assist other local initiatives in planning their own projects.

The restoration of degradation stages 3–5 involves tree planting, so nursery costs should be included in the project budget. Construction of a simple community nursery need not be expensive: for example, the use of locally available materials, such as bamboo and wood, will keep costs down. Tree nurseries last many years, so nursery construction costs represent only a very small component of tree production costs. Reduce the costs of materials by using locally available media, such as rice husk and forest soil, instead of commercially produced potting mixes. Although many such local materials are essentially 'free', don't forget to factor in the labour and transportation costs of collecting them. The only essential nursery items for which there is no effective natural substitute are plastic bags or other containers and a means of delivering water to the plants.

A nursery manager should have overall responsibility for running the nursery and ensuring the production of enough trees of sufficient quality and of the required species. This may be a full-time or part-time salaried position, depending on the numbers of saplings to be produced. Casual labour can be voluntary or paid a daily rate as required. Nursery work is seasonal, with the heaviest workload just before planting and lighter workloads at other times of the year. Nursery staff should also be responsible for seed collection. For a typical nursery, the production rate should be 6,000–8,000 trees produced per nursery staff member per year.

Budget lines for tree production should therefore include:
- construction of a nursery (including a watering system);
- nursery staff;
- tools;
- supplies, e.g. germination trays, containers, media, fertiliser and pesticides;
- water and electricity;
- transportation (for provisioning, seed collection and delivering trees to the restoration site).

Tree planting, maintenance and monitoring costs can be divided into i) labour, ii) materials and iii) transportation. Labour is by far the largest budget item, with fire prevention being the largest labour cost. Therefore, the financial viability of forest restoration often depends on the extent to which paid labour can be replaced with volunteers. It is usually very easy to find people from local schools and businesses to help out on planting day. Fire prevention is also an activity that is usually organised by village committees as a 'community activity'. Therefore, weeding and fertiliser application are the two activities most likely to require paid labour.

To calculate labour costs, begin with the estimated labour inputs suggested in **Table 4.3**. Select those tasks that have been included in your task schedule and remove any for which voluntary labour is assured. Sum up the total person-days labour required for all tasks for year 1 and multiply the sum by the number of hectares to be restored and by the acceptable daily payment for labour. Next, consider how many tasks must be repeated in year 2 and repeat the calculation of labour costs, except add a percentage increase to the daily payment to account for inflation. By year 3, the amount of labour required for weeding and fertiliser application should fall considerably as canopy closure begins to take effect. Therefore, delay calculating labour costs for subsequent years until the progress achieved in years 1 and 2 is assessed.

Materials for planting include glyphosate (a herbicide), fertiliser, and a bamboo pole and possibly a mulch mat for each tree to be planted. Calculate the cost of applying 155 kg of fertiliser per hectare (assumes 50 g per tree × 3,100 (both planted and natural regenerants)) four times in the first year and three times in the second year. If using glyphosate to clear weeds, calculate the cost of 6 litres of concentrate per hectare.

4.6 Fundraising

Having drafted a plan and calculated a budget, the next stage is fundraising. Funding for forest restoration projects can come from many different sources, including governments, NGOs and the private sector, both local and international. A vigorous fundraising campaign should target several potential funding sources.

Corporate social responsibility (CSR) schemes have traditionally been a large source of sponsorship for tree planting events, in return for promoting a 'green image' for the sponsors. Contact local companies involved in the energy industry (e.g. oil companies), in the transportation industry (e.g. airlines, shipping agencies or car manufacturers), or in industries that benefit from a greener environment (e.g. the tourist industry or food and drink manufacturers), as well as companies that have adopted trees or wildlife as their logos.

Application procedures for private-sector grants and the administration of them are usually straightforward. However, before accepting corporate sponsorship, consider ethical issues, such as the use of your project to promote a green image for a company that might be engaged in environmentally damaging activities. To avoid such dilemmas, make sure that the project is supported by a company's social responsibility fund, not by its advertising budget, and check the contract thoroughly.

The recent surge of interest in tropical forests as carbon sinks should increase the corporate sponsorship of restoration projects. It could, however, be having the reverse effect because many companies now only sponsor tree planting projects in return for voluntary carbon credits. This requires that projects register with one of a plethora of organisations[2] that have recently set up standardisation schemes, which monitor projects to verify the additional amount of carbon stored and to ensure that they have no adverse effects. Such services currently cost from US$5,000–40,000 and registration can take up to 18 months. Having to find such hefty start-up costs is now effectively excluding smaller projects from corporate sponsorship and the lengthy and complicated registration process delays project implementation.

---

[2] Such as Carbon Fix Standard (CFS, www.carbonfix.info/), Verified Carbon Standard (VCS, www.v-c-s.org/), Plan Vivo (www.planvivo.org/), and The Climate Community and Biodiversity Standard (CCBS, www.climate-standards.org/).

For smaller projects, charities and foundations are often a good source of funding. They generally provide small grants with uncomplicated reporting and accounting procedures. Domestic government organisations, especially those involved in implementing a country's obligations under the Convention on Biological Diversity (CBD), should also be approached. Local government organisations might also provide small grants for environmental conservation.

If you find applying to grant-awarding organizations a bit daunting, then consider running your own fundraising campaign. For small projects, traditional fundraising events (sponsored runs, raffles and so on) may be sufficient to raise the required funds. But such events require a lot of organisation and usually some upfront payments (such as renting venues). The internet now makes it possible to reach out to more people than ever before with minimum effort. Publicising your project over social networks or through a dedicated project website can generate both interest and funding.

A common approach is the 'sponsor-a-tree' campaign. Calculate your total project costs (see **Section 4.5**) and divide that amount by the number of trees that you intend to plant (to get the cost per tree), then ask visitors to your website or Facebook page to sponsor one or more trees. Many websites currently offer such schemes from US$ 4 to US$ 100 per tree. Internet payment systems such as PayPal can be used to transfer the funds. To overcome the impersonal nature of the internet, show your appreciation of donors by providing personalized feedback. Invite sponsors to join tree-planting events and/or provide them with individual pictures of 'their' tree as it grows. One website even directs sponsors to Google Earth images of the planted sites. Learning the ins and outs of website construction and internet payment schemes will take time at the start, but will pay dividends as the project becomes better known.

On its dedicated website, "Plant a Tree Today" offers sponsorship of tree planting in one of many restoration projects from around US$ 4 per tree.

A comprehensive resource for finding funding for restoration projects agencies is the Collaborative Partnership on Forests (CPF) *Sourcebook on Funding for Sustainable Forest Management* (www.cpfweb.org/73034/en/). This excellent website includes a downloadable database of funding sources for sustainable forest management, a discussion forum and a newsletter on funding issues, as well as useful tips on preparing grant applications.

CHAPTER 5

TOOLS FOR RESTORING TROPICAL FORESTS

With a project plan in place and funding approved, it's time to start work. In this chapter, we discuss how to implement the five main tools for forest restoration: protection, ANR, planting framework tree species, the maximum diversity approach and nurse plantations (or plantations as catalysts). In Chapter 3, we established that these five basic tools are rarely used in isolation. The greater the degree of degradation, the more tools must be combined to achieve a satisfactory result. In Chapters 6 and 7, we go on to provide more details on growing and planting native forest tree species.

*"**The successful restoration of a disturbed ecosystem is the acid test of our understanding of that ecosystem**." Bradshaw (1987).*

5.1 Protection

There is no point in restoring sites that cannot be protected against the damaging activities that destroyed the original forest. Thus, preventing degradation is fundamental to all forest restoration projects, regardless of the degradation stage being tackled. Protection has two basic elements: i) preventing additional encroachment and ii) removing existing barriers to natural forest regeneration. The former involves the prevention of new harmful human activities at the restoration site, whereas the latter engages existing resident communities in fire prevention, stock exclusion and protecting seed-dispersing animals against hunters.

Prevention of encroachment

Unoccupied forest land has always been a magnet for landless people with low incomes. In the past, forest clearance amounted to a legal claim of land ownership and a way out of poverty. But in modern civil societies, and as populations have grown exponentially, 'ownership by deforestation' is no longer acceptable. The vast majority of tropical forest land is now under state control, and there are laws to prevent its exploitation for personal gain. Unfortunately, law enforcement to exclude forest encroachers often targets rural poor people and is therefore heavily criticised by human rights groups, especially where corporations and wealthy landowners can get away with encroachment unpunished. Ultimately, these problems can only be solved by better forest governance[1], but several practical measures can be taken at the local level to prevent further encroachment.

Impoverished villagers, many of whom are poorly educated, are often unaware of the law. Therefore, simply ensuring that everyone is aware of the law and the penalties imposed for encroachment can sometimes be enough to deter it (Thira & Sopheary, 2004). Clearly defined boundaries, with conspicuous signs along them explaining the protected status of the area, also help to ensure that all are aware of legal restrictions and where they apply.

Encroachment tends to occur along roads, so preventing road construction and/ or improvement in protected forest is perhaps the most effective way to prevent it (Cropper et al., 2001), especially in remote areas. Check-points where existing roads enter and leave protected sites can also deter encroachment.

A human presence, such as random patrols, is perhaps the ultimate deterrent to forest encroachment. Maintaining a patrol system is expensive, but forest guardians can have multiple tasks. While on patrol, they may also collect seeds from fruiting trees to supply a tree nursery, or record observations of wildlife, including seed dispersers and pollinators. GPS technology can be used to record the position of seed trees and wildlife, as well as the patrol coverage and signs of encroachment. When integrated into geographic information systems (GIS), the data can be shared and used to predict which areas are most threatened by encroachment. This is the 'smart patrol' concept advocated by the Wildlife Conservation Society (Stokes, 2010).

Preventing further encroachment by communities that are already established within a forest landscape depends on building a strong 'sense of community stewardship'

---

[1] www.iucn.org/about/work/programmes/forest/fp_our_work/fp_our_work_thematic/fp_our_work_flg

for both remaining and restored forest. Local villagers will work together to exclude outside encroachers if they feel that encroachment threatens their community's interests. Community forestry, whereby a village committee (rather than a state agency) becomes responsible for managing a restored forest, provides a strong shared 'sense of stewardship', because the village committee deals with anyone damaging the community's forest resources using self-imposed rules and regulations. Peer pressure replaces the need to involve state law-enforcement agencies. Community forestry is of course impossible where there is no forest. So the prospect of community control over forest resources (once forest has been re-established) provides a powerful motivation for local villagers to contribute to forest restoration projects.

Communities near restoration sites can also benefit from direct employment by restoration projects. Livelihood development schemes can also be provided. These capitalise on the benefits of forest restoration (e.g. the development of ecotourism), reduce the need to clear forest (e.g. by intensifying agriculture) or reduce the exploitation of forest resources (e.g. introducing biogas as a substitute for firewood). If such rewards are offered only to communities in protected areas, however, their effect might be to actually attract outside encroachers who seek to access to the benefits of such development programs.

When protected areas systems were first introduced, the general view was that human settlers should be removed in order to maintain 'pristine' nature. This view disregarded the fact that most areas had in fact been occupied by humans, to a greater or lesser extent, long before they were declared protected. The forced relocation of settlers from protected areas has a sorry history. In most cases, inadequate compensation was paid (if any), the relocation sites were of poor quality and the support promised for agriculture, education and health care at the relocation sites often failed to materialise (Danaiya Usher, 2009). Furthermore, the vacuum left behind when people are moved out of protected areas is often quickly invaded by new encroachers.

Local people who have a long history of living in forest landscapes are a great asset to forest restoration programs. They are a valuable source of local knowledge, especially in regard to the selection of tree species and seed collection. They can provide most of the labour needed for restoration tasks, both in the nursery and in the field, and they can also implement protective measures, such as patrolling and manning road check points, as a civic duty.

Prevention of fire damage

Protecting forest restoration sites from fire is essential for success. In the seasonally dry tropics, fire prevention is an annual activity, and even in the wet tropics, it is necessary during times of drought. Most fires are started by humans, so the best way to prevent them is to make sure that everyone in the vicinity supports the restoration program and understands the need not to start fires. But no matter how much effort is put into raising awareness of fire prevention amongst local communities, fire remains a common cause of failure for forest restoration projects. Most local forest authorities have fire-suppression units, but they cannot be everywhere, so local, community-based fire prevention initiatives are often the most effective way to tackle the problem. Preventative measures include cutting fire breaks and organising fire patrols to detect and extinguish approaching fires before they can spread to restoration sites.

Box 5.1. **Extractive reserves**

Box 5.1. **Extractive reserves.**

Extractive reserves provide local communities with a direct interest in protecting tropical forests by allowing them to exploit non-timber forest products (NTFPs) in a sustainable way. It links villagers' incomes to the maintenance of intact forest ecosystems. The survival of the forest and the livelihoods of the villagers become interdependent.

The concept was pioneered in Brazil in the mid 1980s, when rubber-tappers and local rural workers' unions asked for the designation of areas in the Amazon where they could tap forest rubber trees to support the sustainable development of local communities. Extractive reserves were proposed as conservation areas in which local communities could harvest NTFPs such as nuts and latex. Essentially, the designation of such areas aimed to reconcile issues that policy makers traditionally thought of as incompatible, i.e. protecting forests as conservation areas and allowing local people to exploit them sustainably.

In 1989, the Brazilian Government formally included extractive reserves in its national policy. The land was to be taken into Government ownership for the dual purposes of safeguarding the rights of local people and conserving biodiversity. It was decided that extractive reserves would only be established if requested by local people and where a long tradition of forest use was evident. This has now become a major federal strategy for forest conservation and economic development amongst local peoples. In the case of the rubber tapper unions, under the leadership of Chico Mendes, it was envisaged that the forest would remain standing for use by both rubber-tappers and local people who wished to harvest NTFPs.

Map of Acre showing the location of the Chico Mendes Extractive Reserve (© IUCN).

Box 5.1. continued.

Chico Mendes demonstrating the process of tapping a rubber tree to produce latex in 1988. (Photo: M. Smith, Miranda Productions Inc.)

The best-known extractive reserve in South America is the 980,000 ha Chico Mendes Reserve in the state of Acre in western Amazonia. Chico Mendes himself was assassinated in 1988, but his legacy lives on in over 20 extractive reserves covering approximately 32,000 km². In the Chico Medes Reserve, the rights of local people, who are dependent on the forest, are protected. But in this and other extractive reserves, the IUCN recognises that the use of "economically, environmentally and socially viable forest production as a driver for local development" remains a challenge.

Despite these efforts to protect the Amazon forests, the rate of deforestation in the Amazon increased dramatically in 2010 and 2011, and the Brazilian Parliament had to decide whether to relax environmental laws that protect the forest in favour of farmers seeking more space to raise cattle. It was proposed, for example, that farmers should be allowed to clear 50% of the forest on their land, whereas the existing law allowed them to clear only 20%.

Edinaldo Flor da Silva and his family are benefiting from new rubber production units, which mean they can earn more from their sustainable product. (Photo: © Sarah Hutchison/WWF/Sky Rainforest Rescue)

Fire breaks

Fire breaks are strips of land that are cleared of combustible vegetation to prevent the spread of fire. They are effective at blocking moderate, ground-cover burns. More intense fires throw up burning debris, which can be blown across fire breaks to start new fires far away from where the original fire ignited.

Cut firebreaks at least 8 m wide around restoration sites just before the onset of a dry season. The quickest method is to slash all grasses, herbs and shrubs (trees need not be cut) along the two edges of the firebreak. Pile the cut vegetation at the centre of the firebreak, leave it for a few days to dry out and then burn it. Obviously, using fire to prevent fire can be risky. Make sure plenty of people are available with beaters and water sprayers to prevent accidental escape of the fire into surrounding areas. The risk of fire escaping is considerably reduced by burning fire breaks just before the beginning of a hot, dry season when the surrounding vegetation is too moist to burn easily. Roads and streams act as natural fire breaks. There is usually no need to make fire breaks along streams, but they should be made alongside roads, as fires are often started by drivers throwing cigarettes out of vehicles.

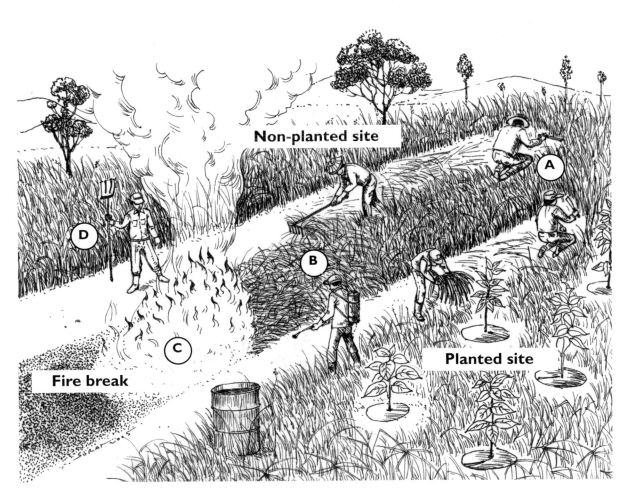

Using fire to fight fire. (A) Slash two strips of vegetation at least 8 m apart. (B) Drag the cut vegetation into the centre. (C) Allow a few days for the cut material to dry out, and then (D) burn it, taking extreme care not to allow the fire to spread outside the firebreak.

Suppressing fires

Organise teams of fire watchers to alert local people when fires are spotted. Try to involve the whole community in the fire prevention programme, so that each household contributes one family member every few weeks to fire prevention duties. Fire watchers must remain alert night and day throughout the dry season.

Place fire-fighting tools and oil drums full of water at strategic places around the planted site. Fire-fighting tools include back-pack water tanks with sprayers, beaters to smother the fire, rakes to remove combustible vegetation from the fire front and a first aid kit. Green tree branches can be used as fire beaters. If a permanent stream runs nearby, above the restoration site, consider laying pipes into the restoration sites. This can greatly increase the efficiency of fire-fighting activities but is very expensive.

Small fires can be controlled with (A) back-pack sprayers, the use of (B) simple tools such as rakes to remove fuel in the fires path, and (C) beaters to extinguish small fires. Oil drums full of water can be placed strategically across the site in advance as refill points for the back-pack sprayers.

Only low-intensity, slow-moving, ground fires can be controlled with hand tools. More serious fires, especially those that move up into tree crowns, must be controlled by professional fire fighters with aerial support. Be ready to contact local fire-fighting authorities if the fire gets out of control, and take extra care, as serious fires move very quickly and can easily cost lives. The forest fire control units of local forest authorities often provide training in fire prevention and fire-fighting techniques to local people. They may be able to supply fire-fighting equipment to community-based fire prevention initiatives, so contact your local forest fire control unit for assistance.

What can be done if restoration sites do burn?

All is not lost. Some tree species can re-sprout (or coppice) from rootstock after having been burnt (see **Section 2.2**). Burnt, dead branches allow the entry of pests and pathogens, so cutting them off can speed recovery after burning. Prune dead branches right back, leaving a stump no longer than 5 mm. After fire, the blackened soil surface absorbs more heat, causing more rapid evaporation of soil moisture. This can subsequently kill young trees that have survived the initial fire. Therefore, laying a mulch of cut vegetation or corrugated cardboard around young, burnt trees can increase their chances of survival and re-growth.

Livestock management

Cattle, goats, sheep and other livestock can completely prevent forest regeneration by browsing on young trees. Ultimately, the decision to reduce the number of livestock or to remove them altogether depends on careful consideration of their economic value to the community and their potential to play a useful role in forest restoration, balanced against any damaging effects they have on young trees. The severity of the damage obviously increases with stocking density.

In Area de Conservacion Guanacaste (ACG), Costa Rica, cattle played a positive role in the early stages of a forest restoration project by grazing on an exotic grass species that fuelled wildfires, but as the developing tree crowns began to shade out the grass, the cattle were gradually removed (see **Case Study 3**, p. 149). Similarly in the montane pastures of Columbia, where grasses are a major barrier to forest regeneration, livestock grazing encouraged shrub establishment, which created a micro-climate that was more suitable for the establishment of tropical montane forest tree species (Posada *et al.*, 2000).

Livestock can also facilitate natural forest regeneration by dispersing tree seeds, especially in forests where wild ungulate species have become extirpated (Janzen, 1981). Free-ranging cattle often consume tree fruits in forests and deposit them in open areas when grazing. Cattle-dispersed tree species mostly grow in dry tropical forests and commonly have dry, indehiscent, brown-black fruits with hard seeds, averaging 7.0 mm in diameter. The family Leguminosae contains many tree species with seeds that are dispersed by cattle; other families with fewer potentially cattle-dispersed tree species include the Caprifoliaceae, Moraceae, Myrtaceae, Rosaceae, Sapotaceae and Malvaceae. Ranchers in the Central Valleys of Chiapas, Mexico purposefully use cattle to plant tree seeds (Ferguson, 2007).

Careful livestock management can, therefore, have beneficial effects for forest restoration if the stocking density is low and the foliage of the desired tree species is unpalatable. But even in such circumstances, livestock can reduce the richness of tree species in restored forest sites by selective browsing.

The impact of livestock can be managed by tethering animals in the field to restrict their movements or by removing them altogether. Stock fences can be erected to exclude livestock during the early stages of forest restoration, but such fences must be maintained until the trees crowns have grown beyond the reach of livestock.

Cattle can act as 'living grass mowers' and can disperse seeds, but dense populations suppress forest regeneration.

In Nepal, villagers often do not allow their cows to roam freely in their community forests. To promote rapid forest regeneration, villagers keep their cows outside the forest. They cut grass and fodder from the forests and carry it to their cows. This feeds the cows without damaging the young, regenerating trees and also encourages the effective weeding of forest plots (Ghimire, 2005).

Protecting seed dispersers

For forest restoration to be successful, with acceptable biodiversity recovery, the protection of trees must be complemented with the protection of seed-dispersing animals. Seed dispersal from intact forest into restoration sites is essential for the return of climax forest tree species. The hunting of seed-dispersing animals can therefore substantially reduce tree species recruitment. There is no point in restoring forest habitat to attract seed dispersers if there are no seed dispersers left to attract.

Simple education campaigns can be effective in turning hunters into conservationists. At Ban Mae Sa Mai in northern Thailand, hill tribe children were the main hunters, capturing birds in traps and killing them with catapults, sometimes to eat but mostly for fun. They particularly targeted bulbuls, the main seed-dispersers from forest into open areas. An effective education campaign (sponsored by the Eden Project, UK) introduced the children to the hobby of bird watching, with the potential for further training some of them to become guides for ecotourists. The project provided binoculars and bird identification books and ran regular bird watching trips. The children established their own small bird reserve and the 'bird police', using peer pressure to dissuade their classmates from hunting. They also took the conservation message home to their parents. Both bird traps and catapults are now a rare sight around the village.

S.O.S. 'Save Our Seed-dispersers': simple education campaigns can turn bird hunters into bird guides. (Photos: T. Toktang).

5.2 'Assisted' or 'accelerated' natural regeneration (ANR)

What is ANR?

ANR is any set of activities, short of tree planting, that enhance the natural processes of forest regeneration. It includes the protective measures that remove barriers to natural forest regeneration (e.g. fire and livestock) already described in **Section 5.1**, along with additional actions to i) 'assist' or 'accelerate' the growth of natural regenerants that are already established in the restoration site (i.e. seedlings, saplings and live stumps of indigenous forest tree species) and ii) encourage seed dispersal into the restoration site.

The UN Food and Agriculture Organisation (FAO), in partnership with the Philippines' government and community-based NGOs, supported much of the research that helped to transform ANR from a concept into an effective and practicable technique (see **Box 5.2**, p. 119). The FAO now promotes ANR as a method for enhancing the establishment of secondary forests by protecting and nurturing seed trees and wildlings already present in the area. With ANR, secondary and degraded forests grow faster than they would naturally. Because this method merely enhances already existing natural

BOX 5.2. ORIGINS OF ANR

Box 5.2. Origins of ANR.

Although humans have long manipulated natural forest regeneration, the concept of actively promoting it to restore forest ecosystems is relatively recent. The formal concept of ANR — 'accelerated' or 'assisted' natural regeneration — first emerged in the Philippines in the 1980s (Dalmacio, 1989). A long-standing partnership between the UN Food and Agriculture Organisation (FAO)'s Regional Office for Asia and the Pacific and the Bagong Pagasa (New Hope) Foundation (BPF), a small NGO in the Philippines, has since played a crucial role in propelling this simple concept from obscurity to the fore-front of tropical forest restoration technology.

Sponsored by the Japan Overseas Forestry Consultants Association (JOFCA), BPF established an early ANR project at Kandis village, Puerto Princesa, on Palawan Island, Philippines, with the aim of restoring 250 ha of degraded water catchment that was dominated by grasses. ANR was tested both as a restoration technique and as a development tool for improving the livelihoods of 51 families. The project combined ANR to restore forest with the establishment of fruit orchards. Treatments included fire prevention, ring weeding of tree saplings and grass pressing. The pioneer trees, which grew up rapidly after the weeding treatments, fostered the regeneration of 89 forest tree species (representing 37 tree families), including many climax forest species. The forest trees were inter-planted with coffee and domestic fruit trees to provide the villagers with income. After three years, a self-sustained forest ecosystem began to develop. Systematic monitoring revealed significant biodiversity recovery and soil improvement (Dugan, 2000).

Although there are now many successful ANR projects in the Philippines, very little information was initially published to enable others to learn from the experiences of organisations such as Bagong Pagasa. Therefore, the FAO has funded several projects to promote ANR for forest restoration in several countries. Launched in 2006, the project "Advancing the application of assisted natural regeneration for effective low-cost forest restoration"[2] created demonstration sites on three geographically different Philippines islands. The project focused on restoring forest to degraded *Imperata cylindrica* grasslands, using weed pressing to liberate shaded tree seedlings. More than 200 foresters, NGO members and community representatives have been trained in ANR methods at these demonstration sites. The project concluded that the costs of ANR are approximately half those of conventional tree planting. As a result, the Philippines Department of Environment and Natural Resources (DENR) allocated US$32 million to support the implementation of ANR practices on approximately 9,000 hectares. The project has generated interest and funding from the mining industry and local municipalities seeking to offset their carbon footprints. The FAO, in collaboration with BPF, is now funding similar ANR trials in Thailand, Indonesia, Lao PDR and Cambodia.

Patrick Dugan, founding chairman of Bagong Pagasa. By building partnerships with the Philippines Government (Department of Environment and Natural Resources) and the FAO, the foundation has promoted the concept of ANR well beyond its origins in the Philippines.

[2] www.fao.org/forestry/anr/59224/en/

processes, it requires less labour than tree planting and there are no tree nursery costs. It can, therefore, be a low-cost way to restore forest ecosystems. Shono *et al.* (2007) provide a comprehensive review of ANR techniques.

ANR and tree planting should not be regarded as mutually exclusive alternatives to forest restoration. More often than not, forest restoration combines protection and ANR together with some tree planting. The site survey technique, detailed in **Sections 3.2** and **3.3**, can be used to determine whether protection + ANR are sufficient to achieve restoration goals, whether they should be complemented with tree planting and, if so, how many trees should be planted.

Where is ANR appropriate?

Protection + ANR may be sufficient to bring about rapid and substantial forest restoration and biodiversity recovery where the forest degradation is at stage-2. At this degradation stage, the density of natural regenerants exceeds 3,100 per hectare, and more than 30 common tree species typical of the target climax forest (or roughly 10% of the estimated number of tree species in the target forest, if known) are present. Where the density of natural regenerants is lower or fewer tree species are represented, ANR should be used in combination with tree planting (see **Section 3.3**). Furthermore, intact forest should remain within a few kilometres of the proposed restoration site, providing a seed source for the re-establishment of climax forest tree species, and seed-dispersing animals should remain fairly common (see **Section 3.1**).

Some advocates of ANR propose its use on highly degraded grasslands, where the density of regenerants (>15 cm tall) is only 200–800 per hectare (i.e. degradation stage 3 or higher, see **Section 3.1**) (Shono *et al.*, 2007). The application of ANR alone under such circumstances usually results in forest of low productive and ecological value because of the dominance of a few ubiquitous pioneer tree species. But even a species-poor secondary forest is a considerable improvement, in terms of biodiversity recovery, on the degraded grasslands that it replaces; and further forest recovery is possible as long as seed trees and seed-dispersers remain in the landscape.

(A) About a year before this photograph was taken (photo May 2007), encroachers illegally cleared lowland, evergreen forest from this Reserved Forest site in southern Thailand to establish a rubber tree plantation. Plenty of sources of natural regeneration remained, including both pioneer and climax tree species, making the site ideal for restoration by ANR. Cardboard mulch mats were placed around remaining tree saplings and seedlings, weeds were cleared and fertiliser was applied three times during the rainy season. (B) Just 6 months later, canopy closure had been achieved (photo November 2007). Most of the canopy tree species were pioneers, and so the understory was enriched by planting nursery-raised saplings of climax forest tree species.

ANR techniques

Reducing competition from weeds

Weeding reduces competition between trees and herbaceous vegetation, increases tree survival and accelerates growth. Before weeding, clearly mark the tree seedlings or saplings with brightly coloured poles to make them more visible. This prevents their being accidentally trampled or cut during weeding.

First, mark sources of natural forest regeneration.

Ring weeding

Remove all weeds, including their roots, using hand tools within a circle of 50 cm radius around the base of all natural seedlings and saplings. Hand-pull weeds (wear gloves) close to small seedlings and saplings, as digging out weed roots with hand tools can damage their root system. Then, lay a thick mulch of the cut weeds around each seedling and sapling, leaving a gap of at least 3 cm between the mulch and the stem to help prevent fungal infection. Where the cut weeds do not yield a sufficient volume of mulching material, use corrugated cardboard as mulch.

Weed pressing or lodging

Remove shade by flattening all remaining herbaceous vegetation between the exposed natural regenerants using a wooden board (130 × 15 cm). Attach a sturdy rope to both ends of the plank, making a loop long enough to pass over your shoulders (attach shoulder pads for comfort). Lift the plank onto the weed canopy and step on it with full body weight to fold over the stems of grasses and herbs near the base. Repeat this action, moving forward in short steps[3]. The weight of the plants should keep them bent down. This is particularly effective where the vegetation is dominated by soft grasses such as *Imperata*. Old, robust or cane-stemmed grasses (e.g. *Phragmites*, *Saccharum*, *Thysanolaena* spp.) should not be pressed, because they can readily re-sprout from nodes along their stems. Pressing weeds is much easier than slashing them; one experienced person can press about 1,000 m² per day.

Grass pressing with a wooden board is particularly suitable for suppressing the growth of *Imperata* grass and releasing natural regenerants from competition.

Pressing is best carried out when the weeds are about 1 m tall or taller: shorter plants tend to spring back up shortly after pressing. The best time to press grass is usually about two months after the rains start when grass stems easily crimp (fold). Before pressing on a large scale, conduct a simple test on a small area. Press the grass and wait overnight. If the grass is starts to spring back up by the morning, then wait a few more weeks before trying again. Always press the weeds in the same direction. On slopes, press grasses downhill. If plants are pressed when they are wet, water on the leaves helps them to stick together, so that they are less likely to spring back up.

Pressing effectively uses the weeds' own biomass to shade and kill them. Plants in the lower layers of the pressed mass of vegetation die because of lack of light. Some plants may

---

[3] www.fs.fed.us/psw/publications/documents/others/5.pdf and www.fao.org/forestry/anr/59221/en/

survive and grow back, but they do so much more slowly than if they had been slashed. Therefore, pressing does not have to be repeated as often as slashing. The pressed vegetation suppresses the germination of weed seeds by blocking light. It also protects the soil surface from erosion and adds nutrients to the soil as the lower layers begin to decompose. Weed pressing opens up the restoration site, making it easier to move around and work with the young trees. It also helps to reduce the severity of fires. Pressed plants are a lot less flammable than erect ones because of the lack of air circulation within the pressed mass of vegetation. They do burn, but the flame height is lower and so tree crowns are less likely to be scorched.

Where the density of natural regenerants is high, the use of herbicides to clear weeds is not recommended because it is very difficult to prevent the spray drifting onto the foliage of the natural regenerants.

Use of fertilisers

Most tree seedlings and saplings of up to about 1.5 m tall will respond well to fertiliser applications, regardless of the soil fertility. Fertiliser application both increases survival and accelerates growth and crown development. This brings about canopy closure and shading out of weeds sooner than if no fertiliser were applied, and thus reduces labour costs for ring weeding and weed pressing. So although chemical fertilisers can be expensive, the costs are partly offset in the long term by the savings in weeding costs. Organic fertilisers, such as manure, can be used as a cheaper alternative to chemical fertilisers. It is probably a waste of effort and expense to apply fertiliser to older saplings and tree stumps, which have already developed deep root systems.

Encouraging tree stumps to sprout

The importance of coppicing tree stumps in accelerating canopy closure and contributing to the richness of tree species in restoration sites was discussed in **Section 2.2**. But besides the general recommendation that tree stumps should be protected from further chopping, burning or browsing, almost no treatments to enhance their potential role in ANR have been tested. Experiments on 'tree stump cultivation' could test the effectiveness of i) applying chemicals to prevent fungal decay or attack by termites, ii) applying plant hormones to stimulate bud growth and coppicing, and iii) pruning back weak coppicing shoots to free up more of the plant's resources for the remaining ones.

Thinning out naturally regenerating trees

Where dense stands of a single species dominate, self-thinning will occur naturally as the taller trees shade out the shorter ones. This process can be accelerated by selectively cutting some of the smaller trees (instead of waiting for them to die back naturally). This provides light gaps in which other, less common, tree species can establish and should increase the overall tree species richness.

Assisting the seed rain

The importance of seed dispersal as a vital and free ecological service that ensures the re-colonisation of restoration sites by climax forest tree species has been emphasised throughout this book (see **Sections 2.2, 3.1** and **5.1**). So how can it be enhanced?

Artificial bird perches are, in theory, a rapid and cheap way of attracting birds and increasing the seed rain in restoration sites. Perches are usually 2–3 m posts that have cross-bars pointing in different directions. Although the seed rain is increased beneath perches (Scott *et al.*, 2000; Holl *et al.*, 2000; Vicente *et al.*, 2010), seedling establishment increases only if the conditions for germination and seedling growth are favourable beneath the perches. Seeds can be predated or young seedlings can be out-competed by herbaceous weeds (Holl 1998; Shiels & Walker 2003). So weeding beneath perches is necessary if they are not on sites with low weed density.

Although artificial perches attract birds, they do so less effectively than actual trees and shrubs, which provide the added benefit of shading out weeds and thus improve conditions for seedling establishment. Establishing structurally diverse vegetation, including fruiting shrubs or remnant trees, is the best way to attract seed-dispersing birds and animals, but it takes time. In the meantime, artificial bird perches can provide a stop-gap measure.

In disturbed areas, the natural seed rain is dominated by secondary forest tree species, often from trees fruiting within the degraded site itself (Scott *et al.*, 2000). Therefore, perches can increase the density of regenerants without increasing species richness. Under such circumstances, the seed rain brought in by birds should be complemented with direct seeding of less common and climax forest trees.

Artificial bird perches can be used to increase the dispersal of tree seeds from intact forest to restoration sites.

Limitations of ANR

ANR acts solely on natural regenerants that are already present in deforested sites. It can achieve canopy cover rapidly, but only where regenerants are present at sufficiently high densities. Most of the trees that colonise degraded areas are of relatively few, common, light-demanding, pioneer species (see **Section 2.2**), which produce seeds that are dispersed by the wind or small birds. They represent only a small fraction of the tree species that grow in the target forest. Where wildlife remains common, the 'assisted' trees will attract seed-dispersing animals, resulting in tree species recruitment. But where large seed-dispersing animal species have become extirpated, planting large-seeded climax forest tree species can be the only way to transform the secondary forest, created by ANR, into climax forest.

5.3 The framework species method

Tree planting should be used to complement protection and ANR wherever fewer than 3,100 natural regenerants can be found per hectare and/or fewer than 30 tree species (or roughly 10% of the estimated number of tree species in the target forest, if known) are represented. The framework species method is the least intensive of the tree planting options: it exploits natural (and cost-free) seed dispersal mechanisms to bring about the recovery of biodiversity. This method involves planting the smallest number of trees necessary to shade out weeds (i.e. to provide site 'recapture') and attract seed-dispersing animals.

For the method to work, remnants of the target forest type that can act as a seed source must survive within a few kilometres of the restoration site. Animals (mostly birds and bats) that are capable of dispersing seeds from remnant forest patches or isolated trees to the restoration site must also remain fairly common (see **Section 3.1**). The framework species method enhances the capacity of natural seed-dispersal to achieve rapid tree species recruitment in restoration plots. Consequently, biodiversity levels recover towards those typical of climax forest ecosystems without the need to plant all of the tree species that comprise the target forest ecosystem. In addition, the planted trees rapidly re-establish forest structure and functioning, and create conditions on the forest floor that are conducive to the germination of tree seeds and seedling establishment. The method was first conceived in Australia, where it was initially used to restore degraded sites within Queensland's Wet Tropics World Heritage Area (see **Box 3.1**, p. 80). It has since been adapted for use in several Southeast Asian countries.

What are framework tree species?

The framework species method involves planting mixtures of 20–30 (or roughly 10% of the estimated number of tree species in the target forest, if known) indigenous forest tree species that are typical of the target forest ecosystem and share the following ecological characteristics:

- high survival rates when planted out in deforested sites;
- rapid growth;
- dense, spreading crowns that shade out herbaceous weeds;
- the provision, at a young age, of flowers, fruits or other resources that attract seed-dispersing wildlife.

In the seasonally dry tropics, where wild fires in the dry season are an annual hazard, an additional desired characteristic of framework species is resilience to burning. When fire prevention measures fail, the success of forest restoration plantings can depend on the ability of the planted trees to re-sprout from their rootstock after fire has burnt their above-ground parts (i.e. coppicing, see **Section 2.2**).

A practical consideration is that framework species should be easy to propagate and, ideally, their seeds should germinate rapidly and synchronously, with subsequent growth of vigorous saplings to a plantable size (30–50 cm tall) in less than 1 year. Furthermore, where forest restoration must yield benefits to local communities, economic criteria such as the productivity and value of the products and ecological services provided by each species can be taken into account.

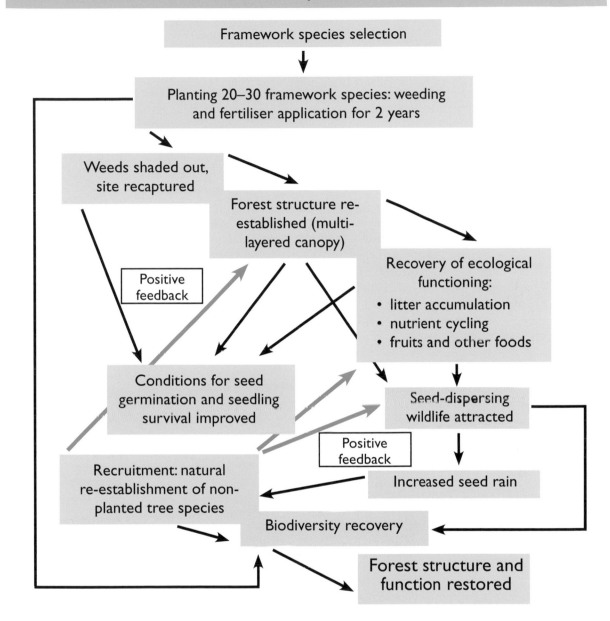

How the framework species method works

Framework species selection

Planting 20–30 framework species: weeding and fertiliser application for 2 years

Weeds shaded out, site recaptured

Forest structure re-established (multi-layered canopy)

Recovery of ecological functioning:
- litter accumulation
- nutrient cycling
- fruits and other foods

Positive feedback

Conditions for seed germination and seedling survival improved

Seed-dispersing wildlife attracted

Positive feedback

Recruitment: natural re-establishment of non-planted tree species

Increased seed rain

Biodiversity recovery

Forest structure and function restored

Are framework trees pioneer or climax species?

The mixtures of framework tree species chosen for planting should include both pioneer and climax species (or species that represent all the successional 'guilds' explained in **Section 2.2**, if known). Forest succession can be 'short-circuited' by planting both pioneer and climax trees in a single step. But in order to achieve rapid canopy closure, Goosem and Tucker (1995) recommend that at least 30% of the planted trees should be pioneers.

Many climax forest tree species perform well in the open, sunny conditions of deforested areas but they fail to colonise such areas naturally because of a lack of seed dispersal. Climax tree species often have large, animal-dispersed seeds and the decline of large mammals over wide areas prevents the dispersal these trees into deforested sites. By including some of them amongst the trees that are planted, it is possible to overcome this limitation and to accelerate the recovery of climax forest.

The planted pioneer trees make the greatest contribution to early canopy closure and shading out of herbaceous weeds. The point at which the tree crowns dominate over the herbaceous sward is called 'site recapture'. Pioneer tree species mature early and some can begin to flower and fruit just 2–3 years after planting. Nectar from flowers, fleshy fruits, and the perching, nesting and roosting sites created within the tree crowns all attract wildlife from nearby forest. Animal diversity increases dramatically as the new trees become established and, most importantly, many of the animals that visit the restoration sites carry with them tree seeds from climax forest. Furthermore, the cool, shady, moist, humus-rich and weed-free forest floor created beneath the canopy of planted trees provides ideal conditions for seed germination.

Pioneer species begin to die back after 15–20 years, creating light gaps. These allow the saplings of in-coming tree species to grow up and replace the planted pioneers in the forest canopy. If just short-lived, pioneer tree species were planted, they might die back before sufficient numbers of incoming tree species had established, leading to the possibility of re-invasion of the site by herbaceous weeds (Lamb, 2011). Planted climax tree species form an understorey that prevents this. They also add diversity and some of the structural features and niches of climax forest right from the start of the restoration project.

Rare or endangered tree species

Rare or endangered tree species are unlikely to be recruited into restoration sites on their own because their seed source is probably limited and they may have lost their primary seed dispersal mechanisms. Including such species in forest restoration plantings can help prevent their extinction, even if they lack some framework characteristics. Information on the world's endangered tree species is collated by the World Conservation Monitoring Centre of the United Nations Environment Program[4].

Selection of framework tree species

There are two stages to the selection of framework species: i) preliminary screening, based on current knowledge, to identify 'candidate' framework species for testing; and ii) nursery and field experiments to confirm framework traits. At the beginning of a project, detailed information about each species is likely to be sparse. Preliminary screening must be based on existing information sources and the target forest survey. As the results of nursery experiments and field trials gradually accumulate, the list of acceptable framework tree species can be gradually refined (see **Section 8.5**). The choice of framework species gradually improves at each planting as poor-performing species are dropped and new species are tried.

Sources of information for preliminary screenings include: i) floras, ii) the results of the target forest survey (see **Section 3.2**), iii) indigenous local knowledge and iv) scientific papers and/or project reports describing any previous work in the area (**Table 5.1**).

---

[4] www.earthsendangered.com/plant_list.asp

In the framework species method, both pioneer tree species (coloured in blue) and climax species (red) are planted 1.8 m apart in a single step, thereby 'short circuiting' succession while also preserving any naturally occurring trees and saplings (green).

The planted pioneer trees grow rapidly and dominate the upper canopy. They begin to flower and fruit a few years after planting. This attracts seed-dispersing animals. The planted climax tree species form an understorey, while seedlings of 'recruited' (i.e. non-planted) species (brought in by the attracted wildlife) grow on the forest floor.

Within 10–20 years, some of the planted pioneer trees begin to die back, providing light gaps in which recruited species can flourish. Climax tree species rise to dominate the forest canopy and forest structure, ecological functioning and biodiversity levels move towards those of climax forest.

Table 5.1. The preliminary screening and final selection of framework tree species draws on a diverse range of different information sources*.

Framework characteristic	Preliminary screening				Final selection	
	Floras	Target forest survey	Indigenous knowledge	Papers and previous project reports	Nursery research (see Section 6.6)	Field trials (see Sections 7.5 and 7.6)
Indigenous, non-domesticated, suited to habitat or elevation	Often indicated in plant descriptions in botanical literature	List of tree species from target forest survey	Unreliable: villagers often fail to distinguish between native and exotic species	EIAs and previous surveys for conservation management plans often list local tree species	–	–
High survival and growth	–	–	Ask local people which tree species survive well and grow rapidly in fallow fields	Unlikely except for economic species in previous forestry projects	Assess survival and growth of seedlings growing in nurseries	Monitor a sample of planted trees of each species for survival and growth (**Section 7.5**)
Dense broad crown shades out weeds	Few texts cover tree crown structure	Observe crown structure of trees in the target forest	–	–	Leaf size and crown architecture can be indicated by saplings in the nursery	Monitor a sample of planted trees of each species for crown breadth and reduced weed cover beneath
Attractive to wildlife	Fleshy fruits or nectar-rich flowers indicated in taxonomic descriptions	Observe fruit type and animals eating fruits or flowers in target forest	Villagers often know which tree species attract birds	–	–	Phenology studies of trees after planting
Resilient to fire	–	Survey trees in recently burnt areas	Villagers often know which tree species recover after burning in fallow fields	–	–	Where fire-prevention measures fail, survey trees in burnt plots immediately after a fire and 1 year later
Easy to propagate	–	–	–	Unlikely except for economic species in forestry projects	Germination experiments and seedling monitoring	–
Climax or large-seeded species	Often indicated in plant descriptions in botanical literature	Observe fruits and seeds of trees in target forest	–	–	–	–

*The organisation and integration of this information is discussed in **Section 8.5**.

Floras can provide basic taxonomic data on species under consideration as well as their suitability to site-specific requirements such as the target forest type being restored or the elevation range. They also indicate if a species produces fleshy fruits or nectar-rich flowers that are likely to attract wildlife.

The target forest survey (see **Section 3.2**) provides a great deal of original information that is useful for the selection of candidate framework tree species, including a list of indigenous tree species, and lists of species that have nectar-rich flowers, fleshy fruits or dense spreading crowns that are capable of shading out weeds. Phenology studies yield information on which trees will attract seed-dispersing wildlife. Studies of the botanical knowledge of local people (ethnobotany) can also provide an insight into the potential of trees to act as framework species. When carrying out such studies, it is important to work with communities that have a long history of living close to the forest, especially those that practice swidden (slash and burn) agriculture. Farmers from such communities usually know which tree species readily colonise fallow fields and grow rapidly and which tree species attract wildlife. The results of such studies must, however, be critically scrutinised. Local people sometimes provide information that they think will please the researcher rather than that based on actual experience. Superstition and traditional beliefs can also distort the objective assessment of a tree species' capabilities. Consequently, ethnobotanical information is reliable only if it is provided independently by members of several different communities with different cultural backgrounds. To design effective ethnobotany surveys, please refer to Martin (1995).

Local people also know if other researchers have been active in the area and which organisations or institutions they come from. Forestry departments and protected areas authorities often carry out biodiversity surveys, although the results might be in unpublished reports. Contact such organisations and ask for access to such reports. The local or national herbarium might also have tree specimens from your project site. Browsing through herbarium labels can reveal a lot of useful information. If any development projects have been carried out near your project site, it is likely that an environmental impact assessment (EIA), including a vegetation survey, was carried out. So it is worth contacting the agency that carried out the EIA. If research students have been active in the area, then universities can also be a source of more detailed information. Finally, there is always the internet. Simply typing the name of your project site into a search engine might reveal major additional sources of information.

Lists of tested framework tree species currently exist only for Australia (Goosem & Tucker, 1995) and Thailand (FORRU, 2006). But trees species in the same genera as those listed for Australia and Thailand might also perform well in other countries, so including some of them in initial framework species trials is well worth a try. Two pan-tropical tree taxa deserve special mention, namely fig trees (*Ficus* spp.) and legumes (Leguminosae). Indigenous species within these two taxa nearly always perform well as framework species. Fig trees have dense and robust root systems, which enable them to survive even the harshest of site conditions. The figs they produce are an irresistible food source for a wide range of seed-dispersing animal species. Leguminous trees often grow rapidly and have the capacity to fix atmospheric nitrogen in root nodules containing symbiotic bacteria, resulting in rapid improvement of soil conditions.

Site management

First, implement the usual protective measures described in **Section 5.1**, particularly measures to prevent both fire and the hunting of seed-dispersing animals. Second, protect and nurture any existing natural regenerants using the ANR techniques described in **Section 5.2**. Third, plant enough framework tree species to bring the total species on-site (including natural regenerants) up to around 30 (or roughly 10% of the estimated number of tree species in the target forest, if known), spaced about 1.8 m apart or the same distance from natural regenerants: this will bring the total density of trees on site up to around 3,100/ha.

Frequent weeding and application of fertiliser to both planted trees and naturally regenerating saplings is recommended during the first two rainy seasons. Weeding prevents herbs and grasses, particularly vines, from smothering the planted trees, enabling the tree crowns to grow above the weed canopy. Fertiliser application accelerates tree growth, resulting in rapid canopy closure. Finally, monitor the survival and growth of the planted trees and biodiversity recovery in restoration sites, so that the choice of framework tree species for future plantings can be continuously improved.

For further information on planting, and the post-planting management and monitoring of framework trees, see **Chapter 7**.

Direct seeding as an alternative to tree planting

Some framework tree species can be established in the field directly from seed. Direct seeding involves:

- collecting seeds from native trees in the target forest ecosystem and if necessary storing them until sowing;
- sowing them in the restoration site at the optimal time of year for seed germination;
- manipulating field conditions to maximise germination.

Direct seeding is relatively inexpensive because there are no nursery and planting costs (Doust *et al.*, 2006; Engel & Parrotta, 2001). Transporting seeds to the restoration site is obviously easier and cheaper than trucking in seedlings, so this method is particularly suitable for less accessible sites. Trees that are established by direct seeding usually have better root development and grow faster than nursery-raised saplings (Tunjai, 2011) because their roots are not constrained within a container. Direct seeding can be implemented in combination with ANR methods and conventional tree planting to increase both the density and species richness of regenerants. In addition to establishing framework tree species, direct seeding can be used with the maximum diversity method or to establish 'nurse tree' plantations, but it does not work with all tree species. Experiments are needed to determine which species can be established by direct seeding and which cannot.

Potential obstacles to direct seeding

In nature, a very low percentage of dispersed tree seeds germinate and even fewer seedlings survive to become mature trees. The same is true of direct seed sowing (Bonilla-Moheno & Holl, 2010; Cole *et al.*, 2011). The biggest threats to sown seeds and seedlings are: i) desiccation, ii) seed predation, particularly by ants and rodents (Hau,

1997) and iii) competition from herbaceous weeds (see **Section 2.2**). By counteracting these factors, it is possible to improve the rates of germination and seedling survival above those for naturally dispersed seeds.

The problem of desiccation can be overcome by selecting tree species whose seeds are tolerant of or resistant to desiccation (i.e. those with thick seed coats) and by burying the seeds or laying mulch over the seeding points (Woods & Elliott, 2004).

Burying can also reduce seed predation by making the seeds more difficult to find. Pre-sowing seed treatments that accelerate germination can reduce the time available for seed predators to find the seeds. Once germination commences, the nutritional value of seeds and their attractiveness to predators decline rapidly. But treatments that break the seed coat and expose the cotyledons sometimes increase the risk of desiccation or make seeds more attractive to ants (Woods & Elliott, 2004). It could also be worth exploring the possibility of using chemicals to repel seed predators. Any carnivores that prey on rodents (e.g. raptors or wild cats) should be regarded as valuable assets on ANR sites. Preventing the hunting of such animals can help to control rodent populations and reduce seed predation.

Seedlings that germinate from seeds are tiny compared with planted, nursery-raised saplings, so weeding around the seedlings is especially important and it must be carried out with extra care. Such meticulous weeding can greatly increase the cost of direct seeding (Tunjai, 2011).

Carnivores, such as this leopard cat (*Felis bengalensis*), can help to control populations of seed-predating rodents, so capturing or killing them at restoration sites should be strongly discouraged.

Species suitable for direct seeding

Species that tend to be successfully established by direct seeding are generally those that have large (>0.1 g dry mass), spherical seeds with medium moisture content (36–70%) (Tunjai, 2012). Large seeds have large food reserves, so they can survive longer than smaller seeds and produce more robust seedlings. Seed predators find it difficult to handle large, round or spherical seeds, especially if such seeds also have a tough and smooth seed coat.

Tree species in the family Leguminosae are most commonly reported as being suitable for direct seeding. Legume seeds typically have tough, smooth seed coats, making them resistant to desiccation and predation. The nitrogen-fixing capability of many legume species can give them a competitive advantage over weeds. Tree species in many other families have also shown promise and are listed in **Table 5.2** (Tunjai, 2011).

Published accounts of direct seeding have tended to concentrate on pioneer tree species (Engel & Parrotta, 2001) because their seedlings grow rapidly, but climax forest tree species can also be successfully established by direct seeding. In fact, because they generally have large seeds and energy reserves, the seeds of climax forest trees may be particularly suited to seeding (Hardwick, 1999; Cole et al., 2011; Sansevero et al., 2011). With the disappearance of large, vertebrate seed-dispersers over much of their former ranges, direct seeding might be the only way that the large seeds of some climax tree species can reach restoration sites (effectively substituting human labour for the roles formerly played by such animals).

Table 5.2. Reports of species and techniques for successful direct seeding from around the tropics. (Prepared by Panitnard Tunjai.)

Location	Optimal sowing time	Forest type	Elevation (m)	Successful species	Recommended methods	Reference
S. Thailand	Early rainy season	Lowland evergreen	<100	*Artocarpus dadah* (Moraceae), *Callerya atropurpurea* (Leguminosae), *Vitex pinnata* (Lamiaceae), *Palaquium obovatum* (Sapotaceae) and *Diospyros oblonga* (Ebenaceae)	Tube to prevent seed movement, no mulching and fertiliser in first two years	Tunjai, 2012
		Dry dipterocarp	300–400	*Afzelia xylocarpa* (Leguminosae) and *Schleichera oleosa* (Sapindaceae)	No weeding after sowing in the first year; scarification to accelerate or maximise germination for both species with hard seed coat	
N. Thailand	Early rainy season	Hill evergreen	1,200–1,300	*Balakata baccata* (Euphorbiaceae), *Syzygium fruticosum* (Myrtaceae), *Aquilaria crassna* (Thymelaeaceae), *Sarcosperma arboreum* (Sapotaceae) and *Choerospondias axillaris* (Anacardiaceae)	No weeding after sowing in the first year	Tunjai, 2012
N. Thailand	Early rainy season	Hill evergreen	1,200–1,300	*Choerospondias axillaris* (Anacardiaceae), *Sapindus rarak* (Sapindaceae) and *Lithocarpus elegans* (Fagaceae)	Burying; pre-sowing seed treatments to accelerate or maximise germination	Woods & Elliott, 2004
Cambodia	Wet season	Deciduous	85	*Afzelia xylocarpa* (Leguminosae), *Albizia lebbeck* (Leguminosae) and *Leucaena leucocephala* (Leguminosae)	Soil ploughing by tractor and applying cow manure before sowing	Cambodia Tree Seed Project, 2004
Hong Kong	Early rainy season	Tropical semi-evergreen	200–550	*Triadica cochinchinensis* (Euphorbiaceae), *Microcos paniculata* (Malvaceae) and *Choerospondias axillaris* (Anacardiaceae)	Burying seeds at 1–2 cm below the soil surface	Hau, 1999
Australia	Rainy season	Complex mesophyll and notophyll vines	121–1,027	*Acacia celsa* (Leguminosae), *Acacia aulacocarpa* (Leguminosae), *Alphitonia petriei* (Rhamnaceae), *Aleurites rockinghamensis* (Euphorbiaceae), *Cryptocarya oblata* (Lauraceae) and *Homalanthus novoguineensis* (Euphorbiaceae)	Burying seeds; mechanical and chemical weeding prior to sowing and two subsequent applications of herbicide (glyphosate) 1 month apart; more consistent establishment when using species with large seeds	Doust et al., 2006

Table 5.2. continued.

Location	Optimal sowing time	Forest type	Elevation (m)	Successful species	Recommended methods	Reference
Brazil	Early rainy season	Seasonal semi-deciduous	464–775	*Enterolobium contortsiliquum* (Leguminosae) and *Schizolobium parahyba* (Leguminosae)	Herbicide (glyphosate) prior to sowing; additional spot application and manual weeding around seedlings	Engel & Parrotta, 2001
Brazil	Late rainy season	Seasonal semi-deciduous	574	*Enterolobium contortisiliquum* (Leguminosae) and *Schizolobium parahyba* (Leguminosae)	Soil ripper to prepare sowing lines at 40 cm depth	Siddique et al., 2008
Brazil	Late rainy season	Terra firme	N/A	*Caryocar villosum* (Caryocaraceae) and *Parkia multijuga* (Leguminosae)	Sowing large-seeded non-pioneer species	Camargo et al., 2002
Brazil	Early rainy season	Evergreen equatorial moist	–	*Spondias mombin* (Anacardaceae), *Parkia gigantacarpa* (Leguminosae), *Caryocar glabrum* (Caryocaraceae), *Caryocar villosum* (Caryocaraceae), *Couepia* sp. (Chrysobalanaceae), *Bertholletia excelsa* (Lecythidaceae), *Carapa guianensis* (Meliaceae) and 27 other species	On opencast mine: deep ripped to 90 cm, 15 cm top soil added; sow seeds along alternate rip lines, 2 × 2 m.	Knowles & Parrotta, 1995
Costa Rica	Early rainy season	Montane	1,110–1,290	*Garcinia intermedia* (Clusiaceae)	Sowing late-successional seeds after establishment of fast-growing and nitrogen-fixing trees	Cole et al., 2011
Mexico	–	Semi-evergreen, seasonal	–	*Brosimum alicastrum* (Moraceae), *Enterolobium cyclocarpum* (Leguminosae) and *Manilkara zapota* (Sapotaceae)	Sowing seeds in young successional forest (8–15 years) or reference forest (>50 years)	Bonilla-Moheno & Holl, 2010
Mexico	Early rainy season	Seasonal tropical	–	*Swietenia macrophylla* (Meliaceae)	Burying seeds 0.5 cm below soil surface; slash and burn to clear sites	Negreros & Hall, 1996
Jamaica	Early rainy season	Dry	140	*Eugenia* sp. (Myrtaceae) and *Calyptranthes pallens* (Myrtaceae)	Sowing seeds under shade with moisture supplementation	McLaren & McDonald, 2003
Uganda	Early rainy season	Moist evergreen semi-deciduous	1,250–1,827	*Strombosia scheffleri* (Olacaceae), *Craterispermum laurinum* (Rubiaceae), *Musanga leo-errerae* (Urticaceae) and *Funtumia africana* (Apocynaceae)	Loosening soil before sowing	Muhanguzi et al., 2005

Aerial seeding

Aerial seeding is a logical extension of direct seeding. It can be useful where direct seeding must be applied to very large areas, for restoring steep inaccessible sites, or where labour is in short supply. Many of the species choices and pre-sowing seed treatments developed for direct seeding can be applied equally well to aerial seeding.

China leads the way with this technology, having carried out dozens of research programs on aerial seeding since the 1980s and having applied the method to millions of hectares to establish plantations of mostly conifers and to reverse desertification. Burying seeds to prevent seed predation is not an option with aerial seeding, and so the Forestry Research Institute of Guangdong Province developed 'R8', a chemical repellent that deters seed predators. Similarly, the Forestry Research Institute of Beipiao, Liaoning Province developed a 'multi-purpose agent' that repels seed predators, prevents seed desiccation, improves rooting, and increases the resistance of seedlings to disease (Nuyun & Jingchun, 1995).

Previous aerial seeding for forestry in America and Australia (usually to establish monocultures of pines or eucalypts) involved dropping seeds, either unprotected or embedded in clay pellets, from planes or helicopters (Hodgson & McGhee, 1992). A more effective delivery system for mixed native tree species might consist of placing seeds in a biodegradable projectile that is capable of penetrating the weed cover and lodging the seeds in the soil surface. In addition to the seed itself, such projectiles could contain polymer gel (to prevent seed desiccation), slow-release fertiliser pellets, predator-repellent chemicals and microbial inoculae (Nair & Babu, 1994), which together would maximise the potential for seed germination, seedling survival and seedling growth. An aerial drone that is capable of accurately delivering up to 4 kg of seeds per flight using GPS technology is currently being investigated (Hobson, pers. comm.). A drone offers low-cost aerial delivery, provides the option of monitoring on a more frequent basis, and makes it possible to monitor hard-to-reach areas.

One of the major obstacles to the success of aerial seeding of large, inaccessible sites is the inability to carry out effective weeding and the consequent failure to protect the germinating seedlings from competition with herbs and grasses. Spraying herbicides from the air is routine in agriculture and could be used to clear restoration sites of weeds initially, provided there are few natural regenerants worth saving. After the tree seeds germinate, however, aerial herbicide sprays would kill the tree seedlings along with the weeds. Specific herbicides are needed that can kill weeds without killing either natural regenerants or seedlings germinating from aerially-delivered seeds.

Limitations of the framework species method

For recovery of tree species richness, the framework species method depends on nearby, remnant forest to provide i) a diverse seed source and ii) habitat for seed-dispersing animals. But how close does the nearest remnant forest need to be? In fragmented evergreen forest sites in northern Thailand, medium-sized mammals such as civets can disperse the seeds of some forest tree species up to 10 km. So the technique can potentially work within a few kilometres of forest remnants, but obviously, the closer the restoration site is to remnant climax forest, the faster biodiversity will recover. If seed sources or seed dispersers are absent from the landscape, recovery of tree species richness will not occur unless nearly all of the tree species from the original forest are replanted, either as seeds or as saplings raised in nurseries. This is the 'maximum diversity' approach to forest restoration.

BOX 5.3. RAINFORESTATION

Box 5.3. Rainforestation.

'Rainforestation' shares many similarities with the framework species method of forest restoration, particularly its emphasis on planting indigenous tree species at high densities to shade out herbaceous weeds and to restore ecological services, forest structure and wildlife habitat. But the Rainforestation method has been adapted to the particular ecological and socio-economic situation of the Philippines. With the most rapidly increasing and densest human population of any country in Southeast Asia (excluding Singapore), growing from 27 million in 1960 to 92 million (or 313 per km²) today, an annual growth rate of 2.1%[5], deforestation has left less than 7% of the country covered in old-growth forest. With so many of the Philippine's species endemic and in imminent danger of extinction because of dwindling primary forest cover, forest restoration clearly has a major role to play in conserving biodiversity. On the other hand, with such intense human pressure, the need is for restoration methods that also generate cash income.

"Introducing the idea of 'let's plant for our forests', the farmers always said we have to think of improving their farming also, so why not include a livelihood component? Rainforestation is a strategy for restoring the forest but at the same time, it can be a way to improve the income of farmers, so you have to enhance it by including crops ... so it becomes a farming system." Paciencia Milan (interview 2011)

Pioneer trees are usually planted in the first year, followed by shade-tolerant, climax tree species (often Dipterocarps) that are under-planted in the second year. Planting densities vary according to project objectives: for example, for timber production, 400 trees/ha (25% pioneers to 75% climax timber trees); for agro-forestry, 600–1000 trees/ha (depending on the canopy of the fruit trees being incorporated); and for wildlife conservation, 2,500 trees/ha. Because wind-dispersed, dipterocarp tree species dominate the Philippines' forests and the remaining primary forest is often reduced to remote fragments, seed dispersal from forest to restoration sites by animals is less evident in Rainforestation than it is with the framework species method.

The concept of Rainforestation was jointly developed by Prof. Paciencia Milan of Visaya State University (VSU, formerly Visayas State College of Agriculture) and Dr Josef Margraf of GTZ (Deutsche Gesellschaft für Internationale Zusammenarbeit) under the ViSCA-GTZ Applied Tropical Ecology Program. The first trial plots were established in 1992 on 2.4 ha of *Imperata* grassland within VSU campus that had patches of coffee, cocoa and banana and portions of pasture.

A 19-year-old original Rainforestation demonstration plot, planted in 1992 in VSU's 625-ha forest reserve on the lower slopes of Mt Pangasugan (50 m elevation). Originally *Imperata* grassland, the site now supports forest that has a complex structure and highly diverse flora and fauna, including the endangered Philippines Tarsier.

[5] 2010 figures at www.prb.org/Publications/Datasheets/2010/2010wpds.aspx

Box 5.3. continued.

Rainforestation quickly evolved from the original concept of an ecological approach to rain forest restoration into 'Rainforestation Farming' or 'closed-canopy and high-diversity forest farming', designed to meet the economic needs of local people by including the cultivation of fruit trees and other crops alongside forest trees. The basic premise is that "the closer the structure of a tropical farming system is to a natural rain forest; the more sustainable it is". The aim of Rainforestation Farming is to sustain food production from tropical forests, while maintaining the forest's biodiversity and ecological functioning. The idea is to replace the more destructive forms of slash-and-burn agriculture with more ecologically sustainable and profitable agricultural systems.

From 1992 to 2005, VSU established 25 Rainforestation demonstration farms on various soil types on Leyte Island, and monitored them in collaboration with local villagers. Rainforestation not only provided farmers with income, it also re-established forest ecosystems with high biodiversity and improved the soil quality. The technique has now diversified into three major types (with 10 sub-types) for different purposes: i) biodiversity conservation and environmental protection (e.g. the introduction of buffer zones and wildlife corridors into protected areas, landslide prevention or riverbank stabilisation); ii) timber production and agro-ecosystems; and iii) projects in urban areas (e.g. road beautification or the introduction of parks). Different tree species and management techniques are recommended to optimise the conservation and/or economic outputs of each project sub-type, but the use of native forest tree species remains central to the Rainforestation concept.

"Rainforestation need not be just for forest restoration. It can be used for other reasons, provided native trees are planted." Paciencia Milan (interview 2011)

A 15-year-old, registered, community-based Rainforestation Farm, established in an over-mature coconut plantation in 1996 by planting 2,123 trees/ha, including 8 species of Dipterocarpaceae and an understorey of shade-tolerant fruit trees (e.g. mangosteen or durian). Benefits are shared amongst community members proportional to their voluntary labour inputs.

Box 5.3. continued.

Rainforestation has been accepted as a national strategy for forest restoration by the Philippine Department of Environment and Natural Resources (Memorandum Circular 2004-06). Native species nurseries and Rainforestation demonstration plots are now being established to further develop the technique at more than 20 state universities and colleges throughout the Philippines, supported by the Philippine Tropical Forest Conservation Foundation and the Philippine Forestry Education Network. The Environmental Leadership & Training Initiative, together with the Rain Forest Restoration Initiative and FORRU-CMU, are working with these institutions to promote the adoption of standardised research and monitoring protocols to facilitate the creation of a national database of native tree species, and the adaptation of Rainforestation to the myriad of social and environmental settings in the Philippines.

Sources: Milan *et al.* (undated and interview 2001); Schulte (2002).
For latest information please log on to the Rainforestation Information Portal at
www.rainforestation.ph/

5.4 Maximum diversity methods

The term 'maximum diversity method' was first coined by Goosem and Tucker (1995), who defined the approach as "attempts to recreate as much as possible of the original (pre-clearance) diversity". The method effectively attempts to recreate the tree species composition of climax forest by intensive site preparation and a single planting event, simultaneously counteracting both habitat and dispersal constraints. For sites in the wet tropics of Queensland, Australia, Goosem and Tucker (1995) recommended intensive site preparation, including deep ripping, mulching and irrigation, as required, followed by the planting of 50–60 cm saplings of up to 60, mostly climax, tree species, spaced 1.5 m apart.

> "*The method is well-suited to smaller plantings, where intensive management is possible and for areas isolated from any native vegetation, which could provide seeds.*" Goosem & Tucker (1995)

The maximum diversity approach becomes applicable wherever natural seed dispersal has declined to such an extent that it is no longer capable of recovering tree species richness in restoration sites at an acceptable rate. This may be because too few individuals or species of seed trees remain within seed-dispersal distances of the restoration sites or because seed-dispersing animals have become rare or extirpated. The absence of this 'free' seed dispersal service must therefore be compensated for by planting most, if not all, of the tree species that comprise the target climax forest, ensuring high tree species richness and the representation of dispersal-limited tree species right from the start of the restoration process.

> "*People planting trees, replace birds dispersing seeds.*"

Consequently, maximum diversity methods of forest restoration are much more intensive and costly than framework species techniques. The difference in costs between the two methods can be viewed as the monetary value of the lost seed dispersal mechanisms.

Expenditure is high at all stages of the process. First, a great deal of research is needed to achieve an effective plantation design, and research is not cheap. Seed collection and propagation of the full range of tree species that comprise the target climax forest ecosystem are both technically difficult and expensive.

Forest patches that are restored by this method tend to be isolated from natural forest, so unfortunately, they are affected by all the problems of fragmentation described in **Section 4.3**. Management efforts may be necessary to i) reduce edge effects (e.g. by densely planting buffer zones with shrubs and small trees as wind breaks, see **Section 4.4**) and ii) retain the small plant and animal populations that might eventually colonise such forest patches.

Planted climax forest trees grow slowly, so trees have to be planted close together to compensate for the delay in canopy closure and shading out of herbaceous weeds (see **Box 5.4**, p. 140). When compared with ANR and the framework species method, the delay in canopy closure means that weed control must continue for longer. Furthermore, climax trees take many years to mature and produce seeds from which an understorey of climax tree saplings can develop. In the meantime, restoration plots can become invaded by undesirable woody weed species (Goosem & Tucker, 1995), which eventually compete with the seedling progeny of the planted climax trees. Eradicating such undesirable undergrowth also adds to costs.

Because of the high costs, the maximum diversity approach has only been implemented by organisations with the financial resources and/or the legal obligation to do so, particularly mining companies, other large corporations and urban authorities.

Mining companies were among the first to experiment with the maximum diversity approach, mainly because of legal requirements to restore opencast mines in tropical forest areas to their original condition. Working at an opencast bauxite mine in Central Amazonia, Knowles and Parrotta (1995) recognised the need to screen the widest possible range of native tree species for possible inclusion in reforestation programs "where natural succession is retarded by physical, chemical and/or biological barriers", in order to "replicate, in an accelerated fashion, natural forest successional processes that lead to complex, self-sustaining forest ecosystems".

> "By including a broad range of tree species in the screening program ... irrespective of their commercial value ... it is far more probable that diversified forests can be re-established that resemble and function as natural forests."
> Knowles & Parrotta (1995)

Even though primary forest grew close to the mine, seed dispersers rarely visited the restoration sites because on-going mining operations created barriers such as desolate open areas and roads with heavy traffic. So the framework species method, which depends on natural seed-dispersal for its success, would not have facilitated tree species recruitment.

Consequently, Knowles and Parrotta systematically screened 160 tree species (around 76%) of the evergreen equatorial moist forest near the mine, to develop a system for selecting species that were suitable for multi-species plantations on an operational scale. They developed a species ranking system (a similar approach is described in **Section 8.5**) that was based on seed germinability, planting stock type and early growth rates. Tree taxa that were recommended for initial plantings were classed as 'highly suitable sun-tolerant' and 'suitable, though prefer shaded conditions initially' (59 taxa (37% of those tested) and 30 taxa (19%), respectively). The remaining 71 shade-demanding taxa represented nearly half of the tree taxa of the target forest ecosystem, and hence Knowles and Parrotta recommended that these taxa should be planted about 5 years later, once the initially planted trees had created the shade and soil conditions conducive to their establishment. Thus, Knowles and Parrotta essentially advocated a two-stage maximum diversity approach, using mostly sun-tolerant pioneers to create the conditions needed for the subsequent addition of all of the other tree species that were representative of the target forest ecosystem.

Restoration sites were levelled and covered with 15 cm of top soil within a year of forest clearance and bauxite extraction. They were deep ripped to 90 cm depth (1 m between rip lines) and tree propagules (direct seeding (**Table 5.2**), wildlings or nursery-raised seedlings) were planted along alternate rip lines at 2 × 2 m spacing (2,500 plants/ha). At least 70 species were planted in a pattern that ensured that trees of the same species were not planted adjacently.

The maximum diversity approach is also particularly suited to urban forestry, adding biodiversity to cityscapes and providing city-dwellings with a rare opportunity to connect with nature. Urban authorities have the responsibility to take care of parks, gardens and roadsides and have budgets that are large enough to pay for intensive landscaping operations. On urban sites, the high costs of maximum diversity techniques are justified by the heavy use and appreciation of urban forests by dense populations and by the high value of the land. When planting trees on urban land, it is important to ensure that they do not disrupt electricity cables or water pipes. Aesthetic considerations, such as the attractiveness of the planted tree species, must also be considered (Goosem & Tucker, 1995).

In summary, the maximum diversity approach can be implemented by single plantings of mostly climax forest tree species or by two-stage plantings, beginning with mostly pioneer trees and following-up, after the pioneers close canopy, by under-planting with shade-dependent climax tree species. The aim is to plant most of the tree species that comprise the target climax forest. However, the difficulties of seed collection and limited nursery capacity have to date limited maximum diversity trials to 60–90 tree species. Most species should be represented by at least 20–30 trees/ha. Greater prominence can be given to i) large-seeded species, ii) 'keystone' species (e.g. *Ficus* spp.) and iii) endangered, vulnerable or rare species to increase the biodiversity conservation value of the operation. Usually, the planting and maintenance methods that are used for the framework species approach (i.e. weeding, mulching and fertiliser application see **Section 7.3**) can also be used for the maximum diversity approach (Lamb, 2011, pp. 342–3), although more intensive site preparation, such as deep ripping, may be necessary at severely degraded sites (Goosem & Tucker, 1995; Knowles & Parrotta, 1995).

5.5 Site amelioration and nurse plantations

At sites with stage-5 degradation, where soil and microclimatic conditions have deteriorated beyond the point at which they can support tree seedling establishment, site amelioration becomes a necessary precursor to forest restoration procedures. Soil compaction and erosion are usually the main problems, but exposure to hot, dry, sunny and windy conditions can also prevent tree establishment, even where soil conditions are not so severely degraded. Site amelioration can involve soil cultivation procedures that are more usually associated with agriculture and commercial forestry (such as those used in the Miyawaki method, see **Box 5.4**, p. 140), and/or establishing plantations of

Box 5.4. The Miyawaki method.

One of the earliest, and perhaps most famous, forms of the maximum diversity approach is the Miyawaki method, invented by Dr Akira Miyawaki, Professor Emeritus of Yokohama National University, Japan and director of the IGES-Japanese Centre for International Studies in Ecology (JISE). Developed in the 1970s, the method is based on 40 years of studies of both natural and disturbed vegetation, all around the world. It was first employed to restore forests at hundreds of sites in Japan, and was subsequently modified successfully for application to tropical forests in Brazil[6], Malaysia[7] and Kenya[8].

The Miyawaki method, or 'Native Forest by Native Trees', is based on the concept of 'potential natural vegetation' (PNV) (synonymous with 'target forest type'): the idea that the climax vegetation of any disturbed site can be predicted from current site conditions, such as existing vegetation, soil, topography and climate. Therefore, restoration begins with detailed soil surveys and vegetation mapping (using phytosociological methods), which are combined to produce a map of PNV units across the restoration site. The PNV map is then used to select tree species for planting and to prepare the project plan (Miyawaki, 1993).

The next stage is to collect seeds, locally, of the trees species representative of the PNV(s). Seedlings of all of the dominant tree species within the PNV(s), and as many associated species (particularly mid- to late-successional species) as possible are grown to 30–50 cm tall in containers in nurseries ready for planting out. Site preparation can involve using earth-moving machines to level or terrace the site and developing a 20–30-cm layer of good top soil, by mixing straw, manure or other kinds of organic compost in with the upper soil layers. On eroded sites, top soil is imported from urban construction sites. The soil is then mounded to increase aeration. Up to 90 tree species are planted, randomly, at very high densities, 2–4 trees/m². After planting, the site is weeded (and the pulled weeds applied as mulch) for up to three years, by which time canopy closure is achieved and maintenance ceases.

"After three years, no management is the best management" (Miyawaki, 1993)

Prof. Akira Miyawaki (in the green hat) poses with children planting trees in Kenya as part of a project using his now famous technique. (Photo: Prof. K. Fujiwara.)

[6] www.mitsubishicorp.com/jp/en/csr/contribution/earth/activities03/activities03-04.html
[7] Currently through a collaborative project involving UPM, Universiti Malaysia Sarawak and JISE, which is sponsored by the Mitsubishi Corporation.
[8] www.mitsubishicorp.com/jp/en/pr/archive/2006/files/0000002237_file1.pdf

Box 5.4. The Miyawaki method

Box 5.4. continued.

The first tropical trials using the Miyawaki method started in 1991 on the Bintulu (Sarawak) Campus of Universiti Pertanian Malaysia (currently known as Universiti Putra Malaysia (UPM))[8]. Eighteen years later, plots restored by the Miyawaki method showed better forest structure and the planted trees were taller, and had wider diameter at breast height (dbh) and greater basal area compared with those of adjacent naturally regenerating secondary forest (Heng *et al.*, 2011). Recovery of the soil fauna is particularly rapid (Miyawaki, 1993). Experiments in northern Brazil, however, were less successful: fast-growing economic pioneers were used in the species mix and these both rapidly over-topped and slowed the growth of the late-successional native species and were more susceptible to wind-throw (Miyawaki & Abe, 2004). Although the high planting density rapidly results in a closed canopy, it can sometimes have undesirable effects. Competition among the closely planted trees can result in high initial mortality and low dbh (more than 70% of trees had a dbh of less than 10 cm when measured 18 years after planting (Heng *et al.*, 2011)).

Sixteen-yr-old plots restored by the Miyawaki method at the Bintulu Campus of Universiti Pertanian Malaysia (UPM). The closely spaced planted trees grew well, creating a multilayered main canopy (left) and completely eliminating weeds (right). (Photos: Mohd Zaki Hamzah.)

The intensive nature of the Miyawaki method (particularly the need for expert-driven site surveys, mechanical site preparation and very high planting densities) means that it is among the most expensive of all forest restoration techniques. As such, it is heavily dependent on the sponsorship of wealthy corporations (e.g. Mitsubishi[9], Yokohama[10], Toyota[11]) and its use is largely confined to 're-greening' small, high-value, industrial or urban sites for recreational and climate amelioration purposes. Benefits to the corporate sponsors include improved public relations, particularly the promotion of a 'green image'. In Japan, the potential of the method for disaster mitigation in urban areas is also being advocated.

[9] www.mitsubishicorp.com/jp/en/csr/contribution/earth/activities03/
[10] yrc-pressroom.jp/english/html/2008916l2mg001.html
[11] www.toyota.co.th/sustainable_plant_end/ecoforest.html

highly resilient tree species to improve the soil and modify the micro-climate — the so-called 'nurse' plantation approach (also known as "plantations as catalysts" (Parrotta et al., 1997a) or "foster ecosystems" (Parrotta, 1993).

Opencast mine sites provide probably the most extreme examples of site degradation. The replacement of top soil and the deep ripping of mine sites have already been mentioned in **Section 5.4** in connection with the maximum diversity method. Deep ripping, sometimes known as sub-soiling, involves slicing thin furrows (up to 90 cm deep, about 1 m apart) through the soil with strong, narrow tines, without inverting the soil. Deep ripping merely opens up soils that have become compacted (e.g. due to machinery or livestock trampling) allowing water and oxygen to penetrate into the subsoil, where the roots of planted trees will subsequently grow. It is carried out by heavy machinery, and so is possible only on relatively flat and accessible sites, and it is very expensive[12]. Mounding is another physical treatment that can improve soil conditions by aerating the soil and reducing the risk of water-logging.

The addition of organic materials such as straw and other organic waste materials (even orange peel from a juice factory was trialled during the ACG project (see **Box 5.2**, p. 119) (Janzen, 2000)) improves soil structure, drainage, aeration and nutrient status and promotes the rapid recovery of soil fauna.

Green mulching (or 'green manure') is a biological approach to soil improvement. It involves sowing the seeds of herbaceous legumes across the restoration site, harvesting their seeds and then mowing the plants. The dead plants are left to decompose on the soil surface or are worked into the upper soil layers with hoes or ploughs. Seeds of commercial legume species can be purchased at agricultural supply stores, but a cheaper and more ecological approach (although more time-consuming) is to select a mix of herbaceous legume species that grow naturally in the area and harvest their seeds for sowing on the restoration site. If seeds are then collected from the plants before mowing them, the seed stock gradually accumulates with each green-mulching cycle, and eventually seeds can be used for other sites. It may be necessary to repeat the procedure for several years before the soil is ready to support tree seedlings. Green mulching can suppress weed growth without the use of herbicide, protect the soil surface from erosion, improve soil structure, drainage, aeration and nutrient status, and facilitate recovery of the soil macro- and micro-fauna.

The application of chemical fertilisers also improves soil nutrient status, but does not provide the benefits to soil structure and fauna offered by organic materials. Several techniques can be employed to determine which soil nutrients are in short supply, including observation of visual symptoms of nutrient deficiency, chemical analyses of soil and/or leaves, and nutrient-omission pot trials (Lamb, 2011, pp. 214–9). However, most of these techniques are expensive and require specialised expertise. If they are considered to be impractical or too expensive, the application of a general-purpose fertiliser (NPK 15:15:15 at 50–100 g per tree) should solve most nutrient-deficiency problems.

Additional opportunities to apply soil treatments arise when holes are dug for tree planting. It is common practice on highly degraded sites to add compost into holes before planting trees (about 50:50 mixed with the backfill from the planting hole). Water-absorbing polymer gels can also be added to planting holes: either 5 g of the dried pellets mixed with the backfill or, in dry soils, two tea-cupfuls of a hydrated gel.

---

[12] www.nynrm.sa.gov.au/Portals/7/pdf/LandAndSoil/10.pdf

Various types of gel are available and the terminology for naming them is confusing and often inconsistent, so discuss options with your agricultural supplier and read the instructions on the product packaging. Laying mulch around the planted trees also helps to preserve soil moisture, adds nutrients and creates conditions that favour soil fauna.

Severely degraded soils probably lack many of the strains of micro-organisms that are required for high performance all of the tree species being planted (particularly the nitrogen-fixing *Rhizobium* or *Frankia* bacteria that form symbiotic relationships with legumes, and the mycorrhizal fungi that improve nutrient absorption for most tropical tree species). Mixing a handful of soil from the target forest ecosystem with compost added to the planting holes is probably the simplest and cheapest way to initiate the recovery of the soil micro-flora.

Another possibility is to inoculate trees in nurseries. Simply including forest soil in the potting medium usually ensures that the trees become infected with beneficial micro-organisms. Research suggests, however, that applying inoculae obtained by culturing micro-organisms collected from adult trees has additional potential to accelerate tree growth. For example, Maia and Scotti (2010) showed that inoculating the leguminous tree *Inga vera*, which is widely used for riparian forest restoration in Brazil, with *Rhizobia* reduced the fertiliser requirement by up to 80% and improved growth. *Rhizobia* inoculae are commercially produced for agricultural legume crops, but they cannot necessarily be used for forest trees because different legume species require different strains of *Rhizobium* for optimum nitrogen fixation (Pagano, 2008). It is unlikely that the specific strains of *Rhizobium* required for the tree species being planted will be commercially available. Making the inoculum entails collecting bacteria from the same tree species and culturing them in a laboratory. The same is true for mycorrhizal fungi. The application of a commercially produced mix of 'ubiquitous' mycorrhizal fungi species to forest tree seedlings grown in a nursery in northern Thailand failed to produce any benefits (Philachanh, 2003).

The planting of 'nurse trees' (Lamb, 2011, pp. 340–1) can improve site conditions, paving the way for subsequent restoration practices to recover biodiversity. By rapidly re-establishing a closed canopy and litter fall, plantations can create cooler, shadier and more humid conditions both above and below the soil surface. This should lead to the accumulation of humus and soil nutrients and, ultimately, to much better conditions for the subsequent seed germination and seedling establishment of less tolerant tree species (Parrotta *et al.*, 1997a)[13]. Such plantations are also capable of producing wood and other forest products at an early stage in the restoration process.

Nurse tree plantations are generally composed of a single (or just a few) fast-growing, pioneer species that is tolerant of the harsh soil and micro-climatic conditions prevalent on sites with stage-5 degradation, but that is also capable of improving the soil. Native tree species are preferred because of their ability to promote biodiversity recovery more rapidly than exotics (Parrotta *et al.*, 1997a). A study of the local tree flora will usually reveal indigenous pioneer tree species that grow just as well as any imported exotics.

---

[13] A special issue of *Forest Ecology and Management* (Vol. 99, Nos. 1–2) published in 1997 was devoted to the potential of tree plantations to 'catalyse' tropical forest restoration. Using 'tree plantations' in its broadest sense (from monocultures to maximum diversity), the 22 papers therein have become essential reading for those involved in tropical forest restoration.

Nevertheless, exotics may be used as nurse trees provided they meet the following conditions:

1) they are incapable of producing viable seedlings and thus becoming woody weeds and ...

2) either, they are short-lived, sun-loving pioneer species that will be shaded out by subsequently introduced climax forest trees or ...

3) they are purposefully killed (e.g. harvested or ring barked and left in place to rot) after they have brought about site improvement and the saplings of replacement trees are well established.

For example, the use of the exotic *Gmelina arborea* in the ACG project (see **Case Study 3**, p. 149) was justified because its sun-loving seedlings could not establish beneath its own canopy and its large, animal-dispersed seeds were not spread outside the plantation. By contrast, the use of the exotic plantation tree *Acacia mangium* in Indonesia is becoming a major problem for future forest restoration because its seedlings rapidly dominate areas around plantations. Their removal from future forest restoration sites will be very expensive. The same is true of *Leucaena leucocephala* in South America and tropical northern Australia. Seedlings of exotic species might be easier to obtain from commercial tree nurseries, but if you are unsure whether the species being considered meets the criteria listed above, it is better to search through the local tree flora for an indigenous alternative.

Plantation species should be light-demanding pioneers (as are many commercial timber trees), extremely hardy and short-lived. In general, better results have been achieved with broadleaved species than with conifers. Planting stock should be of the highest quality (Parrotta *et al.*, 1997a).

Legumes (i.e. members of the Family Leguminosae) and indigenous fig tree species (*Ficus* spp.) nearly always make good nurse plantation species as well as useful framework tree species (see **Section 5.3**). The roots of fig trees are capable of invading and breaking apart compacted soils and even rocks on the most degraded of sites, whereas the nitrogen-fixing capability of many leguminous tree species can rapidly improve soil nutrient status. Planting mixtures of figs and legumes as nurse plantations could, therefore, improve both the physical structure and the fertility of soils, without the need for the intensive and expensive physical soil treatments described above or the application of nitrogen fertiliser.

When establishing a conventional tree plantation, it is tempting to follow conventional forestry production practices. But the design and management of nurse plantations for forest restoration requires a more considered approach. Canopy closure is the first objective of the plantation, and so the trees should be planted closer than is usual for commercial forestry (Parrotta *et al.*, 1997a). If possible, find trees of the same species planted nearby, and try to determine roughly how broad their crowns are after 2–3 years of growth. This will provide the planting distance necessary to close the canopy in 2–3 years. Lamb (2011) recommends a planting density of 1,100 trees per hectare. The canopy should be dense enough to shade out weeds but not so dense as to inhibit the growth of subsequently planted trees or to prevent colonisation of the site by naturally dispersed, incoming tree species.

Conventional forestry demands intensive weeding or 'cleaning up' of plantations. Provided herbaceous weeds do not threaten the early survival of the planted nurse tree saplings (on stage-5 sites, degradation usually limits even weed growth), then weeding is not necessary. Even where it is required, weeding should cease as soon as the crowns

of planted saplings have grown above the weed canopy. On sites where incoming seed dispersal could still be possible, over-vigorous weeding will knock back any tree seedlings that do manage to become established.

As site conditions improve, the nurse trees can be thinned out and replaced by planting a wider range of native forest tree species. This should be done gradually to prevent invasion of the now-fertile soil by light-loving herbaceous weeds. If the nurse trees are of a commercial species, the felled trees can provide income to project participants over several years. When carrying out thinning, precautions must be taken so as not to disturb the understory and thus damage the accumulated biodiversity. Hauling logs out from a plantation without damaging the undergrowth is not easy to say the least, but various 'minimum impact' or 'reduced-impact' logging (RIL) techniques (e.g. using animals instead of machinery) are now being promoted (Putz *et al.*, 2008).

Where seed-dispersal into a restoration site might still be possible, framework tree species should be planted as the nurse trees are gradually cleared: pioneer framework species to replace the nurse trees and climax framework species to build up the understory. But in most restoration sites with stage-5 degradation, seed sources and/ or seed-dispersing animals will have been eliminated from the surrounding area, so biodiversity recovery requires the maximum diversity approach.

The use of nurse plantations is not necessarily restricted to stage-5 degradation with limiting soil conditions. They have often been used on less severely degraded sites, where natural seed dispersal still operates, as a simpler and cheaper alternative to

If fig trees can germinate in and subsequently tear apart the building blocks of Angkor Wat, Cambodia, they will have no difficulty in penetrating even the most severely degraded soils.

the framework species method. The use of plantations of exotic tree species, such as *Gmelina arborea*, adjacent to surviving forest in Costa Rica is described in **Case study 3**. A native species, *Homalanthus novoguineensis*, was used with similar success in Australia to attract seed-dispersing birds from nearby forest into restoration sites (Tucker, pers. comm.). Plantations of the exotic *Eucalyptus camaldulensis* did not, however, facilitate regeneration of native Miombo woodlands in the Ulumba Mountains of Malawi (Bone *et al.*, 1997).

In Costa Rica, a nurse crop of the exotic *Gmelina arborea* stimulated native tree establishment and generated income from felling after 8 years

5.6 Costs and benefits

Practitioners of forest restoration are often asked: "Why don't you just plant economic species?" The answer to this question is: "There is no such thing as a *non*-economic tree species". All trees sequester carbon and produce oxygen, all contribute to watershed stability and all are made of a highly combustible fuel. The question is not whether forest restoration is economic, but whether economic benefits can be converted into cash flows.

How much does it cost?

Very few accounts of the costs of forest restoration have been published (**Table 5.3**). This reflects both the difficulty of carrying out meaningful cost comparisons and perhaps also poor record-keeping among forest restoration practitioners and/or their unwillingness to disclose financial information. Comparing costs among methods and locations is confounded by fluctuations in exchange rates, inflation and huge variations in the costs of labour and materials. Costs are highly location- and time-specific. But precise cost calculations are not needed to show the obvious: the costs of restoration increase from stage-1 degradation to stage-5 degradation as the intensity of the methods required increases.

Table 5.3. Examples of published costs for various forest restoration methods.

Degradation stage	Method	Country	Published cost (US$/ha)	Date	Reference	Present-day costs US$/ha*[14]
Stage 1	Protection	Thailand	–	–	Estimated	300–350
Stage 2	ANR (**Box 5.2**)	Philippines	579	2006–09	Bagong Pagasa Foundation, 2009	638–739
	ANR (Castillo, 1986)	Philippines	500–1,000	1983–85	Castillo, 1986	1,777–3,920
Stage 3	Framework species method (**Section 5.3**)	Thailand	1,623	2006	FORRU, 2006	2,071
Stage 4	Maximum diversity with mine site amelioration (**Section 5.4**)	Brazil	2,500	1985	Parrotta et al., 1997b	8,890
	Miyawaki method (**Box 5.4**)	Thailand	9,000	2009	Toyota, pers. comm.	9,922
Stage 5	Site amelioration and nurse plantation		–	–	–	None found ?

*total costs for whole period needed to achieve a self-sustaining system

Potential value of the benefits

The potential economic value of the benefits of achieving a climax forest ecosystem, in terms of ecological services and diversity of forest products, is the same, regardless of the starting point. The Economics of Ecosystems and Biodiversity study (TEEB, 2009)[15] put the average annual value of fully restored tropical forests at US$6,120/ha/yr in 2009 (**Table 1.2**), equivalent to US$6,747 today, allowing for inflation. Even the most expensive forest restoration methods do not cost more than US$10,000/ha in total, so the value of potential benefits from a restored forest far outweighs the costs of establishment within very few years after the climax forest condition is achieved.

The speed of delivery of those benefits depends, however, on the initial degradation stage and on the restoration methods used. As the degradation stage increases, the time required to realise the full range of potential benefits increases from a few years to several decades. Therefore, the return on investment is delayed. The full potential benefits of forest restoration, in cash terms, can only be realised if they are marketed and people are prepared to pay for them. Schemes to market forest products and ecotourism or to sell carbon credits and 'payments for environmental services' (PES) all require a great deal of development and upfront investment before the full cash potential of restored forests can be realised (see **Chapter 1**).

[14] estimated by applying a constant 5% annual inflation rate.
[15] www.teebweb.org/

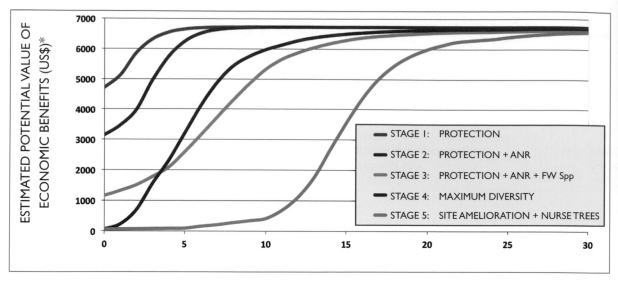

YEARS AFTER STARTING RESTORATION

Hypothetical curves representing the increase in potential economic benefits over time of the five major approaches to forest restoration. The restoration of stage-1 degradation yields considerable benefits from the start, whereas projects to restores sites with degradation at stages 4 and 5 start off by yielding zero benefits. At such sites, the initial increase in potential economic benefits is slow, until canopy closure promotes an influx of recruited species, which increase the rate of accumulation of benefits (e.g. more biodiversity leads to more forest products, or leaf litter improves soil water-holding capacity). As the accumulation of benefits nears the maximum, the rate of increase slows because the final few benefits take a long time to achieve (due to their dependence on slow environmental processes or the return of rare species). Note that the maximum diversity method achieves more rapid economic benefits because more tree species are planted at the start. With stage-5 degradation, site amelioration and the cultivation of nurse trees yield low economic benefits until a more diverse tree community is established.

Degradation	Restoration Costs	Incremental increase in benefits	Delivery of full benefits
Stage 1	LOW	SMALL	RAPID
Stage 2			
Stage 3	↓	↓	↓
Stage 4			
Stage 5	HIGH	LARGE	DELAYED

Summary of the economic costs and benefits of restoring the different stages of forest degradation.

CASE STUDY 3   Area de Conservacion Guanacaste (ACG)

Country: Costa Rica

Forest type: A mosaic of dry tropical forest, rain forest and cloud forest fragments surrounded by pasture.

Ownership: The Guanacaste Dry Forest Conservation Fund (GDFCF) has funded the purchase of 13,500 hectares of forest from private owners.

Management and community use: Cattle grazing and potential to harvest the 'nurse plantation' of *Gmelina arborea*.

Level of degradation: Cleared of all but fragments of forest for livestock and crop agriculture.

One of the first large, scientifically based, forest restoration projects in Central America is continuing in the Area de Conservacion Guanacaste (ACG) in north-western Costa Rica (www.gdfcf.org/). Largely the brain-child of American biologist Daniel Janzen and his wife Winnie Hallwachs, the project has become a classic example of how landscape-level forest restoration can be achieved, mostly through the protective measures described in **Section 5.1** and then by letting nature take its course.

The project site, Hacienda Santa Rosa (the second Spanish ranch founded in Costa Rica) was cleared of all but fragments of its dry tropical forest, beginning in the late 1500s, and was used mainly for mule and cattle ranching, wild meat, water for irrigation, and croplands. The Inter-American Highway was carved through its centre in the 1940s and jaragua pasture grass (*Hyparrhenia rufa*) was introduced from East Africa. This grass provided much of the fuel for human-caused, annual dry-season fires, which effectively blocked forest succession, because the ranchers wanted 'clean' pasture. The result was a mosaic of dry forest, rain forest and cloud forest fragments surrounded by pasture.

In 1971, the 10,000 ha Santa Rosa National Park was designated. In the 1990s, the 165,000-ha ACG expansion became part of the new Sistema National de Areas de Conservacion (SINAC), one of 11 conservation units that cover about 25% of Costa Rica. Cattle and horses were removed, but this allowed the jaragua grass to grow up to 2 m tall, fuelling ravenous fires that annually consumed trees and forest remnants. If the fires could not be stopped, there would soon be no forest remnants left to supply the tree seeds needed for restoration.

In September 1985, Janzen and Hallwachs wrote an unsolicited plan for the long-term survival of Santa Rosa's dry forest, which became the Projecto Parque Nacional Guanacaste (PPNG). The project's mission included: i) allowing seeds from forest remnants to restore 700 km² of the original dry forest to "maintain in perpetuity all animal and plant species and their habitats originally known to occupy the site"; ii) "offering a menu of material goods" to society; and iii) providing a study site for ecological research and a "re-awakening to the intellectual and cultural offerings of the natural world".

"The technological recipe for the restoration of this large dry forest ecosystem was obvious: purchase large tracts of marginal ranch and farmland, adjacent to Santa Rosa, and connect it with the wetter forests to the east, stop the fires, farming, and the occasional hunting and logging, and let nature take back its original terrain" (Janzen, 2002).

Guanacaste Province residents were hired to prevent fires, but with the grass growing so high, the fires were difficult to control with hand tools. A major part of the solution was to bring back the cattle. During the first five years of the project, ACG's to-be-restored-to-forest pastures were rented out as grazing land for up to 7,000 head of cattle at any one time. The cattle acted as 'biotic mowing machines', keeping fuel loads so low that the fire-control program could manage the less-severe fires. As naturally established trees grew up and began to shade out the grass, the cattle were removed.

Tree planting was also tried in a few select sites for a couple of years, but this was abandoned because natural forest regeneration from seeds, which were dispersed by the wind and vertebrates into the restoration sites from the interspersed patches of secondary forest, far outweighed the effort and expense of planting trees.

In the rain forested part of ACG, however, the natural regeneration of abandoned pastures was much slower. Compared with dry forest, fewer plant species were wind-dispersed, fewer animal seed-dispersers ventured out from the forest into rain forest pastures, and the survival of tree seedlings was hindered by the hot, dry and sunny conditions of the pastures. In such areas, a 'nurse plantation' (see **Section 5.5**) approach was employed, using abandoned plantations of the exotic timber tree species, *Gmelina arborea*. The dense canopies of *G. arborea* plantations shaded out grasses in 3–5 years and the understorey filled with a diverse community of rain forest trees, shrubs and vines, which were brought in as seeds by small vertebrates from neighbouring rain forest. After one rotation of 8–12 years, the *G. arborea* logs could have been harvested and the stumps killed with herbicide, generating income to support the project, but owing to a lack of purchasers, ACG elected to let the trees die of old age at 15–20 years. Such trials demonstrated that, provided forest seed sources and seed-dispersing animals remain nearby, rain forest pastures could easily be transformed into young rain forest by planting them with *G. arborea* and then abandoning the forest (rather than pruning and cleaning as is normal with a plantation).

In the 1980s, when Janzen and Hallwachs initiated the project, forest restoration was a new idea, a departure from the classical notion that national parks were created only to protect existing forest. The project was disapproved of by several conservation NGOs, which were surviving largely on the fund-raising slogan of "once tropical forest is cut, it is gone forever." Today, attitudes have changed. ACG and Janzen's publications are regarded as milestones of tropical forest restoration science. Having firmly established many of the practices needed to restore tropical forest in Costa Rica, the need now is to determine how to bring about and maintain the stable political and sociological conditions that will enable such techniques to be implemented elsewhere on a sustainable and long-term basis, and how to maintain the normal annual funding to support the staff and operations needed by any large conserved wildland:

"The key management practice was to stop the assault — fire, hunting, logging, farming — and let the biota re-invade the ACG. The key sociological practice was to gain social acceptability for the project locally, nationally and internationally … The question is not whether a tropical forest can be restored, but rather whether society will allow it to occur" (Janzen, 2002)

Abridged from Janzen (2000, 2002) www.gdfcf.org/articles/Janzen_2000_longmarchfor ACG.pdf

(A) Jaragua–forest boundaries were characteristic of tens of thousands of hectares of the ACG at the beginning of the restoration process (photo December 1980). At least 200 years old, the pasture was formerly occupied by native grasses and had been burned every 1–3 years. The old secondary oak forest retained more than 100 tree species. (B) The same view (photo November 2000) after 17 years of fire prevention. The oak forest canopy is still visible and Winnie Hallwach's hand is 2 m above the ground. The regeneration is dominated by *Rehdera trinervis* (Verbenaceae), a medium-sized wind-dispersed tree, intermixed with 70 other woody species. Such invasion of pastures by forest as a result of fire prevention is now characteristic of tens of thousands of hectares of the ACG. (Photos: Daniel Janzen.)

CHAPTER 6

GROW YOUR OWN TREES

High-quality planting stock is essential for the success of all forest restoration projects that involve tree planting (i.e. for the restoration of degradation stages 3–5). Saplings of all tree species must be grown to a suitable size and must be robust, growing vigorously and disease-free when the season is optimum for tree planting. This is difficult to achieve when growing a large number of different native forest tree species, which will fruit at different times of the year and vary greatly in their germination and seedling growth rates, especially if those species have never been mass-produced in nurseries before. In this chapter, we provide standard tips that are generally applicable for a first attempt to grow native forest trees for a forest restoration program. We also include research protocols that can be used to improve your tree propagation methods, leading to the development of detailed production schedules for each species being propagated.

6.1 Building a nursery

A nursery must provide ideal conditions for the growth of tree seedlings and must protecting them from stresses. It must also be a comfortable and safe place for nursery workers.

Choosing a location

A nursery site should be protected from extremes of climate. It should be:

- flat or slightly sloping, with good drainage (steeper slopes require terracing);
- sheltered and partially shaded (a site protected by existing trees is ideal);
- close to a permanent supply of clean water (but free from the risk of flooding);
- large enough to produce the number of trees required and to allow for future expansion;
- close to a supply of suitable soil;
- accessible enough to allow the convenient transportation of young trees and supplies.

If an exposed site cannot be avoided, a shelter belt of trees or shrubs could be planted, or large containerised trees could provide shelter.

How much space is needed?

The size of the nursery ultimately depends on the size of the area to be restored, which in turn determines how many trees must be produced each year. Other considerations include seedling survival rates and growth rates (which determine how long plants must be kept in the nursery).

Table 6.1 relates the area to be restored each year to the minimum size of the nursery required. These calculations are based on the germination of seeds in trays and their subsequent transplantation into containers, with relatively high survival rates. For example, if the area to be restored is 1 hectare per year, up to 3,100 trees will be needed, requiring a nursery of approximately 80 m².

Essential features of a tree nursery

Building a tree nursery need not be costly. Locally available materials, such as recycled wood, bamboo and palm leaves, can all be used to build a simple inexpensive nursery. The essential requirements include:

- a shaded area with benches for seed germination that is protected from seed predators by wire mesh; shade can be provided by commercial materials, but alternatives include palm leaves, coarse grasses and bamboo slats;
- a shaded area where potted seedlings can be grown until ready for planting (the shading should be removable if the young trees are to be hardened here prior to planting);
- a work area for seed preparation, pricking-out etc.;
- a reliable water supply;
- a lockable store for materials and tools;
- a fence to keep out stray animals;
- a shelter and toilet for staff and visitors.

Table 6.1. Relation between the space needed for a nursery and the size of the restoration site.

Area to be restored (ha/year)	Maximum number of trees needed[a]	Seed germination area (m²)	Standing-down area[b] (m²)	Storage, shelter, toilet etc. (m²)	Total nursery area needed (m²)
0.25	775	3	11	15	29
0.5	1,550	6	22	15	43
1	3,100	13	44	15	72
5	15,500	63	220	15	298
10	31,000	125	440	15	580

[a] Assuming absence of natural regenerants
[b] An additional area of similar size might be required for hardening-off seedlings if it is not possible to remove the shading from the containerised seedlings.

Designing a nursery

A carefully considered nursery layout can greatly increase efficiency. Think about the various activities to be carried out and the movement of materials around the nursery. For example, position the container beds and hardening-off areas near to the main access point, i.e. close to where the trees will eventually be loaded onto vehicles for transport to the restoration site; place the lockable store and media store near the potting area.

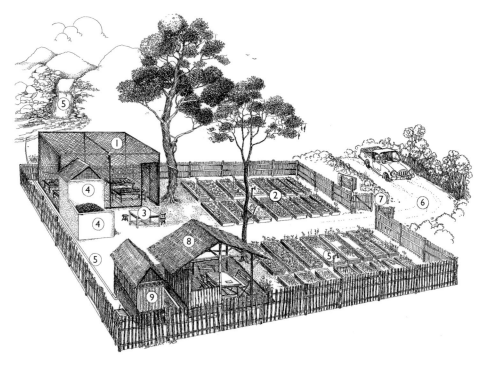

An ideal nursery layout: (1) germination shelter that is protected from seed predators; (2) standing-down area (shade removed); (3) potting work area; (4) media store and lockable equipment store; (5) reliable water supply; (6) easy access; (7) fence to exclude stray animals; (8) shelter from the sun and rain; and (9) toilet.

Nursery tools

Growing trees requires simple, inexpensive equipment. Many of the items illustrated here are readily available in an average agricultural community and could be borrowed for nursery work:

- shovel (1) and buckets (2) for collecting, moving and mixing potting media;
- trowels (3) or bamboo scoops (4) for filling containers with potting medium;
- watering cans (5) and a hose, both fitted with a fine rose;
- spatulas or spoons for pricking-out seedlings;
- sieves (6) for preparing the potting medium;
- wheelbarrows (7) for moving plants and materials around the nursery;
- hoes (8) for weeding and maintaining the standing-down area;
- secateurs (9) for pruning seedlings;
- a ladder and basic construction tools for erecting shade netting etc.

A lockable store for the safe storage of equipment and a media store are essential parts of a tree nursery.

Essential nursery equipment.

6.2 Collecting and handling tree seeds

What are fruits and seeds?

The structure that is sown in a germination tray is not always just the seed. For tree species such oaks and beeches (northern hemisphere Fagaceae and southern hemisphere Nothofagaceae), the whole fruit is sown. For, other species, we sow the pyrene, which consists of one or several seeds enclosed within the hard inner wall of the fruit (i.e. the endocarp, which can delay the penetration of the seed embryo by water). So a basic understanding of fruit and seed morphology can be helpful in deciding which pre-sowing seed treatments (if any) are appropriate.

A seed develops from a fertilised egg cell (ovule) that is contained within the ovary of a flower, usually after pollination and fertilisation. Being the products of sexual reproduction, during which the genes of the two parents are combined, seeds are an essential source of genetic diversity within tree populations.

Seeds consist of three main parts: a covering, a food store and the embryo. The seed coat or testa protects seeds from harsh environmental conditions and plays an important role in dormancy. Food reserves, which sustain metabolism during and immediately after germination, are stored in the endosperm or the cotyledons. The embryo consists of a rudimentary shoot (plumule), a rudimentary root (radicle) and seed leaves (cotyledons).

Fruits are derived from the ovary wall. They may be broadly classified as 'simple' (formed from the ovary of a single flower); 'aggregate' (formed from the ovary of a single flower, but with several fruits fused into a larger structure) or 'multiple' (formed from ovaries of several flowers fusing). Each broad category contains several fruit types.

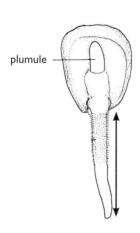

plumule

At germination, the radical (first root) and plumule (shoot bud) burst through the outer coat (testa) of the seed fueled by food reserves from the endosperm.

A

B

C

D

Simple fruits can have either a fleshy pericarp, like that of tomato, or A) a dry covering, such as the pods of legumes. B) Custard apple (*Annona reticulata*) produces aggregate fruits whereas C) jackfruit trees (*Artocarpus heterophyllus*) produce multiple fruits. D) The multiple fruit of fig trees essentially consist of an enclosed infructescence (syconia).

When should seeds be collected?

In all tropical forests, different tree species fruit in every month of the year, so at least one seed collection trip is needed every month. In seasonal tropical forests, fruiting peaks at the end of the dry season and at the end of the rainy season. Reduced numbers of fruiting tree species in the early rainy season means that fewer seed collection trips are needed then.

In parts of Southeast Asia and Central America, the fruiting months of many tree species are well-known, but for many regions, phenology studies are needed to provide this information (see **Section 6.6**). Find seed trees in the forest and monitor them frequently, from flowering onwards, to judge the best time to collect fruits. Collect fruits once they are fully ripe, but just before they are dispersed or consumed by animals. Seeds that are collected too early will be undeveloped and will fail to germinate, whereas those collected too late may have lost viability.

For fleshy fruits, ripeness is usually indicated by a change in the colour of the fruit, usually from green to a brighter colour that attracts seed-dispersing animals. Animals' grazing on the fruits is a sure sign that the seeds are ready for collection. Dehiscent fruits, such as those of some legumes, start to split open when they are ripe. It is usually better to cut fruits from the tree branches rather than to pick them up from the ground.

If you have received appropriate training, climb the tree to cut down ripe fruit. Use a safety harness and never do this alone. A more convenient method of seed collection for shorter trees is to use a cutter mounted on the end of a long pole. Fruits can also be dislodged by shaking smaller trees or some of the lower branches.

The collection of fruits from the forest floor may be the only option for very tall trees. If this is the case, make sure that the seeds are not rotten by cutting them open and looking for a well-developed embryo and/or a solid endosperm (if present). Do not collect fruits or seeds that have signs of fungal infection, teeth marks from animals or small holes made by seed-boring insects. Collect fruits or seeds from the forest floor when the first truly ripe fruits begin to fall.

Seed collection trips require planning and liaison with the people responsible for treating and sowing the seeds because seeds are vulnerable to desiccation and/or fungal attack if they are not processed quickly. Sow the seeds as soon as possible after collection or prepare them for storage as described later in this chapter. Before sowing, do not leave them in damp places, where they might rot or germinate prematurely. If they are sensitive to desiccation, do not leave them in full sun.

Choosing seeds for collection

Genetic variability is essential to enable a species to survive in a changeable environment. Maintaining genetic diversity is therefore one of the most important considerations in any restoration program aiming to conserve biodiversity. It is therefore crucial that the planted trees are not all closely related. The best way to prevent this is to collect seeds from at least 25 to 50 high-quality parent trees locally, and preferably to augment this with some seed from trees located in more distant, eco-geographically matched areas (see **Box 6.1**, p. 159). If seeds are collected from just a few local trees, their genetic diversity may be low, reducing their capacity to adapt to environmental change. Equal numbers of seeds from each seed tree should be mixed together (known as bulking-up) before sowing, to ensure that all the seed trees are represented equally. Once the trees mature within the restored plots, they may inbreed with each other, further reducing genetic variability in subsequent generations. Cross-pollination with unrelated trees can restore genetic diversity, but only where such trees grow close to restoration sites.

The number of seeds collected depends on the number of trees required, seed germination percentage and seedling survival rates. Keep accurate records to determine the numbers required in future collections.

Information that should be recorded when collecting seeds

Species number: Batch number:

SEED COLLECTION RECORD SHEET

Family:

Species: **Common name:**

Date collected: **Collector's name:**

Tree label no.:   **Tree girth:**

Collected from ground [ ] or from tree [ ]

Location: **Elevation:**

Forest type:

Approximate no. seeds collected:

Storage/transport details:

Pre-sowing treatment: **Sowing date:**

Voucher specimen collected [ ]

Notes for herbarium label:

Each time you collect seeds from a new species, give that species a unique species number. Nail a numbered, metal tag onto the tree, so that you can find it again. Collect a specimen of leaves and fruits for species identification. Place the specimen in a plant press, dry it and ask a botanist to identify the species. Use a pencil to write the species name (if known), date and species number on a label and place the label inside the bag with the seeds.

On a data sheet (example below), record essential details about the seed batches collected and what happened to them from collection time until they are sown in germination trays. This information will help to determine why some seed batches germinate well whereas others fail, and thus will improve seed-collection methods in the future. A more detailed seed collection data sheet that might be used for research purposes is provided in the **Appendix** (**A1.3**).

Box 6.1. Gene flow, adaptive genetic diversity and sourcing seed

Box 6.1. Gene flow, adaptive genetic diversity and sourcing seed.

Global climate change has profound consequences for tropical forest ecosystems. The evolutionary adaptability of a species, its capacity to survive environmental change, depends on the genetic diversity present among the individuals of the species. Populations of trees that have a wide range of adaptive genetic variation have the best chance of surviving climate change or changes in other environmental factors, such as increased salinity, the use of fertilisers and vegetation redistribution resulting from habitat conversion.

Consider individual trees of a species, each of which might possess different versions or 'alleles' of a gene that encodes a certain protein. If one of these alleles functions better in drier conditions, then the individuals carrying this allele might survive better if rainfall declines and so would be more likely to pass on their version of the gene to subsequent generations. Conversely, trees that carry a different allele of the same gene or of a different gene might survive better if conditions were to become wetter. Consequently, maintaining genetic variability among individual trees that comprise a species population is one of the most important considerations in any restoration program for biodiversity conservation.

Adaptive genetic variation depends on rates of gene mutation, gene flow and other factors. Natural selection increases the frequency of traits, which confer advantages to individuals, at a particular time or place. In the case of tropical trees, it may act at the seedling stage, when saplings have an opportunity to replace a fallen tree in the canopy. It may help tree populations to cope with future climate-induced stress.

Adaptive genetic diversity increases as a result of gene flow, that is when different genes are introduced into a population by pollen or seed from another tree, or population of trees. Gene flow can occur over distances of up to hundreds of kilometres (Broadhurst et al., 2008). Habitat fragmentation hinders the dispersal of both pollen and seeds. Furthermore, tree populations that are adapted to the current environmental conditions might not have sufficient adaptive genetic diversity to enable sufficient numbers of their offspring to survive climate change. Forest restoration practitioners must consider whether local gene pools have sufficient adaptive genetic diversity and resilience to meet the challenges of climate change, and to adapt quickly enough as the environment changes. Consequently, there could be a strong case for sourcing a proportion of the seed for restoration projects from areas that are not local in an attempt to mimic natural gene flow.

It has been recommended that seeds for forest restoration projects are collected locally, from 'high quality' parent trees, because local trees are the product of a long history of natural selection that has adapted them genetically to survive and reproduce in the local prevailing conditions. Given the need to maintain high levels of genetic diversity so as to ensure adaptability to climate change, however, locally sourced seed could be supplemented with a small percentage of seeds collected from other areas that have similar environmental and climatic conditions to the planting site. 'Composite provenancing' has been proposed as a way of enhancing natural gene flow (Broadhurst et al., 2008). For example, the majority of seeds could be collected from as many local parent trees as is practicable, but also incorporate nearby and eco-geographically matched sources (Sgró et al., 2011). A smaller proportion (10–30%) could be sourced from much further afield (Lowe, 2010). The resulting new gene combinations might enable the tree populations to respond to environmental change, which is crucial if natural selection is to act on restoration plantings.

Box 6.1. continued.

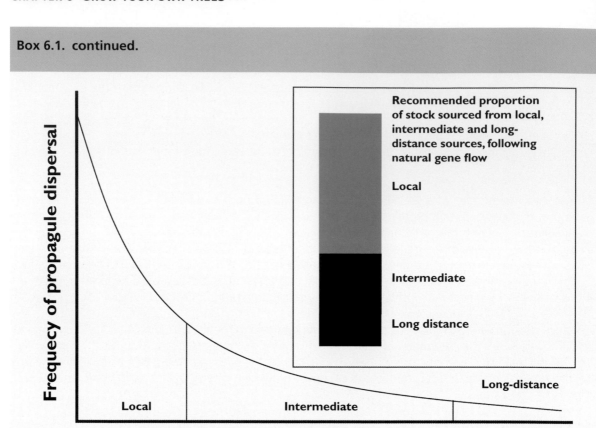

Distance from parent plant

Figure reproduced with kind permission from Sgró *et al.* (2011).

For most species, seed dispersal is local, with much lower proportions of seed being dispersed over intermediate and longer distances. Composite provenancing mimics this dispersal pattern, using a high proportion of locally adapted seed, and lower proportions of seed collected at intermediate (mimicking intermediate gene flow) and distant locations. The seed collected some distance from the restoration site could introduce novel genes into the population.

Extracting seeds from fruits

For most species, the seeds should be removed from the fruits and cleaned before sowing.

With fleshy fruits, remove as much of the fruit pulp as possible with a knife and wash off remaining pulp with water. Soak firm fruits in water for 2–3 days to soften the pulp sufficiently to ease seed extraction. Once the fruit pulp has been removed, the seeds might germinate quickly, so either sow them immediately or process them for storage. Failure to remove fruit pulp encourages fungal infection. In some species, removal of the pulp reveals a woody or stony pyrene containing one or more seeds. If the seeds are to be planted immediately, crack open the tough endocarp to allow water to penetrate into the embryo and trigger germination. Use a vice, hammer or knife to crack the endocarp gently without damaging the seed(s) inside.

Dry dehiscent fruits, such as the pods of trees in the Leguminosae family, often split open naturally, so lay them out in a dry, sunny place until they open and the seeds either fall out on their own or can be easily shaken out.

For dry indehiscent fruits that do not split open naturally, cut pods open or prise them apart with secateurs or other tools. The seeds of some indehiscent fruits, such as samaras and nuts, are not usually extracted and the whole fruits should be placed in germination trays. Fruit appendages, such as the wings of samaras (e.g. *Acer*, *Dipterocarpus*) or the cupules of nuts including acorns or chestnuts, should be removed for easier handling. The germination of seeds that are covered by an aril is nearly always accelerated by scraping the aril.

Ensuring seed quality

It is very important to sow only the highest-quality seeds available. They should have no signs of fungal growth, teeth marks from animals or small holes made by seed-boring insects such as weevils. For larger seeds, dead seeds can be rapidly identified by immersing the seeds in water and waiting for 2–3 hours. Skim off those seeds that remain floating as they have air inside instead of dense cotyledons and a functioning embryo. Sowing poor-quality seed is a waste of time and space, and could encourage the spread of diseases.

Sorting good seed from bad: the good seed sinks (left) the bad seed floats (right).

Seed storage

Although it is usually best to germinate seeds as soon as possible after collection, seed storage can be a useful in streamlining tree production, sharing seeds among nurseries and accumulating seeds for direct seeding. Depending on their physiological storage potential, seeds may be classified as orthodox, recalcitrant or intermediate. The storage behaviour of many species can be found at http://data.kew.org/sid/search.html.

Orthodox and recalcitrant seeds

Orthodox seeds remain viable when dried to low moisture contents (2–8%) and chilled to low temperatures (usually a few degrees above freezing), so they can usually be stored for many months or even years.

Recalcitrant seeds are more common in species from moist tropical habitats and tend to be large and to have thin seed coats or fruit walls. They are very sensitive to desiccation and cannot be dried to moisture contents lower than 60–70%. Furthermore, they cannot be chilled and are relatively short-lived. Therefore, it is very difficult to store recalcitrant seeds for longer than a few days without losing viability.

There is also a sub-group of species that have 'intermediate' seeds. These can be dried to low moisture contents, approaching those tolerated by orthodox seed, but they are sensitive to chilling when dried.

Drying and storing orthodox seed

First, determine whether the majority of the seeds are ripe or immature, because individual trees may disperse their seeds at slightly different times. Ripe seeds, which are ready for dispersal, respond best to drying. Immature seeds are generally more difficult to dry.

Immature seeds, either fresh or after drying, do not germinate. They can, however, be ripened and their viability increased markedly by storing them at a controlled humidity and temperature. A relative humidity of 65% is low enough to reduce the chances of mould. Alternatively, store the fruits under as natural conditions as possible, i.e. with the fruit left on stems and seed left in the fruit. Examine a few seeds occasionally to determine when the batch reaches maturity.

Ripe seeds must be handled carefully between collection in the forest and storage or sowing in the nursery. Once the seeds are harvested, they begin to age, particularly if kept at high moisture contents. They may be attacked by insects, mites and/or fungi (if not kept well aerated) or they may germinate.

Development of seed quality with maturation time. (Reproduced with kind permission of the Board of Trustees of the Royal Botanic Gardens, Kew)

Development of seed quality

	Immature Seeds may not be fully dessication tolerant. Seeds will not have attained maximum storage potential.	Optimum time to collect	Post-Harvest Seeds may lose viability rapidly in hot, humid conditions.
Seed formation differentiation	Maturation reserve accumulation	Post-abscission ripening	Dispersal/post-harvest ageing or repair

→

Seed development — time after flowering

Measuring moisture content

To retain their viability during storage, orthodox seeds must be dried, but how dry is dry enough? To determine if seeds are dry enough for storage, half fill a glass jar with seeds and add a small hygrometer or moisture indicator strip to the jar (Bertenshaw & Adams, 2009a). Wait for the air in the jar to reach a stable humidity in the shade. This is called the equilibrium relative humidity (eRH). **Table 6.2** shows that an eRH% of 10–30% is recommended for the long-term storage of orthodox seeds.

Either a sophisticated and expensive digital hygrometer (left) or simple and cheap dial hygrometers (right) can be used to assess the moisture content of seeds.

Table 6.2. Relation between eRH%, seed moisture content and the survival of seeds in storage.

eRH%	Approximate moisture content (varies with seed oil content and temperature)		Seed survival
	Non-oily seed (2% oil)	Oily seed (25% oil)	
85–100%	>18.5%	> 16%	High risk of mould, pests and disease
70–85%	12.5–18.5%	9.5–16%	Seeds at risk of rapid loss of viability
50–70%	9–12.5%	6–9.5%	Rate of deterioration slower; seeds may survive for 1–2 yrs
30–50%	7.5–9%	5.5–6%	Seeds could survive for several years
10–30%	4.5–7.5%	3–5.5%	Seeds can be kept alive for decades
< 10%	< 4.5%	< 3%	Risk of damage, therefore best avoided

Crude salt test for seed moisture content. The small jam jar (front), the third vial from the right and the far right vial all contain free-flowing salt indicating seeds that are dry enough for storage.

Salt can also be used in a crude test for moisture content. Quarter fill a small glass jar with very dry table salt, add about an equal volume of seeds and shake. If the salt forms clumps, the eRH is higher than 70%. If the salt remains free flowing, then the seeds can be stored, at least in the short term.

An alternative method of determining the moisture content of dried seeds is to weigh a sub-sample of the sun-dried seeds, then put them in an oven at 120–150°C for an hour before reweighing them. If the following calculation gives a values of <10%, the seeds are ready for storage:

$$\frac{(\text{Seeds mass after sun-drying} - \text{Seed mass after oven-drying}) \times 100\%}{\text{Seed mass after sun-drying}}$$

Throw away the sub-sample of seeds used for this test.

Drying seeds

The simplest way to dry seeds is to clean them and leave them in the sun for a few days. Spread the seeds in thin layers on a mat, and turn them regularly with a rake so that they dry quickly and evenly without overheating. Direct sunlight for extended periods reduces seed viability. Shade the seeds during the hottest part of the day and protect the seeds at night or after rain to prevent moisture re-absorption. If practical, transfer the seeds to sealed containers overnight. Once every 24–48 hours, test the moisture content of a sample of the seeds, and continue drying until the eRH falls to 10–30% (equivalent to 5–10% seed moisture content). Drying time will depend on the size of the seed, the structure and thickness of the seed coat, ventilation and temperature.

Seeds drying on a mat in Tanzania. (Photo: K. Gold)

Desiccants

Desiccants are substances that absorb moisture from air. A wide range of desiccants can be used to dry seeds in sealed containers. Silica gel is perhaps the best known, but local products, such as toasted rice and charcoal, are cheaper alternatives. The Royal Botanic Gardens, Kew has developed a seed-drying technique that uses natural or 'lump-wood' charcoal, which is universally available in rural tropical communities (Bertenshaw & Adams, 2009b). First, dry the seeds for 2–3 days under ambient conditions; meanwhile, dry small lumps of charcoal in an oven or in direct sunshine. Place the charcoal in the bottom of a sealable container, then cover it with newspaper and place the seeds on top of the paper. Add a hygrometer or a moisture strip, seal the container and store it in a cool place. Alternatively, put the seeds in cloth bags and hang them in larger containers, such as plastic drums, with charcoal at the bottom. To attain an eRH of 30% use a charcoal:seed weight ratio of 3:1; for an eRH% of 15% use a ratio of 7.5:1.

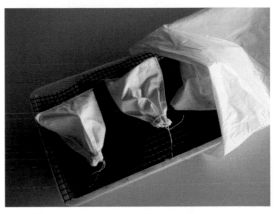

Charcoal is a cheap desiccant that is widely available in tropical rural communities.

Once the seeds are dried, store them in air-tight containers under conditions that reduce seed metabolism and prevent the entry (or growth) of pests and pathogens. Containers can be plastic, glass or metal and their seals might be improved with the use of rubber inner-tubes. Fill the containers to the top to minimise the volume of air (and moisture) inside. Efficient sealing of containers is crucial, to prevent the entry of moisture or fungal spores. Even a rise of just 10% eRH can reduce the storage life of the seeds by half. If containers are likely to be opened frequently, store the seeds in small sealed packets within larger containers to minimise the exposure of the remaining seeds to air and moisture. Putting a small sachet of coloured silica gel into the containers will indicate if any moisture is getting into the container.

Charcoal in a sealed container or in sealed bags can be used as a natural dessicant.
(Photo: K. Mistry)

Storing the containers at ambient temperatures should be sufficient to maintain viability for 12–24 months. Keeping seeds for longer periods may require storage at low temperatures, but this can be expensive and is not usually necessary for forest restoration projects.

Storing recalcitrant and intermediate seed

The storage tolerance of recalcitrant and intermediate seeds varies enormously. Some species have no dormancy at all. Highly recalcitrant seeds die when their moisture content drops below 50–70%, whereas less sensitive ones can remain viable down to 12% moisture content. Chilling tolerance also varies. Keep the storage duration for recalcitrant seeds to an absolute minimum. When storage is unavoidable, prevent desiccation and microbial contamination and maintain an adequate air supply.

For a comprehensive account of seed collection and handling, the reference text *"A Guide to Handling Tropical and Subtropical Forest Seed"*, by Lars Schmidt (published by the DANIDA Forest Seed Centre, Denmark, 2000) is highly recommended.

6.3 Germinating seeds

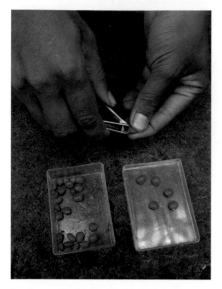

For larger seeds that have hard seed coats, dormancy may be broken by manually cutting the seed coat.

In the nursery, dormancy prolongs tree production time (see **Box 6.2**, p. 168). Therefore, various treatments are commonly applied to shorten dormancy. The treatment used for each species depends on the particular dormancy mechanism(s) present.

A thick, impervious seed coat can prevent water or oxygen reaching the embryo, so one of the simplest techniques to break dormancy is to cut away a small piece of the seed coat with a sharp knife or nail clippers. For smaller seeds, gently rubbing them with sandpaper can be equally effective. These techniques are called scarification. During scarification, care must be taken not to damage the embryo within the seed.

For species with mechanical dormancy, acid treatment is recommended. Acid can kill the embryo, so seeds must be soaked in acid long enough to soften the seed coat but not for long enough to allow the acid to reach the embryo.

When germination is inhibited by chemicals, simply ensuring complete removal of fruit pulp can solve this problem. But if the chemical inhibitors are present within the seed, they must be washed out by repeated soaking. For more on pre-sowing seed treatments, see **Section 6.6**.

Sowing seeds

Germination is the most vulnerable time in the long life of a tree.

Sow seeds in germination trays filled with a suitable medium. Large seeds can be sown directly into plastic bags or other containers. The advantage of using trays is that they can be easily moved around the nursery, but remember that they can dry out quickly if neglected. Seed trays should be 6–10 cm deep, with plenty of drainage holes in the bottom.

The germination medium must have good aeration and drainage and must provide adequate support for germinating seedlings until they are ready for pricking-out. Seedling roots need to breathe, so the germination medium must be porous. Too much water fills the air spaces in the medium and suffocates seedling roots. It also encourages disease. Compacted soil inhibits both germination and seedling growth.

Mix forest soil with organic materials to create a well-structured medium. Chiang Mai University's Forest Restoration Research Unit (FORRU-CMU) recommends a mixture of two-thirds forest top-soil to one-third coconut husk. A mix of 50% forest soil with 50% coarse sand is more suitable for small seeds, especially those (e.g. *Ficus* spp.) susceptible to damping-off fungi. Include some forest soil in the medium to provide a source of mycorrhizal fungi, which are required by most tropical forest tree species. If forest top-soil is not available, use a mix that includes coarse sand (to encourage good drainage and aeration) and sieved organic matter (to provide texture, nutrients and water retention). Do not add fertiliser to the seed germination medium (except when germinating seeds of *Ficus* spp.), as the seedlings will not require it.

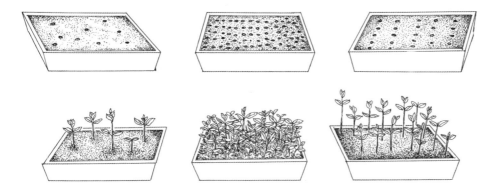

Sowing seeds too far apart (left) is a waste of space, but sowing them too close together (centre) increases the risk of disease.

Sow small to medium-sized seeds on the surface of the medium and then cover them with a thin layer of germination medium (with a depth of approximately 2–3 times the seeds' diameter), which should come to within 1 cm of the tray's rim. Seeds larger than 5 mm in diameter require an equivalent depth of germination medium. This protects the seeds from predators and drying out and prevents them from being washed away during watering. If rats or squirrels are a problem, then cover the germination trays in wire mesh. Place the trays in shade to reduce the drying out and scorching of leaves.

Space the seeds at least 1–2 cm apart (further if the seeds are large) to prevent over-crowding. If the seeds are sown too closely together, the seedlings may be weakened and hence more susceptible to diseases such as damping-off. Water the germination trays lightly, immediately after sowing the seeds and regularly thereafter, using a spray bottle or a watering can with a fine rose to prevent compaction of the medium. Too frequent watering encourages damping-off diseases.

A perfect germination room at Lake Eacham National Park in Queensland, Australia with the germination trays on wire grid benches. The trays at the back are protected by wire cages, which are lowered at night to exclude rats and birds. Note that all of the germination trays are clearly labelled with the species and date of sowing.

Box 6.2. Dormancy and germination.

Dormancy is the period during which viable seeds fail to germinate, despite having conditions (moisture, light, temperature etc.) that are normally favourable for the later stages of germination and seedling establishment. It is a survival mechanism that prevents seeds from germinating during seasons when the seedlings are likely to die.

Dormancy can originate in the embryo or in the tissues that surround it (i.e. the endosperm, testa or pericarp). Dormancy that originates in the embryo can be due to i) a need for further embryonic development (after-ripening); ii) chemical inhibition of metabolism; iii) a block on the mobilisation of food reserves; or iv) insufficient plant growth hormones. Dormancy that is due to the seed coverings can be caused by i) restriction of transport of water or oxygen into the embryo; ii) mechanical restriction of embryo expansion; or iii) chemicals that inhibit germination (most commonly abscisic acid). In many plant species, dormancy results from a combination of several such mechanisms.

Germination consists of three overlapping processes. i) The absorption of water causes swelling of the seed and splitting of the seed coat. ii) Food reserves in the endosperm are mobilised and transported to the embryonic root (radicle) and shoot (plumule), which begin to grow and push against the seed coat. iii) The final stage (and the most precise definition of germination) is the emergence of the embryonic root through the seed coat. In germination trials, this can be difficult to observe as the seeds are buried, so emergence of the embryonic shoot can also be used to indicate germination.

Seed germination is influenced by moisture, temperature and light. Seedlings are at their most vulnerable to disease, mechanical damage, physiological stress and predation just after germination, so take care to protect germinating seeds from infection, drying winds, heavy rain and strong sunshine.

Species number: **Batch number:**

GERMINATION RECORD SHEET

Species:

Date sown: **Number of seeds sown:**

Germinated	Date	Days since sowing
First seed		
Median seed		
Final seed		

Number germinated: **% Germination:**

Pricking-out date:

No. of seedlings pricked out:

Date	No. Germinated	Date	No. Germinated

Keeping track of germination gradually improves nursery efficiency over time.

Damping-off diseases

The term 'damping-off' refers to diseases that are caused by several genera of soil fungi, including *Pythium*, *Phytopthera*, *Rhizoctonia* and *Fusarium*, which can attack seeds, pre-emergent sprouts and young seedlings. Pre-emergence damping-off softens the seeds and turns them brown or black. Post-emergence damping-off attacks the soft tissue of recently germinated seedlings just above the soil surface. Infected seedlings appear to be 'pinched' at the base of the stem, which turns brown.

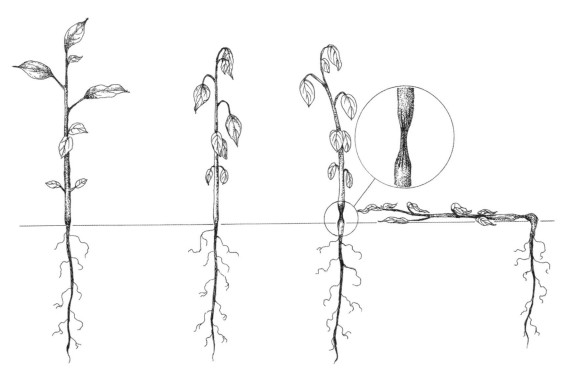

Damping-off diseases, which are caused by various fungi, start with brown lesions appearing on the stem, at or just above the soil surface. The lesions spread and the leaves wilt. Finally, the stem collapses and the seedling dies.

If they become a serious problem, damping-off diseases can be controlled with fungicides such as Captan. The use of chemicals is undesirable, but prompt application of fungicide at the outbreak of disease can mean the difference between saving the tree crop and having to wait another year to collect seeds again.

Remove infected seedlings immediately and destroy them to prevent the disease from spreading. Basic hygiene measures can significantly reduce the incidence of damping-off diseases and are preferable to spraying with a fungicide. These include not sowing seeds too densely, maintaining a well-structured germination medium, not over-watering, ensuring free air movement around the seedlings and disinfecting any nursery tools that have come into contact with soil.

Box 6.3. Propagation of *Ficus* species.

Ficus species play a vital role in tropical forest restoration (see **Box 2.2**) and several species should always be growing in a restoration tree nursery. But propagating them requires a few special techniques. *Ficus* planting stock is best grown from seed: although propagation from cuttings is efficient, the planting stock derived from seed is usually healthier and more vigorous. Stock that has been raised from seed is also more genetically diverse, a crucial consideration for biodiversity conservation projects. But growing fig trees that are large enough for planting from seed can take 18–22 months, so if planting stock is required more urgently, try cuttings.

First make sure that there are seeds in the fig as there are in this female fig of *Ficus hispida*.
(Photo: C. Kuaraksa)

Collect ripe figs, break them open and see if they contain seeds. Figs of monoecious *Ficus* species contain both male and female flowers, so all of their figs have the potential to produce seeds if they have been visited by pollinating fig wasps (see **Box 2.2**, p. 31). Dioecious figs have separate male and female trees. Obviously, the figs on male trees never contain seeds, so consult a flora to find out if the species that you want to propagate are monoecious or dioecious.

A single fig can contain hundreds or even thousands of miniscule, light brown and hard seeds. Scrape out the mush that contains the seeds from inside the figs with a spoon. Press the mush through a piece of mosquito netting over a bowl of water. Viable seeds will pass through the net and sink. Pour off most of the water and pour the remaining water, along with the seeds (which have sunk to the bottom of the bowl), through a fine tea strainer. Wash the seeds thoroughly and leave them to dry slowly over 1–2 days.

Separate the seeds from the mush inside the fruit and air dry them for a few days.

Sprinkle the seeds evenly (aiming for gaps of 1–2 cm) over the surface of a germination medium comprising a 50:50 mix of sand and charred rice husk or similar materials (do not include forest soil in the medium). Do not cover the seeds. Water the trays by hand using a fine spray bottle.

Most species will start to germinate within 3–4 weeks and germination will be complete within 7–8 weeks. Fig seedlings are tiny and grow slowly at first. Adding a few granules of slow-release fertiliser (e.g. Osmocote) just below the surface of the germination medium can accelerate seedling growth, but it can also increase seedling mortality. *Ficus* seedlings are particularly susceptible to damping-off, so remove infected seedlings immediately and apply a fungicide such as Captan if an outbreak occurs. Prick out the seedlings after the second pair of true leaves has expanded (4–10 months after germination) and pot them into standard containers and media.

To produce planting stock from cuttings, follow the method in **Box 6.5**. If propagating a dioecious species, collect an equal number of cuttings from both male and female trees. Apply synthetic auxins to stimulate rooting (Vongkamjan, 2003).

By Cherdsak Kuaraksa

Shade

Germinate all seeds, whether they are from light-demanding or shade-tolerant species, under shade. If practical, provide more shade to shade-tolerant species. As the time for pricking out approaches, reduce the level of shade to that of the growing-on area. If several layers of plastic shade cloth have been used, remove them one at a time.

6.4 Potting

Containers or soil beds?

Trees that have been grown in beds of soil are termed 'bare rooted' because they retain very little soil with the roots when they are dug out for planting. Their exposed roots quickly lose water and are easily damaged. When the root system is reduced but the leaf area remains the same, the roots are unable to deliver enough water to the shoots to maintain transpiration and to retain turgidity of the leaf cells, resulting in wilting and increased mortality. Thus, bare-rooted planting stock often suffers from 'transplantation shock' when planted out in deforested sites, resulting in mortality rates that are much greater than those for trees that have been grown in containers.

With a containerised system, seedlings are transplanted from germination trays into containers in which they grow until they are large enough to be planted out. The containers protect the trees during transportation to the planting site where the whole root ball can be removed from the container, thus minimising transplantation stress

Choosing your containers

The containers must be large enough to allow the development of a good root system and to support adequate shoot growth. They must have sufficient holes to permit good drainage, and be lightweight, inexpensive, durable and readily available. Containers can be made out of a variety of materials, such as polythene, clay and biodegradable materials. When funding is insufficient to allow for the purchase of containers, try improvising by converting cartons, plastic bottles or old cans (don't forget to add drainage holes); even banana leaves can be folded to make an adequate container.

Plastic bags are probably the most commonly used containers. They come in a range of sizes and are strong, lightweight and cheap, and they have been used successfully with a very wide range of species. Large plastic bags are difficult to transport and require a lot of medium, whereas small ones restrict root development. The optimum size is 23 × 6.5 cm, which allows tap roots to reasonable length before they reach the bottom of the bag and begin to spiral.

Plastic bags (23 × 6.5 cm) are cheap but not reusable and can cause root curling of fast-growing tree species.

Plastic bags do have some disadvantages. They can bend easily, particularly during transportation; this can damage the root ball, causing it to crumble during planting. The roots of fast-growing tree species can fill the bags rapidly and begin to spiral around at the bottom. This poor root formation can increase the vulnerability of trees to wind-throw later in life. Roots can grow through the drainage holes into soil beneath, so that roots are severed when the tree is lifted just before planting, causing transplantation shock. Root trainers can reduce this problem.

Root trainers

Root trainers are rigid plastic pots with grooves down the sides that direct root growth downwards, thus preventing root spiralling. Large holes in the bottom allow air pruning (see **Section 6.5**). Although initially more expensive than many other types of container, they can be re-used many times and their rigidity protects the root ball during transportation.

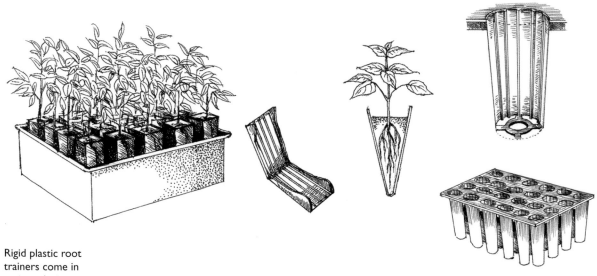

Rigid plastic root trainers come in various designs and sizes.

What makes a good potting medium?

A potting medium consists of coarse and fine soil particles with pores between them that allow aeration and drainage. The medium must provide growing trees with support, moisture, oxygen, nutrients and symbiotic micro-organisms.

Tree roots that are growing in containers have access to only a limited volume of medium. Soil alone is an unsuitable medium because it is easily compacted and the container prevents free drainage, causing water-logging that suffocates the roots. Good drainage is essential, but the medium must also have an organic matter content that is adequate to ensure that the medium remains adequately moist between waterings.

Various materials can be included in potting media, including coarse sand or gravel (washed to remove salts) and forest topsoil. Organic matter can be added in the form of rice husk charcoal, coconut husk, peanut husks and even waste products from agricultural production, such as coffee fruit pulp or pressed sugar cane. Alternatively, try making compost from domestic organic waste. Adding cow dung to the mixture can dramatically increase seedling growth rates because of its rich nutrient content.

Although forest topsoil alone is a poor potting medium, it is an important component of potting mixes because it carries the spores of soil micro-organisms that help trees to grow, such as *Rhizobium* bacteria and mycorrhizal fungi. To prevent compaction, mix forest soil with bulky organic matter or coarse sand. Mixing forest soil with these ingredients 'opens out' the medium and improves drainage and aeration. Whichever materials you choose, they should be cheap and locally available throughout the year.

Table 6.3. Standard potting mix.

Ingredient	Proportion	Beneficial properties	Examples
Forest soil	50%	Nutrients, soil micro-organisms, structural support	Top 15 cm of black forest soil
Coarse organic matter	25%	Air spaces	Peanut husks, leaf litter, domestic compost, tree bark
Fine organic matter	25%	Moisture retention, nutrients	Coconut fibre, charcoal made from rice husks, dried cattle dung

A standard, general purpose medium consists of 50% forest top soil mixed with 25% fine organic matter and 25% coarse organic matter (**Table 6.3**).

Store the potting medium in a moist condition but protected from rain. To prevent the spread of diseases, never re-cycle the potting medium. When disposing of weak or diseased trees, dispose of the potting medium in which they grew well away from the nursery.

When making a potting medium, sieve the materials to remove stones or large clumps and mix them together on a hard, flat surface using a shovel. Large nurseries use electric cement mixers to mix their potting media.

Box 6.4. Wildlings as alternatives to seeds.

Growing a mixed crop of framework tree species from seeds can take 18 months or more because you have to wait for the parent trees to fruit and for the seeds to germinate. So, is there a faster way to produce framework tree saplings? Wildlings are seedlings that are dug up from the forest and cultivated in a nursery. Forest trees usually produce vast numbers of surplus seedlings, most of which die, so digging up a few of them for transfer into a nursery does not harm the forest ecosystem. If wildlings are transplanted from a cool, shady forest directly into an open deforested site they usually die of transplantation shock. So wildlings must first be potted, cared for in a nursery, and hardened off before they are planted out. Researchers at Chiang Mai University's Forest Restoration Research Unit (FORRU-CMU) have determined how to use wildlings to produce framework trees for planting (Kuarak, 2002).

In the forest, locate several suitable parent trees of the required species that fruited heavily the previous fruiting season. It is best to collect seedlings from around many parent trees to maintain genetic diversity. Collect seedlings that are no taller than 20 cm (larger ones have high mortality because of severe transplantation shock) within a 5 m radius of the parent tree (which would otherwise die as the result of competition from the parent tree). The primary consideration when collecting wildlings is to minimise root damage, so dig them up during the rainy season, when the soil is soft. Lever out very young, small seedlings carefully with a spoon or dig up larger seedlings with a trowel, retaining a plug of soil around the roots. Place the seedlings in a bucket with a little water, or use containers made from banana stems.

In the Philippines, containers made from sections of banana plant stems make cheap containers for the transfer of wildlings from forest to nursery.

If wildlings are more than 20 cm tall, consider pruning them just after digging them up to reduce mortality and increase growth rate. Cut back the stem by one-third to one-half, but remember that not all species tolerate pruning, so you might need to carry out some experiments. Make a 45° cut about 5 mm above an axillary bud. Alternatively cut back the larger leaves by about 50%. Secondary roots might need to be trimmed to enable seedlings to be potted easily into 23 × 6.5 cm plastic bags filled with standard potting mix without bending the tap root. Keep the potted wildlings under deep shade (20% of normal sunlight) for about 6 weeks or construct a recovery chamber. Thereafter, follow the same procedures used for care and hardening-off of saplings that have been grown from seed. When compared with growing planting stock from seed, these techniques can shorten the time needed to grow trees to a plantable size by several months to a year and can reduce production costs considerably.

BOX 6.4. WILDINGS AS ALTERNATIVES TO SEEDS

Box 6.4. continued.

A recovery chamber of 1 × 4 m is large enough for 1,225 plants. This example is constructed in a shaded area of the nursery from a simple split bamboo frame. The frame is covered with a polythene sheet, the edges of which are buried in a shallow trench around the structure, thereby sealing the chamber. Humidity builds up in the chamber, preventing transplantation shock. After a few weeks, the chamber is partially opened so that the plants can acclimatise to ambient conditions and eventually the cover is removed completely.

By Cherdsak Kuaraksa

How much potting medium is needed?

To calculate the volume of media needed to fill your containers, measure their radius and height and apply the following formula:

Total volume of medium required = (container radius)^2 × container height × 3.14 × number of containers

For example, for 2,000 plastic bags of 23 × 6.5 cm, you will need $(6.5/2)^2$ × 23 × 3.14 × 2,000 = 1,525,648 cm^3 or approximately 1.5 m^3 of medium.

Filling the containers

First, make sure the medium is moist but not too wet: spray it with water if necessary. When pricking-out small seedlings, fill the containers to the brim with medium using a trowel or bamboo scoop. Bang each container on the ground a few times to allow the medium to settle, before topping-up the containers with more medium to 1–2 cm below the container's rim. The medium should not be so compact as to inhibit root growth and drainage, but neither should it be too loose. The consistency of medium within plastic bags can be checked by firmly grasping the bag. The impression of your hand should remain after you let go and the bag should stand up straight, unsupported.

'Pricking-out'

'Pricking-out' (potting) is transferring seedlings from germination trays into containers, a task that should be carried out in shade, late in the day. Seedlings are ready for pricking out when the first 1–3 pairs of true leaves have fully expanded. Fill the containers as previously described. Then use a spoon to make a hole in the medium that is large enough to take the seedling's roots without bending them. Handle the fragile seedlings with care. Gently grasp a leaf (not the stem) of a seedling and slowly prise it out of its germination tray with a spoon. Place the seedling's root into the hole in the potting medium and fill the hole with more medium. Bang the container on the

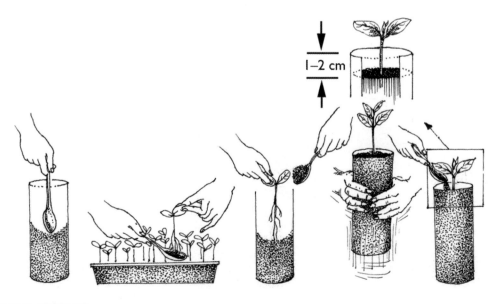

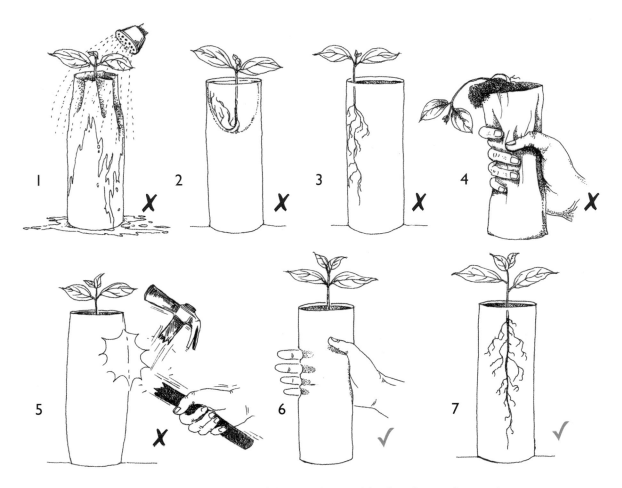

Potential problems with potting: (1) the medium has settled causing the rim of the plastic bag to collapse and blocking watering; (2) curled roots will make the adult tree susceptible to wind throw; (3) the seedling is not placed centrally; (4) the medium is too soft; (5) the medium is compacted; (6) excellent medium consistency; and (7) the perfectly potted seedling!

ground to settle the medium. Top up until the medium surface is 1–2 cm below the container's rim and the seedling's root collar (the junction between the root and shoot) is at the medium's surface. Then, press the medium to make sure the plant is upright and centrally placed. For larger plants, suspend the roots in a partly filled container and then carefully add medium around the roots.

'Standing down'

'Standing-down' refers to the time that the containerised trees are kept in the nursery from potting until transportation to the planting site. After potting the seedlings, place the containers in a shaded area and water the seedlings. Make sure that plastic bags remain upright and are not squeezed together. At first, the containers can be touching each other (i.e. 'pot thick'), but as the seedlings grow, space the containers a few centimetres apart, to prevent neighbouring seedlings from shading each other.

The containers can be stood down on bare ground, on ground covered by various materials or on raised wire grids. If the containers are stood down on bare earth, tree roots can grow through the holes in the base of the containers into the underlying soil.

Box 6.5. Cuttings as an alternative to seeds.

Vegetative propagation is not normally recommended for producing planting stock for forest restoration projects because it tends to reduce adaptive genetic diversity (see **Box 6.1**, p. 159). It may be appropriate, however, for highly desirable, rare, framework tree species whose seeds are difficult to find or germinate. For such species, propagation by cuttings is acceptable provided that the cuttings are collected from as many parent trees as possible.

Trees that are grown from cuttings often mature early — a desirable characteristic for a framework tree species. Low-tech methods can be used to root cuttings. Longman and Wilson (1993) report that most tropical tree species tested to date can be rooted as leafy stem cuttings in low-technology 'poly-propagators' and/or under mist. These authors also provide a comprehensive review of techniques, but bear in mind that little work has been done on the vegetative propagation of the vast majority of tropical tree species that are to be useful for the restoration of tropical forest ecosystems.

A study of vegetative propagation at Chiang Mai University's Forest Restoration Research Unit (FORRU-CMU) provided the following recommendations for rooting cuttings of framework species, using a simple method based on plastic bags (Vongkamjan, 2003).

Cut medium-sized, vigorous juvenile shoots (leafy shoots can often be found on stumps after chopping or burning), from as many parent trees as possible, with a sharp, clean pair of secateurs or a knife. Place the cuttings in plastic bags with a little water and take them to a nursery immediately. In the nursery, trim the cuttings into 10–20 cm lengths. Remove the lower, woody parts and the fragile apical section. If each node has a leaf or bud, single nodes can be used, but for shoots that lack buds and have short internodes, cuttings can include 2–3 nodes.

Cut back the leaves transversely by 30–50%. Cut the bases of the cuttings with a sharp propagation knife into a heel shape just below a node.

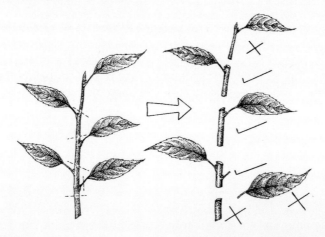

Hormone treatments are usually required to stimulate the cuttings to root. Each species responds differently to the various hormone preparations that are available, so some experimentation will be necessary. Products that contain auxins, either indole-3-butyric acid (IBA) or naphthalene-1-acetic acid (NAA), in various concentrations are most likely to be effective. These products are usually powders, which should be dusted lightly onto the bases of the cuttings. Some rooting powders also contain a fungicide such as 'Thiram' or 'Captan' which helps to discourage disease. Follow the instructions on the packet. For more advice on rooting cuttings of tropical trees see: www.fao.org/docrep/006/AD231E/AD231E00. htm#TOC

BOX 6.5. CUTTINGS AS AN ALTERNATIVE TO SEEDS

Box 6.5. continued.

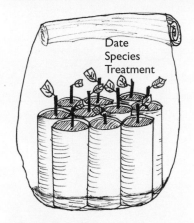

Bags within bags can be used to maintain 100% humidity while the cuttings grow roots.

Mix 50% sand with 50% rice husk charcoal to make a rooting medium and place it in small, black, plastic bags. Push the bases of the cuttings into the medium. Water the medium and press it to make it firm around each cutting. Put groups of 10 small bags into larger plastic bags (20 × 30 cm). Add one litre of water and seal the larger bag, resulting in an atmosphere of 100% humidity that will keep the cuttings alive until the roots grow and are able to supply sufficient water to the cuttings' shoots. Label each bag with the species name and starting date. Keep records of how many cuttings develop roots and shoots. Top up the bags with water weekly and remove dead cuttings and dried leaves. When the cuttings show vigorous root and shoot development, transplant them into 23 × 6.5 cm plastic bags and care for them as described in **Section 6.5**.

By Suphawan Vongkamjan

When the trees are lifted for planting, these roots will break, suddenly reducing the supply of water from the root to the shoot. This can cause the plant to go into shock before it even reaches the planting site. Therefore, the containers must be lifted every few weeks, and any protruding roots pruned back before they can penetrate the soil. Covering the standing down beds with very coarse gravel can help to prevent this problem. Roots growing into the gravel find no nutrients and little moisture, and are gradually killed by exposure to air. Covering the standing down area in plastic sheets also prevents roots from penetrating the soil, but non-porous plastic can obviously create drainage problems.

(A) Standing down on bare earth works well, but the young trees require constant attention to prevent roots growing into the soil under the pots. In this nursery, bamboo guard rails are used to keep the plants upright. (B) Covering the ground with gravel and then a porous sheet ('weed-mat') prevents roots from growing into the underlying soil. In this nursery, an automatic sprinkler system waters the plants, which are grown in square, rigid, reusable, plastic pots.

The ultimate (and most expensive) solution is to stand down containers on raised wire grids. Roots that grow out from containers are exposed to air and either stop growing or die. This is called air pruning (see **Section 6.5**). It encourages root branching within containers and the formation of a dense root ball, which increases the survival chances of trees after planting out.

Five-star accommodation for trees. The trees sit in wire grid trays on a frame that is raised off the ground, allowing the air pruning of roots. Removable shade netting permits control of lighting conditions.

6.5 Caring for trees in the nursery

Shade requirements

After pricking-out, place the seedlings under about 50% shade to prevent scorching of the leaves and wilting. Shade netting, graded according to the percent shade cast, can be bought at most agricultural supplies stores. Hang it on a frame 0.5–2.5 m above the ground. If shade netting is unavailable or too costly, local materials such as coconut palm leaves, thin strips of bamboo or even dried grass are also effective, but take care not to provide too much shade with these materials. More than about 50% shade will produce tall, weak trees that are susceptible to diseases. Even when well-established in containers, trees remain vulnerable to high temperatures and full sunlight. Consequently, they are usually grown under light shade until they are ready for hardening-off.

Watering

Each container holds a relatively small amount of water, so seedlings can dry out rapidly if watering is interrupted for more than a day, especially in a dry season. By contrast, over-watering can saturate the potting medium, which suffocates the roots, and this can be just as damaging to plant growth as dehydration.

Water the trees early in the morning and/or late in the afternoon to avoid the heat of the day. If there is any doubt about the reliability of the water supply, install a system of water tanks as a reserve supply. Nursery workers who are responsible for watering should make a record on a calendar each time watering is carried out.

Large commercial nurseries often use a system of sprayers that are inter-connected with pipes, allowing effortless watering whenever the tap is turned on, but such systems are expensive. In nurseries producing many different tree species with different water requirements, watering by hand using a watering can or a hose with a fine rose attached is recommended. This allows nursery workers to assess the dryness of each batch of trees and to adjust the amount of water delivered accordingly.

Some training is usually required to enable the person responsible for watering to judge how much water to provide. During the rainy season, it may be possible to go several days without watering the saplings in an open nursery. By contrast, in a dry season, it may be necessary to water the saplings twice in a day. The saplings are ready for water when the soil surface is

Hand watering allows more control than is provided by automated sprinklers.

starting to dry out. The presence of mosses, algae or liverworts on the surface of the potting medium indicates that the seedlings are being given too much water; they should be removed and watering reduced. Weeds can compete aggressively for water, so should be removed from containers.

Extra care is required when watering germination trays: a very fine rose must be used and the watering should be carried out in a sweeping motion to avoid damage to the seedlings.

Fertiliser

Trees require large amounts of nitrogen (N), phosphorus (P) and potassium (K), moderate amounts of magnesium, calcium and sulphur and trace amounts of iron, copper and boron and other mineral nutrients to sustain optimal growth. The potting medium might supply adequate quantities of these nutrients, especially if rich forest soil is being used, but the application of additional fertiliser can accelerate growth. Your local agricultural extension service or agriculture college might be able to analyse the nutrient content of the medium you use and advise you on fertiliser requirements.

The decision to apply fertiliser depends not only on the availability of nutrients in the potting medium but also on the growth rate required, or the appearance of the seedlings. Plants that have symptoms of nutrient deficiency, such as yellowing leaves, should receive fertiliser. Fertiliser should also be applied when it is necessary to accelerate growth to ensure that the plants are ready for transplantation by the planting season.

The application of 10 granules of slow-release fertiliser every 3–6 months can mean the difference between seedlings growing tall enough by planting day and having to keep them in the nursery for another year.

The use of slow-release fertiliser granules at a rate of about 1.5 g per litre of potting medium is recommended. At FORRU-CMU, good results have been achieved for many species by adding about 10 granules of 'Osmocote' NPK 14:14:14 (about 0.3 g) to the surface of the medium in each container every 3 months. 'Nutricote' is also widely available, and recommended. Although slow-release fertilisers are expensive, only very small quantities are applied every 3–6 months, and hence the labour costs of applying them are very low.

Alternatively, ordinary fertiliser can be used, either as solid fertiliser mixed in the potting medium (a rough guide is 1–5 g per litre of medium) or dissolved in water. Dissolve roughly 3–5 g of fertiliser per litre of water and apply with a watering can. Then, water the saplings again with fresh water to wash any fertiliser solution from the leaves. This treatment must be repeated every 10–14 days, so it requires much more time and labour than using slow-release granules.

Do not apply fertiliser i) to rapidly growing species that will reach a plantable size before the optimal planting time (as they will outgrow their containers), ii) to species in the family Leguminosae, or iii) immediately before hardening-off (as new shoot growth should not be encouraged at that time). The overuse of fertiliser can result in the 'chemical burning' of plants and can kill beneficial soil micro-organisms such as mycorrhizal fungi.

Mycorrhizal fungi

Mycorrhizal fungi (literally 'fungus root') form mutually beneficial 'symbiotic' relationships with plants. They form extensive networks of fine fungal hyphae that radiate out from tree roots into the surrounding soil. The fungi transfer nutrients to trees from a much greater volume of soil than can be exploited by the trees' root systems alone. In return, the fungi gain carbohydrates (as an energy source) from the trees. There are two main types of mycorrhiza that associate with trees: ectomycorrhiza and vesicular arbuscular mycorrhiza (VAM). Ectomycorrhiza form a sheath of fungal threads around the outside of tree roots that extends between the plant's cells but does not penetrate them. All dipterocarps, some legumes, many conifers and a few broadleaved trees (e.g. oaks) have ectomycorrhiza. VAM live within roots and actually penetrate the root cells. They can be found on the vast majority of tropical trees but we know relatively little about the diversity of mycorrhiza in tropical forests or about their role in maintaining the complexity of tropical forest ecosystems. It is therefore impossible to prescribe detailed actions for the use of mycorrhiza in forest tree nurseries.

We do know, however, that mycorrhizal inoculation can increase the survival and growth of nursery grown trees after they are planted out, especially on highly degraded land that has been without native vegetation or top soil for several years (e.g. mined land). When forest soil is included in the potting medium, most native forest tree species become naturally infected with mycorrhizal fungi and application of commercially produced mycorrhizal inoculae has no significant advantage (Philachanh, 2003).

Weed control

Weeds that are present in the nursery can harbour pests, and their seeds may spread into containers. Grasses, herbs and vines should all be removed from the nursery grounds before they can flower. Weeds that colonise containers compete with tree seedlings for water, nutrients and light. If not dealt with when small, weeds can be difficult to remove from containers without damaging the roots of tree seedlings. Check containers frequently and use a blunt spatula to remove weeds while they are still small. Weed in the morning, so that any remnant weed fragments dry out in the heat of the day. Wear gloves when dealing with thorny or noxious weeds. Also remove any mosses or algae that are growing on the medium surface. Obviously, herbicides cannot be used to control weeds in tree nurseries.

Take care to avoid poisonous snakes or insects in the dense foliage of a batch of container-grown tree seedlings.

More weeds than trees? Weed the nursery regularly to prevent the build-up of weeds in the containers.

Diseases

Disease prevention

Diseases can occur even in the best-maintained nurseries and there are three main causes:

- **fungi** — although some species are beneficial, others cause damping-off, root-rots and leaf-spots (blights and rusts);
- **bacteria** — most are harmless, but some cause damping-off, canker and wilts; and
- **viruses** — most do not cause problems, but some cause leaf-spots.

Prevention is better than cure, so keep containers, tools and work surfaces clean by washing them in a solution of domestic bleach. Follow the manufacturer's instructions for dilution, taking care to avoid getting bleach onto your skin or into your eyes. Thoroughly wash rigid plastic containers when re-using them. Do not recycle plastic bags or medium.

Rigid plastic pots can be re-used provided they are properly cleaned, but plastic bags must be disposed of well away from the nursery to prevent the build-up of pathogens.

Detecting and controlling disease

Constant vigilance is needed to prevent disease outbreaks. Ensure that all of the nursery staff learn how to recognise the symptoms of common plant diseases and that all the young trees are inspected at least once a week. To prevent disease spread, make sure that the plants are not being over-watered, that there is adequate drainage within and beneath the containers, and that the plants are well-spaced to allow air movement around them and to prevent direct transfer of pathogens from individual seedlings to their neighbours. Use disinfectant to wash tools or rubber gloves that come into contact with diseased plants.

If a disease outbreak occurs, remove infected leaves or dispose of diseased plants immediately. Burn them well away from the nursery. Do not recycle the medium or plastic bags in which they grew. If using rigid containers, wash them with disinfectant and dry them in the sun for several days before re-using them. Inspect the plants daily until the outbreak is over.

Routine spraying with chemicals should not be necessary. Chemicals are expensive and they are a health hazard if not handled properly. If it is necessary to spray an infected batch of plants, first try to identify the type of disease (fungal, bacterial or viral) and select an appropriate chemical. For example, Iprodione is active against fungal leaf-spots, whereas Captan is particularly effective against damping-off fungi.

When using any fungicides, read the health warnings on the packet and follow all the protective precautions recommended.

Where diseases become prevalent, consider pasteurising the potting medium by heating it in the sun. This will kill most pathogens, pests and weed seeds, but it could also kill beneficial soil micro-organisms so consider re-inoculating the medium with mycorrhiza.

Pest control

Most insects are harmless or indeed beneficial, but some can rapidly defoliate young trees or damage their roots causing death. Not all pests are insects: nematode worms, slugs and snails, and even domestic animals can all cause problems.

The most important pests include leaf-eaters, such as caterpillars, weevils and crickets; shoot borers, particularly beetle and moth larvae; juice-suckers, such as aphids, mealy bugs and scale insects; root-eaters, such as nematode worms; cutworms, the larvae of certain moths; and termites, which also destroy nursery structures. In addition to eating the plants, pests can transmit diseases.

Inspect the trees for pests regularly and carefully to ensure that an infestation cannot develop. Remove harmful animals or their eggs by hand, or spray the saplings with a mild disinfectant. If this fails to prevent infestation, then spray the saplings with an insecticide. Prevention is better than cure as most insecticides are poisonous to humans. It is therefore essential to read the labels on insecticide packaging and follow the instructions carefully, observing all the health precautions recommended by the manufacturer. Select the most appropriate chemical for the particular pest species present. For example, 'Pirimicarb' is active against aphids, 'Aldrin' can be used to control termites, and 'Pyrethrin' is a more general insecticide.

Not all pests are small — this nursery in western Thailand is protected from elephants by an electric fence.

Not all pests are small. Dogs, pigs, chickens, cattle and other animals can wreak havoc in a tree nursery in just a few minutes. So, where such animals occur, make sure that the plants are protected within a sturdy fence.

Quality control by grading

Grading is an effective method of quality control. It involves arranging the growing trees in order of size, while at the same time removing stunted, diseased or weak ones. In this way, only the most vigorous and healthy trees are selected for hardening-off and planting-out and thus post-planting survival is maximised. When the nursery is full, the smallest and weakest plants can be identified easily and removed to make room for new more vigorous plants.

Grading is the best form of quality control.

Carry out grading at least once each month. Root pruning and disease inspection can be carried out at the same time. Wash hands, gloves and secateurs in disinfectant frequently to prevent the spread of diseases from one block of plants to another. Dispose of poor-quality plants by burning them, well away from the nursery, and do not recycle the medium or plastic bags in which they were grown. Nursery workers are sometimes reluctant to dispose of poor quality plants, but keeping them is a false economy, as they waste space, labour, water and other nursery resources that would be more efficiently provided to healthy plants. Poor-quality plants are susceptible to disease, and therefore pose a health risk to the entire nursery stock.

The nursery manager must produce high-quality trees that will perform well when planted out in the harsh conditions typical of deforested sites. Both the shoot and root systems of the young trees should be healthy and in balance with each other. This reduces transplantation stress, tree mortality and the risk of having to replant the following year. It is a false economy and a waste of time to plant poor-quality trees.

1   2   3   4   5   6   7

1. Unbalanced root and shoot growth: the shoot is too long and thin and may well break during handling. Prune back well before planting time.
2. A malformed stem compromises future growth, plants that have such stems need to be disposed of.
3. Plants that have been attacked by insects should be burnt and surviving plants sprayed with insecticide to prevent the infestation from spreading.
4. Dispose of plants whose growth is stunted growth when compared with other plants of same age.
5. This plant is losing its leaves, possibly as a result of disease; it should be burnt.
6. This container was knocked over and spent some time lying on its side, resulting in a non-vertical stem – dispose of such plants.
7. The perfect plant is well balanced, disease-free and straight; with adequate care and rigorous grading, all of the plants in your nursery should look like this.

A healthy root system

Root systems are far more crucial to the survival of trees than shoot systems. A plant can survive and re-sprout after losing its shoot, but not after losing its roots. The root system must constantly supply water and nutrients to the shoots. Root growth is affected by the container, the potting medium, the watering regime and by pests and diseases. By planting time, the root systems of containerised trees must:

- form a compact root ball that does not fall apart when the tree is removed from its container;
- be densely branched with a balance between thick, supporting roots and fine ones that absorb water and nutrients;
- not be spiralling at the base of the container;
- be able to support the shoot system;
- be infected with mycorrhizal fungi and (if the tree is a legume) with nitrogen-fixing bacteria; and
- be free of pests and diseases.

If the containers are stood down on bare earth, lift them frequently and prune back protruding roots using a clean pair of secateurs (do this in the late afternoon to minimise moisture loss). Alternatively, inhibit root growth beyond the containers by standing down the trees on gravel or on raised wire-grid benches, which allows for air pruning of the roots (see **Section 6.4**).

Size of saplings at planting time

The actual height of the saplings at planting is less important than their capacity to produce vigorous new growth. Some fast-growing pioneer tree species can be planted out when only about 30 cm tall; for *Ficus* species, the recommended size is 20 cm tall (Kuaraksa & Elliott, 2012). For slower-growing climax forest tree species, it is better to plant trees of around 40–60 cm tall. Small saplings have much higher post-planting mortality rates than larger ones because of competition with weeds, but very large saplings are much more susceptible to transplantation shock and more difficult to transport.

Root pruning encourages both root-branching within the pot and the formation of a compact root ball, thus increasing the probability of survival after planting out.

Shoot pruning

Shoot pruning is necessary for plants of fast-growing species that must be kept in the nursery for a long time. Such trees can become too large for their roots to support or too cumbersome to handle during transportation and planting. The stems of tall saplings are easily broken when they are being moved. In some species, pruning encourages branching. This is a desirable characteristic because spreading crowns shade out weeds and rapidly close the canopy. Never prune shoots in the month before planting out because pruning will promote the growth of new leaves just as the saplings are about to be stressed by transplantation. Immediately after planting, the root system might not be able to take up enough water to supply new leaves, so anything that stimulates bud break shortly before planting out should be avoided. Some species do not respond well to pruning or become highly susceptible to fungal infections after pruning. So before attempting to prune large numbers of saplings, experiment with a few to test the effects of pruning.

Hardening-off

Weaning, or 'hardening-off', prepares saplings for the difficult transition from the ideal environment in the nursery to the harsh conditions of deforested sites. If they are not hardened to the hot, dry, sunny conditions of planting sites, the planted trees suffer transplantation shock and mortality rates are high.

About 2 months before planting, move all of the saplings to be planted to a separate area in the nursery and gradually reduce the shade and the frequency of watering. Light-demanding trees should stand in full sunlight for their final month in the nursery. Shading should be reduced but not removed for shade-bearing species that will not be planted in full sun.

Gradually reduce watering by approximately 50% to slow shoot growth and to ensure that any newly forming leaves will be relatively small. During hardening, water saplings just once in the late afternoon instead of twice (early morning and late afternoon). Water saplings that are normally watered once a day once every other day. Do not reduce watering to the point at which the leaves wilt, as that would stress and weaken the saplings. Regardless of the normal schedule, water the saplings as soon as any wilting is observed.

Record keeping

Record keeping and labelling seed trays facilitates efficient nursery management.

Learning from experience is only possible if both nursery activities and the performance of each species are recorded accurately. Records are essential to prevent new nursery workers from repeating the mistakes of previous ones. They are also used to assess the productivity and achievements of the nursery (e.g. numbers of species or saplings grown) and for the development of species production schedules.

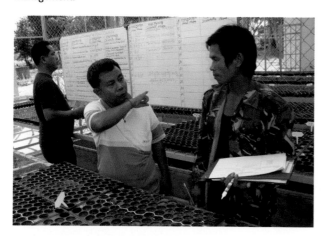

Label the seed trays and plants in the nursery with species names, batch numbers and dates of seed collection and pricking-out. Use the record sheet in the **Appendix** (**A1.6**) to record when and where each batch of seeds was collected and which seed treatments were applied, together with germination rates, growth rates, diseases observed and so on. Finally, record when and to where the saplings were dispatched for planting.

6.6 Research for improving native tree propagation

The standard protocols outlined above are sufficient to get you started on growing a wide range of native forest tree species. But as you gain experience, you will want to refine these techniques and to develop individual production schedules for each species being propagated, thereby improving the efficiency and cost-effectiveness of your nursery. Here, we provide a few basic research procedures to help you produce high-quality, vigorous, and disease-free saplings of the required size, by the optimum planting time, as rapidly and cost effectively as possible. This is achieved by conducting basic controlled experiments to test treatments that either accelerate or slow down seed germination and/or seedling growth.

Selection of species for research

Guidelines for selecting candidate framework species and nurse plantation species were provided in **Sections 5.3** and **5.5**, respectively. It is very likely that propagation protocols will have already been well-researched for any commercially valuable species. So, start by doing a literature search to find out what is already known about the species you want to grow and where the gaps in knowledge lie.

Recognising and identifying trees

At the start of a restoration research program, not all of the scientific names of the tree species to be grown will be known, so it is useful to assign a species number to every tree species from which seeds are collected: the first species to provide seeds becomes S001, the second S002 and so on. Subsequent seed batches that are collected from the same species are labelled with the same 'S' number, but are assigned their own batch number. So 'S001b1' would be the first batch of seeds collected from species no. 1, and 'S001b2' would be the second batch of seeds collected from species no. 1, either from the same tree on a different date or from a different tree of the same species. Nursery staff often remember species numbers more easily than scientific names and, with a little experience, the numbers will be used more consistently than local names. List all species and their 'S' numbers on a board in the nursery and keep it up-to-date. Then label every seed germination tray and block of containerised seedlings with their 'S' and 'b' numbers.

Display a list of 'S' numbers, alongside local and scientific names, so that all nursery staff know which species they are working with.

All species numbers must be matched with scientific names. Local, vernacular names can also be noted, but they cannot be relied upon because local people often group similar species under a single name or use different names to refer to the same tree species. Collect voucher specimens of all trees from which seeds are collected. If there are any subsequent doubts about the tree species in the nursery or those planted in field trials, the voucher specimen of the seed tree can be re-examined to confirm or change the species name. Botanical taxonomists frequently revise plant classifications and change species names, so having a voucher specimen with a species number attached can reduce confusion.

Always collect a voucher specimen that can be used to confirm species identification.

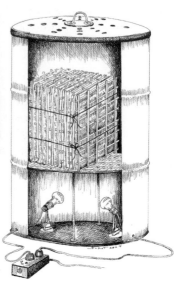

Construct a simple drying box in which light bulbs are used to dry specimens gently.

Use a cutter mounted on a pole to obtain a sample of foliage, fruits and/or flowers. Trim the specimen without losing essential features (e.g. leaf arrangement, infructescence branching etc.) until it fits well in a standard-sized plant press. In the nursery, construct a simple drying box that uses light bulbs to provide gentle heat to dry the specimens. Write a label for every specimen that includes 'S' and 'b' numbers, the local name(s), details of the tree's location, and descriptions of the bark and any features that may change with drying, particularly colours.

Mount the specimens on robust paper using standard herbarium techniques. If there is space and appropriate staff and facilities, start your own herbarium. Store mounted specimens in suitable cabinets and enter the information from the specimen labels into a database. Take precautions to prevent insects or fungi from attacking the specimens. For additional security, make several herbarium sheets for each specimen and lodge duplicates in recognised herbaria. Have the specimens examined and identified by a professional botanical taxonomist. For more detailed information on herbarium techniques, see 'The Herbarium Handbook' published by the Royal Botanic Gardens, Kew, UK (www.kewbooks.com).

Phenology

Phenology is the study of the responses of living organisms to seasonal cycles in environmental conditions. In forestry, phenological studies are used to determine when to collect seeds and to learn how forests function (particularly in regard to tree reproduction and forest dynamics), so that the same functionality can be replicated in restored forest.

The flowering and fruiting of many tropical trees are usually related to seasonal variations in moisture (Borchert et al., 2004) and solar radiation energy (insolation) (Calle et al., 2010). Cycles in reproductive events are most marked in the seasonal tropics, but cycles of flowering and fruiting can be observed even in the less seasonal, equatorial forests. Not all tropical trees reproduce seasonally. Some flower and fruit twice or several times each year, whereas others exhibit 'masting', i.e. mass fruiting at intervals of several years.

Obtaining ripe seeds is the first big challenge in tree-planting projects, so it is worth the effort of carrying out phenology studies to determine optimal seed collection schedules so that the nursery is well stocked with all of the required species. Phenological studies can also be used to predict the length of seed dormancy, and which pre-sowing seed treatments are likely to be successful in breaking or prolonging dormancy. Furthermore, they enable the identification of 'keystone' tree species: those that flower and fruit at times when other food resources for animals are in short supply (Gilbert, 1980). Keystone tree species, such as fig trees (*Ficus* spp.), support whole communities of animal pollinators and seed dispersers upon which other tree species rely for their reproduction. They are obvious candidates for testing as framework tree species. Observations of pollination and seed dispersal mechanisms can also be made during phenological studies. Additional data on the leafing phenology of the trees are usually collected at the same time. These data can help to predict optimal planting sites for individual tree species; for example, deciduous species are more suited to drier habitats and evergreen species to wetter habitats.

Establishing phenology study

Phenology trails are set up as part of the target forest survey according to the procedure described in **Section 4.2**. Label at least five individuals of each tree species that characterise the target forest type. Collect voucher specimens (as described previously) from each labelled tree and get a botanist to identify them. Write a brief note, describing where each tree is located in relation to the trail (e.g. "10 m to the left"; "right 20 m by rocky overhang"). As you repeat the observations month by month, you will soon be able to remember where each individual tree is located.

How often should data be collected?

The trees should be inspected at least once each month. Even with monthly observations, some tree-flowering events might be missed as some trees produce and drop their flowers within a month. Usually, such rapid-turnover flowering events can be inferred when the trees are subsequently observed in fruit. In such cases, the dataset can be adjusted during processing to add the 'estimated' time of a flowering event. If many flowering events are being missed, increase the frequency of data collection to twice each month.

Scoring system for phenology

We recommend the 'crown density' method for recording tree phenology, which was originally devised by Koelmeyer (1959) and subsequently much modified by various authors. This semi-quantitative method uses a linear scale of 0–4, in which a score of 4 represents the maximum intensity of reproductive structures (flower buds (FB), open flowers (FL) and fruits (FR)) in the crown of a single tree. Scores of 3, 2 and 1 represent approximately ¾, ½ and ¼ of the maximum intensity, respectively. The 'maximum intensity' of a flowering or fruiting event varies among species, and judgments of it are bound to be subjective at first but they improve with experience.

The same approach can be used to score leafing. For individual tree crowns, estimate scores of between 0 to 4 for i) bare branches, ii) young leaves, iii) mature leaves and iv) senescent leaves. The sum of these four scores should always equal 4 (which represents the entire tree crown). Scores for flowers + fruits are always less than 4, except when flowering or fruiting is occurring at the maximum intensity typical for the species being observed.

Examples of phenology scores for flowers. (Designed by Khwankhao Sinhaseni.)

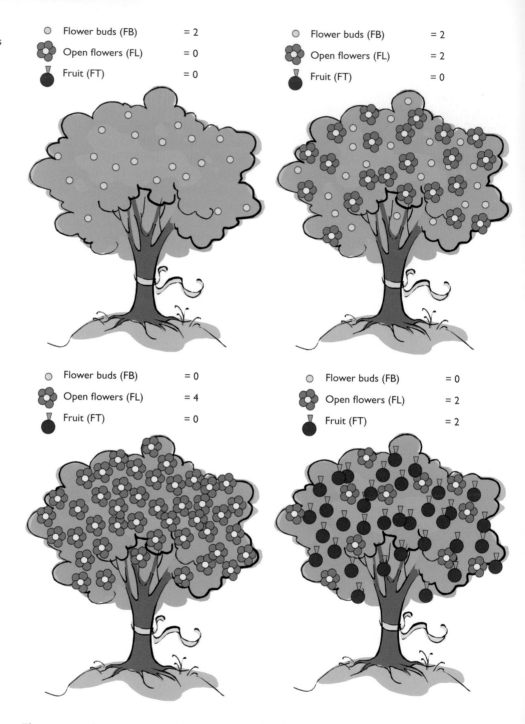

Flower buds (FB)	= 2	
Open flowers (FL)	= 0	
Fruit (FT)	= 0	

Flower buds (FB)	= 2	
Open flowers (FL)	= 2	
Fruit (FT)	= 0	

Flower buds (FB)	= 0	
Open flowers (FL)	= 4	
Fruit (FT)	= 0	

Flower buds (FB)	= 0	
Open flowers (FL)	= 2	
Fruit (FT)	= 2	

The crown density method is a compromise between very time-consuming absolute counts of flowers and fruit (or estimates of their biomass using litter-fall traps) and the very quick qualitative method of recording simple presence or absence. It is quick and it allows quantitative analytical techniques to be applied to the data. At the beginning of a study, however, it is important to train all data collectors to be consistent in their scoring, thereby minimising the subjectivity of the technique.

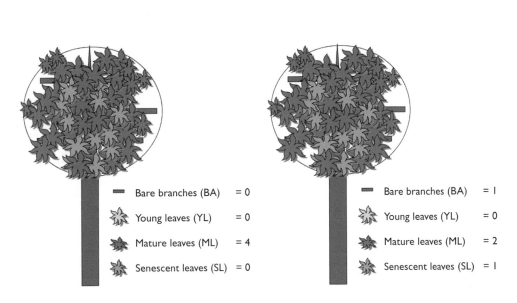

Examples of phenology scores for leaves. (Designed by Khwankhao Sinhaseni.)

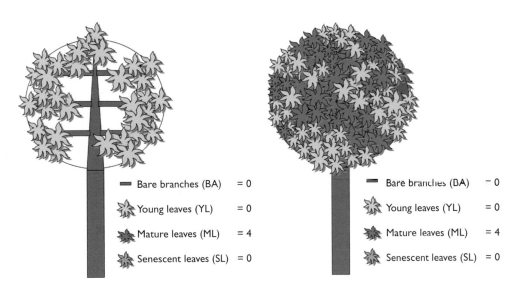

Presenting and analysing phenology data

Microsoft Excel spreadsheets are ideal for storing and manipulating phenology data. Once the study trees have been selected and labelled, prepare a data sheet as shown below. List the trees in the order in which they are encountered along the phenology trail. In the field, carry the previous month's data sheets with you, as well as blank sheets for recording the current month's data.

Month by month, accumulate all of the data into a single spreadsheet. Always enter new data at the bottom of the spreadsheet (rather than to the right). After each data-collection session, paste a copy of the blank data record sheet at the bottom of the spreadsheet and then add the newly collected data.

To analyse the data, use the tools within Excel to sort the data first by 'SPECIES', then by 'LABEL', and finally by 'DATE'. This arranges the data in chronological order, for each individual tree of each species (see overleaf).

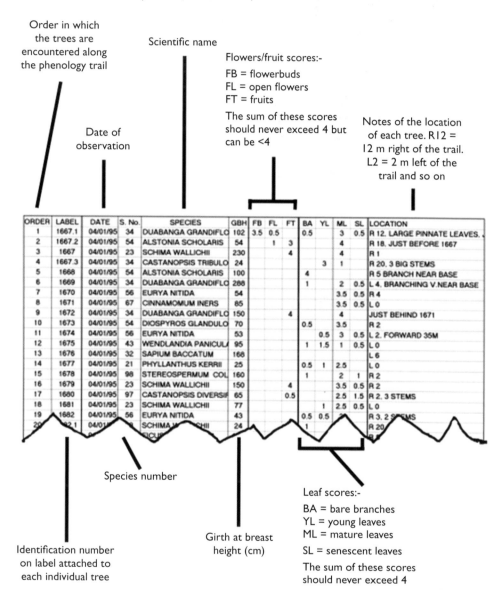

Order in which the trees are encountered along the phenology trail

Scientific name

Flowers/fruit scores:-

FB = flowerbuds
FL = open flowers
FT = fruits

The sum of these scores should never exceed 4 but can be <4

Notes of the location of each tree. R12 = 12 m right of the trail. L2 = 2 m left of the trail and so on

Date of observation

ORDER	LABEL	DATE	S. No.	SPECIES	GBH	FB	FL	FT	BA	YL	ML	SL	LOCATION
1	1667.1	04/01/95	34	DUABANGA GRANDIFLO	102	3.5	0.5		0.5		3	0.5	R 12. LARGE PINNATE LEAVES.
2	1667.2	04/01/95	54	ALSTONIA SCHOLARIS	54		1	3			4		R 18. JUST BEFORE 1667
3	1667	04/01/95	23	SCHIMA WALLICHII	230			4			4		R 1
4	1667.3	04/01/95	34	CASTANOPSIS TRIBULO	24					3	1		R 20. 3 BIG STEMS
5	1668	04/01/95	54	ALSTONIA SCHOLARIS	100				4				R 5 BRANCH NEAR BASE
6	1669	04/01/95	34	DUABANGA GRANDIFLO	288				1		2	0.5	L 4. BRANCHING V.NEAR BASE
7	1670	04/01/95	56	EURYA NITIDA	54						3.5	0.5	R 4
8	1671	04/01/95	67	CINNAMOMUM INERS	85						3.5	0.5	R 0
9	1672	04/01/95	34	DUABANGA GRANDIFLO	150			4			4		JUST BEHIND 1671
10	1673	04/01/95	54	DIOSPYROS GLANDULO	70				0.5		3.5		R 2
11	1674	04/01/95	56	EURYA NITIDA	53					0.5	3	0.5	L 2. FORWARD 35M
12	1675	04/01/95	43	WENDLANDIA PANICULA	95		1	1.5	1		0.5	L 0
13	1676	04/01/95	32	SAPIUM BACCATUM	168								L 6
14	1677	04/01/95	21	PHYLLANTHUS KERRII	25				0.5	1	2.5		L 0
15	1678	04/01/95	98	STEREOSPERMUM COL	160				1		2	1	R 2
16	1679	04/01/95	23	SCHIMA WALLICHII	150			4			3.5	0.5	R 2
17	1680	04/01/95	97	CASTANOPSIS DIVERSIF	65				0.5		2.5	1.5	R 2. 3 STEMS
18	1681	04/01/95	23	SCHIMA WALLICHII	77					1	2.5	0.5	L 0
19	1682	04/01/95	56	EURYA NITIDA	43				0.5	0.5	3		R 3. 2 STEMS
20	82.1	04/01		SCHIMA WALLICHII	24				1				R 20

Species number

Identification number on label attached to each individual tree

Girth at breast height (cm)

Leaf scores:-

BA = bare branches
YL = young leaves
ML = mature leaves
SL = senescent leaves

The sum of these scores should never exceed 4

Then use the MS Excel graph wizard to construct a visual phenological profile like that shown opposite. Start by making a profile for each individual tree of each species. This will give you some idea of the variability in phenological behaviour within each species population and will enable you to assess the synchrony of phenological events. Only then, calculate mean score values across all of the individuals within each species

ORDER	LABEL	DATE	S. No.	SPECIES	GBH	FB	FL	FT	BA	YL	ML	SL	LOCATION -
272	296	05/01/95	34	ACROCARPUS FRAXINIF	222	3	0	0	1.5		1.5	1	L 4, OPP.297
272	296	26/01/95	34	ACROCARPUS FRAXINIF	222	0	4	0	3	1			L 4, OPP.297
272	296	15/02/95	34	ACROCARPUS FRAXINIF	222	0	1	3	1.5	2.5			L 4, OPP.297
272	296	08/03/95	34	ACROCARPUS FRAXINIF	222	0	0.5	3			4		L 4, OPP.297
272	296	30/03/95	34	ACROCARPUS FRAXINIF	222	0	0	3			4		L 4, OPP.297
272	296	20/04/95	34	ACROCARPUS FRAXINIF	222	0	0	3			4		L 4, OPP.297
272	296	12/05/95	34	ACROCARPUS FRAXINIF	222	0	0	3.5			4		L 4, OPP.297
272	296	01/06/95	34	ACROCARPUS FRAXINIF	222	0	0	3.5			4		L 4, OPP.297
272	296	23/06/95	34	ACROCARPUS FRAXINIF	222	0	0	3.5			4		L 4, OPP.297
272	296	14/07/95	34	ACROCARPUS FRAXINIF	222	0	0	1			4		L 4, OPP.297
272	296	06/08/95	34	ACROCARPUS FRAXINIF	222	0	0	0			4		L 4, OPP.297
272	296	30/08/95	34	ACROCARPUS FRAXINIF	222	0	0	0			4		L 4, OPP.297
272	296	21/09/95	34	ACROCARPUS FRAXINIF	222	0	0	0			4		L 4, OPP.297
272	296	13/10/95	34	ACROCARPUS FRAXINIF	222	0	0	0			4		L 4, OPP.297
272	296	02/11/95	34	ACROCARPUS FRAXINIF	222	0	0	0			4		L 4, OPP.297
272	296	25/11/95	34	ACROCARPUS FRAXINIF	222	0	0	0			4		L 4, OPP.297
272	296	16/12/95	34	ACROCARPUS FRAXINIF	222	0	0	0			4		L 4, OPP.297
329	464	05/01/95	34	ACROCARPUS FRAXINIF	575						4		EG 10/5
329	464	26/01/95	34	ACROCARPUS FRAXINIF	575	3	0	0	2.5		1.5		EG 10/5
329	464	15/02/95	34	ACROCARPUS FRAXINIF	575	3.5	0.5	0	3.5	0.5			EG 10/5
329	464	08/03/95	34	ACROCARPUS FRAXINIF	575	0	0	2	1.5	2	0.5		EG 10/5
329	464	30/03/95	34	ACROCARPUS FRAXINIF	575	0	0	0.5		3	1		EG 10/5
329	464	20/04/95	34	ACROCARPUS FRAXINIF	575	0		0			4		EG 10/5

population and construct an 'average' profile for each species. When analyzing flower or fruit data, the most important point to look for is the period during which the fruit scores decline for each species. This indicates the optimal seed collection month for that year, when natural seed dispersal is occurring. For example, the graph below shows that the optimum seed collection time for *Acrocarpus fraxinifolius* is from late June to early July, when maximum seed dispersal occurs. The fruit/seed maturation period is from February to June.

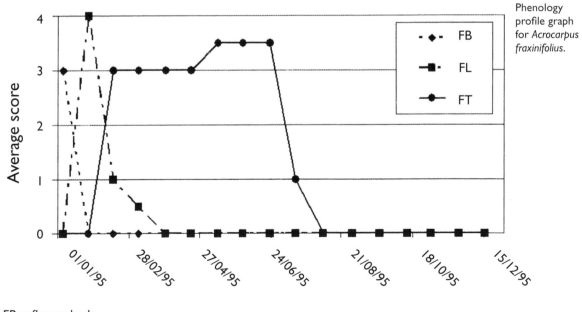

Phenology profile graph for *Acrocarpus fraxinifolius*.

FB = flower buds

FL = open flowers

FT = fruits

After phenology has been studied for several years, various useful indices of seed production can be calculated by extracting data from the spreadsheets (Elliott *et al.*, 1994).

- **Duration** – the mean length of flowering-fruiting episodes (in weeks or months) for each individual tree and averaged across all trees sampled in a species.
- **Frequency** – the total number of flowering-fruiting episodes recorded for each individual divided by the number of years the study has run: then averaged across all individuals of the same species.
- **Intensity** – mean of the maximum flower or fruit scores (for each flowering-fruiting episode) recorded for each individual tree: then averaged for all flowering-fruiting individuals in the species sample.
- **Prevalence** – number of individual trees that flowered and fruited in each year, expressed as a percentage of the total number of individual trees in each species sample, averaged across the total duration of the study (in years).
- **Fruit set index** – for each flowering-fruiting episode, the maximum fruit score observed expressed as a percentage of the maximum flower score: averaged for all flowering-fruiting episodes for all individuals in the species sample.

Germination trials

Don't risk your life to collect a few seeds. If you need to climb trees, wear a safety harness.

Phenology studies provide ideal opportunities to collect seeds for germination trials, but remember that seeds can be collected from any trees bearing ripe fruits, even if they are not included in the phenology studies. Collect fruits when they are fully ripe but just before they are dispersed or consumed by animals. Label each seed tree with a unique number and fill in a seed collection data sheet. If a GPS is available, record the location of each seed tree.

Date collected: 20/03/2005   **Species No.:** 071   **Batch No.:** 1

SEED COLLECTION DATA SHEET

Family: *Rosaceae*   **Botanic name:** *Cerasus cerasoides* (Buch.-Ham. ex. D. Don) S.Y. Sokolov

Common name: Nang Praya Sua Klong

Location: Doi Suthep-Pui National Park, roadside by Cinchona plantation

GPS location: 18 48 23.37 N; 98 54 44.76 E   **Altitude:** 1,040 m

Forest type: primary evergreen forest, disturbed roadside area, granite bedrock

Collected from: ☒ ground   ☒ tree

Tree label no.: 71.1   **Tree girth:** 88 cm   **Tree height:** 6 m

Collector: S. Kopachon   **Date seed sown:** 20/03/2005

Notes: Bulbuls were eating the fruit

☒ Voucher collected?   ✂ ✂

HERBARIUM, BIOLOGY DEPARTMENT, CHIANG MAI UNIVERSITY
FOREST RESTORATION RESEARCH UNIT, VOUCHER
NOTE: all dates are day/month/year

FAMILY: *Rosaceae*

BOTANICAL NAME: *Cerasus cerasoides* (Buch.-Ham. ex D. Don) S.Y. Sokolov

PROVINCE: Chiang Mai   **DATE:** 20/03/2005

DISTRICT: Suthep   **ELEVATION:** 1,040 m

LOCATION: Doi Suthep-Pui National Park, roadside by *Cinchona* plantation

HABITAT: Primary evergreen forest, disturbed roadside area, granite bedrock

NOTE: Height 6 m: DBH 28 cm

    Bark lenticellate, peeling, dark brown

    Fruit 14 mm × 6 mm, pericarps juicy, bright red

    Seed stoney pyrene, about 7–10 diameter, light brown, contains 1 seed

    Leaf blades green above, light green underneath

COLLECTED BY: S. Kopachon   **NUMBER:** S071   **DUPLICATES:** 5

Germination trials can answer two basic questions: i) how many seeds germinate (percent germination) and ii) how quickly or slowly do the seeds germinate? Both of these parameters can be used and even manipulated when planning the growth of sufficient numbers of tree saplings for a specific planting time.

In seasonal tropical forests, the seeds of most tree species tend to germinate at the beginning of the rainy season (Garwood, 1983; FORRU, 2006). Seeds that are produced shortly before the rainy season usually have a short dormancy period; whereas those that are produced earlier have a longer dormancy. For the former, the saplings will be too small to plant in the first planting season, so it may be necessary to delay germination by storing seed as described in **Section 6.2** in order to prevent the saplings from outgrowing their containers before the second planting season. Conversely, it might be necessary to break the dormancy and accelerate the germination of seeds that are produced well before the ideal planting season so as to produce a crop of saplings that are ready to plant in less than 1 year. Failure to break the dormancy of such seeds could mean that plants must be kept in the nursery for 18 months or longer.

The objective of a germination trial is not to test the germination that would occur in nature, but to determine germination rates and periods under nursery conditions. Hence, seeds should be prepared using the standard protocol (see **Section 6.2**): fruit flesh should be removed, seeds should be air-dried and non-viable seeds identified by a flotation test removed.

Testing treatments for overcoming dormancy

To accelerate and maximise germination, seed treatments must overcome any dormancy mechanisms that are present (see **Box 6.2**, p. 168). The most common dormancy mechanisms involve the seeds coverings; treatments that perforate those coverings (scarification) are often effective as they allow water and oxygen to diffuse into the embryo. Use sand paper to roughen the entire seed surface or nail clippers to make small individual holes in the end of the seed opposite to that at which the embryo is located. Try cracking large pyrenes, which are covered by a hard, stony or woody endocarp, open gently in a vice or tapping them with a hammer. Scraping off a soft aril, if present, nearly always increases germination.

Try cracking large hard seeds gently in a vice.

Acid can also be tested as a scarifying agent to break down impermeable seed coats. Soak the seeds in concentrated sulphuric acid for a few minutes to several hours (depending on the size of the seed and the thickness of its coat). You will need to experiment with the time required. This treatment is usually effective with legume tree seeds. Acids are obviously dangerous substances and must be handled with caution, following manufacturer's safety guidelines. If physical dormancy is suspected (i.e. if embryo development is restricted by a hard but permeable seed coat), acid may penetrate rapidly and kill the embryo, so acid treatment is not recommended for such seeds. Freezing and heat treatments (particularly burning) are also not recommended for tropical tree species. If dormancy is caused by chemical inhibitors, experiment with soaking the seeds in water for various lengths of time to dissolve out the inhibitory chemicals. Another option that is worth investigating is to collect seeds at different times of the year, from the same or different individual trees of the same species. Such experiments can be used to determine the optimum seed collection time.

Try to design treatments that change only one factor, even though this can be difficult to achieve in practice. For example, putting seeds into hot water has two simultaneous effects, i.e. soaking and heating.

Experimental design for germination trials

Use a randomised complete block design (RCBD) as described in **Appendix (A2.1)** to test for treatment effects. Place a control germination tray containing seeds that have been prepared in a standard way and several treatment trays, each one containing seeds that have been subjected to a different pre-sowing treatment, adjacent to each other on a nursery bench as a 'block'. Replicate the blocks several times on different benches and represent each treatment equally in every block (i.e. with the same number of seeds subjected to each of the treatments and in the control tray).

Allocate the positions of the control and the treatment replicates randomly within each block. The typical design shown here has four treatments (T1–T4) and a control (C), replicated in four blocks. Using a minimum of 25 seeds per replicate, this design requires 125 seeds per block or 500 seeds in total. If you do not have enough seeds, then reduce the number of treatments tested, but try to keep the number of replicates above three. If you have enough seeds, then increase the number of seeds per replicate to 50–100 (which would require 1000–2000 seeds, respectively, for the whole experiment).

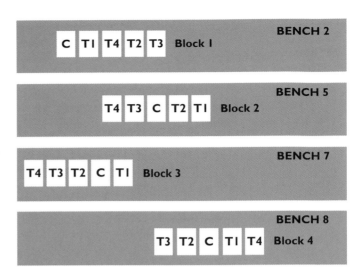

BENCH 2

C T1 T4 T2 T3 Block 1

BENCH 5

T4 T3 C T2 T1 Block 2

BENCH 7

T4 T3 T2 C T1 Block 3

BENCH 8

T3 T2 C T1 T4 Block 4

Fill modular germination trays with the regular germination medium used in the nursery. Then, sow a single seed into each module. Do not bury the seeds too deeply, otherwise it will be difficult to observe when each seed germinates. Clearly label the trays with the species number and treatment applied, and if necessary cover the trays with wire mesh to prevent animals from interfering with the experiments.

Setting up one block of an RCBD germination experiment in Cambodia.

Collecting data in germination trials

Prepare a seed germination data sheet like that illustrated here. Inspect all seed germination trays at least once per week. During periods of very rapid germination, more frequent data collection might be necessary. For each seed that has germinated (see the definition in **Box 6.2**), use a correction fluid pen ('Liquid Paper') to place a waterproof white dot on the rim of the module, always in the same orientation (e.g. always on the top edge of the module). Count the total number of white dots and record the result on the data sheet. White dots indicate all those cells in which a seed has germinated, even if the seedling subsequently dies and disappears. Therefore, counting the white dots provides a better assessment of actual germination than counting the number of visible seedlings.

Early seedling mortality (i.e. death occurring after germination but before the seedlings grow large enough for pricking out) is also a useful parameter when calculating the number of trees that can be generated from a given number of seeds collected. To record early seedling mortality, count the number modules with white dots that contain no visible seedling or an obviously dead one. For additional insurance, draw diagrams of each modular tray, with one square representing each module. Then record in each square the date on which germination or seedling death was first observed.

Species Number: 133 **Batch Number: 10**

SEED GERMINATION DATA COLLECTION SHEET

Species name: *Afzelia xylocarpa* (Kurz) Craib **Family: Leguminosae**

Date seeds collected: 20/8/2010 **Date seeds sown: 24/11/2010** **Seeds sown per replicate: 24**

Description of standard seed preparation procedures applied to all seeds:

TREATMENT DESCRIPTIONS	
T1	Control
T2	Scarification
T3	Soaking in water for 1 night

	BLOCK 1						BLOCK 2						BLOCK 3							
	T1R1		T2R1		T3R1		T1R2		T2R2		T3R2		T1R3		T2R3		T3R3		Total germinated	Total died
Date	G	GD	G	GD	G	GD	G	GD	G	GD	G	GD	G	GD	G	GD	G	GD		
1/12/2010	0	0	0	0	0	0	0	0	0	0	0	0	0	0	0	0	0	0	0	0
8/12/2010	0	0	0	0	0	0	0	0	0	0	0	0	0	0	0	0	0	0	0	0
15/12/2010	0	0	2	0	0	0	1	0	0	0	0	0	0	0	0	0	0	0	3	0
22/12/2010	0	0	5	0	0	0	1	0	0	0	0	0	0	0	0	0	0	0	6	0
29/12/2010	0	0	6	0	0	0	1	0	0	0	0	0	0	0	0	0	0	0	7	0
5/1/2011	0	0	9	0	0	0	1	0	0	0	0	0	0	0	0	0	0	0	10	0
12/1/2011	0	0	9	0	0	0	3	0	0	0	0	0	0	0	5	0	0	0	17	0
19/1/2011	0	0	12	0	0	0	5	0	0	0	0	0	0	0	6	0	0	0	23	0
26/1/2011	0	0	17	1	0	0	7	0	0	0	0	0	0	0	7	0	0	0	31	1
2/2/2011	0	0	17	1	0	0	7	0	0	0	0	0	0	0	7	0	0	0	31	1
9/2/2011	0	0	19	1	0	0	8	0	0	0	0	0	0	0	9	0	0	0	36	1
16/2/2011	0	0	22	1	0	0	12	1	0	0	0	0	0	0	9	0	0	0	43	2
23/2/2011	0	0	22	2	0	0	15	1	0	0	0	0	0	0	11	0	0	0	48	3
2/3/2011	0	0	22	2	0	0	17	1	0	0	0	0	0	0	15	1	0	0	54	4
9/3/2011	0	0	22	2	0	0	17	1	0	0	0	0	0	0	19	1	0	0	58	4

Germination curves

One of the simplest and clearest ways to represent the results of germination trials is a germination curve, with time elapsed since sowing on the horizontal axis and cumulative number (or percentage) of seeds germinated (combined across replicates) on the vertical axis. The germination curve combines into a single graphic all germination parameters, including length of dormancy period, rate and synchronicity of germination, and final percent germination.

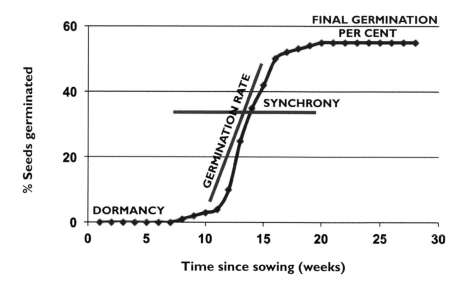

Germination curves can inform decision making without the need for complex statistical tests. In the example illustrated, the pre-sowing seed treatment accelerates germination but reduces the number of seeds that germinate. Getting the seeds to germinate faster may mean the difference between achieving a crop of saplings that are ready to plant by the first rainy season after seed collection and having to maintain saplings in the nursery until the second rainy season after seed collection. So even though the treatment reduces germination, it could have beneficial results.

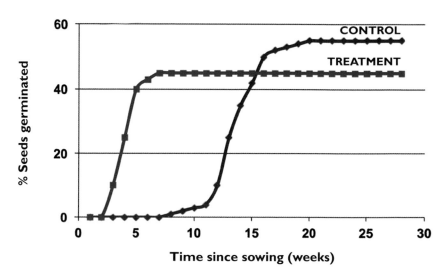

Measuring dormancy

Dormancy length is defined as the number of days between sowing a seed and emergence of the radicle (the embryonic root or the plumule if the radicle cannot be seen). In any batch of seeds, this period of time varies among the seeds. One way to express the dormancy of a batch of seeds is to add together the number of days that each individual seed is dormant and then divide the total by the number of seeds that germinate. This is 'mean dormancy'. Within any batch of seeds, however, a few seeds usually take an exceptionally long time to germinate. This increases the mean dormancy disproportionately and can produce misleading results. For example, if 9 seeds germinate 50 days after sowing and one seed germinates 300 days after sowing, the mean dormancy is ((9×50)+300)/10) = 75 days. Even though germination was complete for 90% of the seeds by day 50, a single outlying seed, increased the recorded mean dormancy by 50%.

Median length of dormancy (MLD) overcomes this problem by defining dormancy as the length of time between sowing and the germination of half the seeds that eventually germinate. In the above example, MLD would be the time between sowing and germination of the 5th seed, i.e. 50 days.

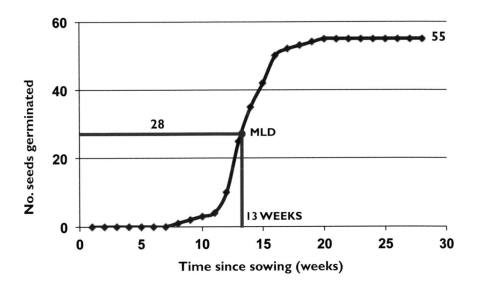

Comparing germination treatments

For each treatment and for the control, sum the final number of seeds that germinate from all replicate blocks and divide the result by the number of blocks to calculate the mean value and then repeat the calculation for the MLD values. Then use an analysis of variance (ANOVA) (see **Appendix A2.2**) to test for significant differences among the means (i.e. among the treatments and control). If the ANOVA shows significant differences, then perform pair-wise comparisons between each treatment mean and the control mean to determine which treatments increase or decrease germination and/or dormancy (see **Appendix A2.3**).

Experimenting with seed storage

If you want to experiment with seed storage, first try to confirm from the literature, or by a pilot study, whether the species you want to work with has orthodox, intermediate or recalcitrant seeds (see **Section 6.2** and http://data.kew.org/sid/search.html). Seed storage is useful for those tree species with orthodox seed whose saplings would otherwise grow rapidly and reach a plantable size well before the optimal planting time. Tending such plants for longer than is necessary wastes nursery space and resources. Furthermore, pruning them becomes an added chore when the plants start to outgrow their containers, and some species do not respond well to pruning.

For such tree species, use records of previously germinated seedlings to calculate how many months are required to grow saplings to a plantable size. Count back that number of months from the optimal planting date to obtain the optimal seed-sowing date. Next, count forward from the fruiting month to the optimal sowing date to arrive at the duration of seed storage necessary to optimise nursery production. Carry out germination trials with some seeds immediately after collection, to determine their original viability (this is the 'control'). Then, store the rest of the seeds for the calculated length of time required. Sample the seeds at intervals to monitor any changes in viability. If there are enough seeds, experiment with different storage conditions (e.g. dry the seeds to different moisture contents or vary the storage temperature). Then, perform germination tests to determine if viability declines when the seeds are stored for the required length of time.

For direct seeding, carry out a germination trial on a sample of seeds immediately after collection. Then store the rest of the seeds for the required length of time (from seed collection to optimal direct seeding date). Remove the seeds from storage and sow samples in the nursery and in the field. Compare germination between these two groups and with the seed sample tested at collection time.

For species that fail to fruit every year, experiment with storing seeds for 1 year or longer to determine if seeds that are collected in fruiting years can be stored to grow seedlings in years when fruits are not produced. Similar experiments are useful for distributing seeds to other locations or if seeds are collected elsewhere to supplement a planting program (see **Box 6.1**, p.159).

When carrying out seed storage experiments, pre-sowing treatments can also be tested, but for a valid comparison, apply the same treatments to both the control batch (sown immediately after collection) and the stored batches.

Seedling growth and survival

Monitoring the performance of tree species in nurseries enables calculation of the time needed to grow trees of each selected species to a plantable size by the planting-out date. It also allows assessment of the susceptibility of each species to pests and diseases and detection of other health problems; thus it also provides a mechanism for quality control.

Comparing species and treatments

Tree species that grow well in nurseries usually perform well in the field. So one of the most useful nursery experiments is to compare survival and growth among species. Adopt a standard production method for all species and use a RCB experimental design (see **Appendix A2.1**) to compare performance among species. In this case, there are no 'control' and 'treatment' replicates. A 'block' consists of one replicate (no less than 15 containers) of each species.

Subsequently, additional experiments can be carried out to develop more efficient production methods for selected high-performing species. These should test different techniques to manipulate growth rates in order to grow saplings that reach a suitable size in time for hardening-off and planting-out. So many factors affect plant growth; the number of potential treatments is bewildering. The best plan is to start with the simplest and most obvious treatments, such as different container types, media composition and fertiliser regimes and test others (e.g. pruning, inoculation with mycorrhizal fungi) later if necessary.

The benefits of each treatment must be weighed against its costs and feasibility. So it is important to record the cost of applying each treatment. The main question being addressed is whether or not improving the quality of the planting stock in the nursery ultimately results in increased survival and growth of trees planted in the field. So, it is also useful to label trees that have been subjected to different nursery treatments and to continue to monitor them after they have been planted out in the field.

Factors that might influence seedling survival and growth

Container type

Experiments should be performed to test which container type is the most cost-effective for the species being grown. Start with a standard container type, such as plastic bags, and carry out simple experiments with different bag sizes to determine the effects of container volume on the size and quality of trees produced by planting-out time. Then, compare plastic bags with other container types that exert more control over root form (with or without air-pruning), such as rigid plastic cells or tubes (see **Section 6.4**).

Media and fertiliser regime

Start with a standard potting medium (see **Section 6.4**) and then experiment with varying its composition by using different forms of organic matter (e.g. coconut husk, rice husk or peanut husk) or by adding nutrient-rich materials such as cattle dung. For slow-growing species, try accelerating growth by experimenting with different fertiliser treatments (fertiliser type, dosage and frequency of application).

Pruning

If trees start to out-grow their containers before planting-out time, experiment with shoot-pruning treatments. Tree species vary in their responses to shoot pruning. Some are killed by pruning whereas others branch, producing a denser crown that enables them to shade out weeds more rapidly after planting out. Compare different shoot pruning intensities, timing and frequencies. In addition to growth and mortality data, also record plant form during pruning experiments.

Saplings that have a dense, fibrous root system are better able to supply their shoots with water. Therefore, a high root:shoot ratio improves the chances of survival after planting out. Large, woody roots are most resistant to desiccation, but they must have a dense network of young, fine roots for efficient water absorption. Experiment with different root pruning schedules. At the end of such experiments, sacrifice a few plants for the recording of root form and root:shoot ratio.

Mycorrhizal fungi

Most tropical tree species develop symbiotic relationships with fungi that infect their roots to form mycorrhiza. Such relationships enable trees to absorb nutrients and water more efficiently than the tree's own root system can (see **Section 6.5**). If forest soil is included in the potting medium, most saplings become naturally infected with mycorrhizal fungi (Nandakwang *et al.*, 2008). So first, survey saplings that are growing in the nursery to confirm the presence of mycorrhiza and assess the frequency of root infection.

For arbuscular mycorrhiza, i) wash a sample of fine roots; ii) treat them with a clearing solution (10% (w/v) KOH at 121°C for 15 minutes) to render the roots transparent; iii) apply 0.05% trypan blue in lactic acid:glycerol:water (1:1:1 v/v) to stain the fungal cells, and finally, iv) examine the roots under a dissecting microscope to estimate the percentage that are infected. Follow the safety precautions recommended for each of the chemicals.

For ectomycorrhiza, estimate the percentage of fine roots that have characteristic swollen ends to the root tips, then observe the roots under a microscope for the presence of fungal hyphae. Mycorrhizal fungus species are identified by examining their spores under a compound microscope. This requires specialist help (for general techniques for the study of mycorrhizae, see Brundrett *et al.*, 1996).

If the tree roots of any species are not colonised by mycorrhizal fungi, or colonised only very sparsely, then consider experiments to assess the effect of artificial inoculation. Commercial preparations containing mixtures of common mycorrhizal fungi spores may be available for testing (but be aware that they may not contain the particular fungus species or strains required by the tree species being grown). Alternatively, it is possible to collect fungal spores from around the roots of forest trees and then culture them in pots on domestic crop plants such as sorghum. Such home-made inoculae might be more specific for the trees being grown, but producing them is time-consuming and requires specialised techniques. The success of inoculation is often reduced if the plants are given fertiliser. So, try experiments that test various combinations of fertiliser treatments with the application of mycorrhizal inoculum. First, determine if artificial inoculation can increase infection rates (and ultimately tree performance) above those achieved naturally by including forest soil in the potting medium. Compare the performance of saplings grown in standard medium (which includes forest soil) with those subjected to supplementary sources of inoculum at various doses. Mycorrhizal fungi can easily be spread from one container to another by water, either by splashing or drainage. So, raise the containers off the ground on a wire grid and separate treatment replicates with plastic shielding to prevent splashing.

Designing experiments to test sapling performance

As with germination experiments, use a randomised complete block design (RCBD; see **Appendix A2.1**) and analyse the results using a two-way ANOVA, followed by paired comparisons (**Appendices A2.2** and **A2.3**). The example experimental design for germination trials can be used equally well for sapling performance experiments (substituting 'beds' for 'benches').

The number of treatments that can be applied and the number of replicates possible (i.e. the number of blocks) depends on the number of seedlings that survive after potting. Decide on the treatments that can be applied. Then, for each block, select a minimum of 15 plants (more is better) to constitute one 'replicate' for each treatment, and the same for the control. Make sure that all treatments (and the control) are represented by the same number of plants in all blocks. Place each block, consisting of one replicate of each treatment + control, in a different bed in the standing down area of the nursery. Within each block, position the treatment and control replicates randomly.

Seedling growth experiments in Cambodia: replicates are 15 seedlings in 23 cm × 6.5 cm plastic bags (3 rows of 5 plants), surrounded by a single guard row of 20 plants.

Select uniform plants for experiments; reject unusually tall or short plants and any showing signs of disease or malformations. Plants at the edge of a replicate may experience a different environment to those within it because treatments, such as watering or fertiliser application, may 'spill over' from one replicate to another. In addition, the plants at the edge of a block experience no competition from neighbours on one side and they may be affected by people brushing up against them. Reduce these 'edge effects' by surrounding each replicate with a 'guard row' of plants that are not assessed in the experiment. A simple experiment testing four treatments + a control in four blocks, would require a minimum of 15 × 5 = 75 uniform, healthy plants in each block, or 300 totally, plus extra plants to make the guard rows.

Assessment of growth

Collect data immediately after the experiment has been set up (as soon as possible after potting) and at intervals of approximately 45 days thereafter. The final data collection session should be just before the trees are removed from the nursery for planting out (even if this occurs earlier than 45 days after the previous data collection session).

Measure the height of each sapling (from root collar (i.e. the point at which the shoot meets the root) to apical meristem) with a ruler. Measure RCD (i.e. diameter at the 'root collar') at the widest point with Vernier-scale callipers (available from most stationery stores). At the zero mark on the lower sliding scale, read number of millimetres diameter from the upper scale. For the decimal point, look for the point at which the division marks on the lower scale are exactly aligned with the division marks on the upper scale. Then, read the decimal point off the lower scale. The Vernier scale in the example

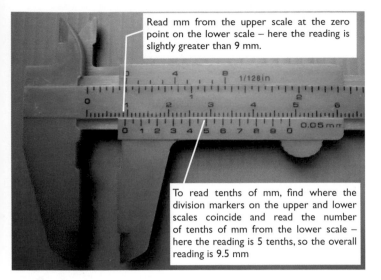

Read mm from the upper scale at the zero point on the lower scale – here the reading is slightly greater than 9 mm.

To read tenths of mm, find where the division markers on the upper and lower scales coincide and read the number of tenths of mm from the lower scale – here the reading is 5 tenths, so the overall reading is 9.5 mm

illustrated here reads 9.5 mm. Because RCD is a small value, it must be measured with high accuracy. For best results, measure RCD twice by turning the callipers at right angles and then use the average reading.

Use a simple scoring system to record plant survival and health (0 = dead; 1 = severe damage or disease; 2 = some damage or disease but otherwise healthy; 3 = good health). Also, record descriptions of any pests and diseases observed, as well as any signs of nutrient deficiency. Note when leaf shedding, bud break or branching occurs and record any unusual climatic events that might affect the experiment.

Determine root:shoot ratio (dry mass) by sacrificing a few plants at the end of the experiment. At the same time, photograph the structure of the root system. Remove sample plants from their containers and wash out the medium, taking care not to break the fine roots. Separate the shoot from the roots at the root collar. Dry them in an oven at 80–100°C. Weigh the dried shoot and dried root systems and calculate root dry weight divided by shoot dry weight for each plant sample.

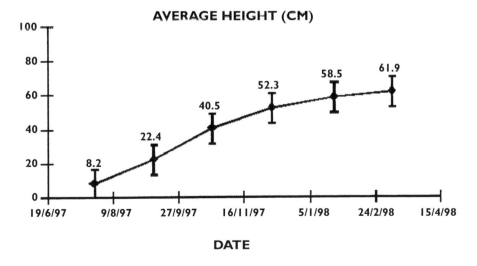

Seedling growth data for a pioneer tree species. Trees reach a size suitable for planting out by January, six months ahead of the optimal planting time. Therefore, seed storage to delay germination is recommended to prevent waste of nursery space and to avoid the need to prune the saplings.

Species: *Cerasus cerasoides*   **S. No.: S71B1**

Pricked out: June 6th 1997   **BLOCK: I**   **TREATMENT: NONE (CONTROL)**

HEIGHT DATA (CM)

							SEEDLING NUMBER										
DATE	DAYS	1	2	3	4	5	6	7	8	9	10	11	12	13	14	15	AVG
7/6/97	1	5.0	4.0	3.5	2.0	4.0	3.0	4.0	3.0	3.5	3.0	5.0	4.0	3.0	4.0	4.5	3.7
25/7/97	49	11.0	12.0	8.0	3.0	8.0	5.5	7.5	5.5	6.5	8.5	12.0	9.0	8.5	9.0	9.5	8.2
8/9/97	94	29.0	38.0	23.0	33.0	x	16.0	19.0	17.0	13.0	14.0	35.0	20.0	25.0	16.0	16.0	22.4
23/10/97	139	67.0	67.0	44.0	34.0	x	32.0	35.0	25.0	32.0	29.0	66.0	27.0	50.0	28.0	31.0	40.5
7/12/97	184	70.0	70.0	55.0	34.0	x	52.0	61.0	36.0	48.0	47.0	71.0	38.0	58.0	40.0	52.0	52.3
23/1/98	231	73.0	70.0	57.0	34.0	x	64.0	67.0	41.0	52.5	53.0	80.0	46.0	72.0	43.0	66.0	58.5
9/3/98	276	73.0	70.0	60.0	34.0	x	64.0	67.0	49.0	58.0	54.0	81.0	55.0	73.0	53.0	75.0	61.9

ROOT COLLAR DIAMETER DATA (MM)

							SEEDLING NUMBER										
DATE	DAYS	1	2	3	4	5	6	7	8	9	10	11	12	13	14	15	AVG
7/6/97	1	0.5	0.7	0.4	0.8	0.4	0.5	0.6	0.7	0.6	0.7	0.7	0.6	1.0	0.6	0.7	0.6
25/7/97	49	1.4	2.2	1.3	1.1	1.3	1.0	1.5	1.6	1.3	1.2	1.4	1.1	2.1	1.3	1.4	1.4
8/9/97	94	2.8	3.2	2.7	1.4	x	1.5	1.6	3.3	2.7	2.5	2.4	2.5	2.2	2.3	1.4	2.3
23/10/97	139	4.2	4.0	3.0	1.7	x	1.8	2.1	3.3	2.7	2.7	3.6	2.5	3.0	2.3	1.6	2.8
7/12/97	184	4.4	4.0	3.0	2.5	x	2.9	2.9	3.3	2.7	3.0	3.7	3.0	3.0	2.3	3.0	3.1
23/1/98	231	4.4	4.0	4.2	2.5	x	4.5	4.5	3.3	3.2	3.5	4.2	3.0	4.0	2.6	4.5	3.7
9/3/98	276	5.2	6.0	4.2	2.6	x	5.0	5.5	3.6	4.0	4.3	4.6	3.5	4.5	3.0	5.0	4.4

HEALTH DATA (0–3)

							SEEDLING NUMBER										
DATE	DAYS	1	2	3	4	5	6	7	8	9	10	11	12	13	14	15	AVG
7/6/97	1	2.5	2.5	2.5	1.5	2.0	1.5	3.0	3.0	2.5	3.0	3.0	2.5	2.0	3.0	3.0	2.5
25/7/97	49	3.0	3.0	3.0	2.0	3.0	2.5	3.0	2.5	3.0	3.0	3.0	3.0	3.0	3.0	3.0	2.9
8/9/97	94	3.0	3.0	3.0	2.0	x	2.5	3.0	3.0	2.5	2.5	3.0	3.0	3.0	3.0	2.5	2.8
23/10/97	139	3.0	2.5	3.0	2.5	x	3.0	3.0	3.0	3.0	3.0	3.0	3.0	1.5	3.0	3.0	2.8
7/12/97	184	3.0	3.0	3.0	3.0	x	3.0	3.0	3.0	3.0	3.0	3.0	3.0	3.0	3.0	3.0	3.0
23/1/98	231	3.0	3.0	3.0	3.0	x	3.0	3.0	3.0	3.0	3.0	3.0	3.0	3.0	3.0	3.0	3.0
9/3/98	276	3.0	3.0	3.0	3.0	x	3.0	3.0	3.0	3.0	3.0	3.0	3.0	3.0	3.0	3.0	3.0

Calculations from growth data

Use a standard data collection sheet for each replicate in each block. After each data collection session, calculate mean values (and standard deviations) for each of the parameters measured.

Also calculate relative growth rates (RGR), thereby removing the effects of differences in the original sizes of seedlings or saplings immediately after potting on subsequent growth. This makes it possible to assess treatment effects despite differences in the initial sizes of the plants at the beginning of the experiment. RGR is defined as the ratio of the growth of a plant to its mean size over the period of measurement, according to the equation below:

$$\frac{(\ln FS - \ln IS) \times 36{,}500}{\text{No. days between measurements}}$$

where ln FS = natural logarithm of final sapling size (either sapling height or RCD) and ln IS = natural logarithm of initial sapling size. The units are per cent per year.

Analysing survival data

For each replicate, count the number of saplings that survive until planting-out time. Then calculate the mean value and the standard deviation for each treatment; repeat for the control. Apply ANOVA (see **Appendix A2.2**) to determine if there are significant differences in mean survival among the treatments. If so, then use paired comparisons (see **Appendix A2.3**) between each treatment mean and the control mean to identify which treatments significantly increase survival. The same approach can be used to make comparisons among species.

Analysing growth data

Represent sapling growth graphically by constructing a growth curve that can be updated after each data collection session. Plot time elapsed since pricking out (horizontal axis) against mean sapling height (or mean RCD), averaged across blocks, for each treatment (vertical axis). By extrapolation, such curves can be used to estimate roughly how long it will take saplings growing in the nursery to reach the optimum planting size.

Just before optimum planting-out time, calculate the mean sapling height and RCD for each replicate and average these mean values across all blocks to arrive at treatment means. Carry out an ANOVA (see **Appendix A2.2**) to determine if there are significant differences among treatment means and, if so, use paired comparisons (see **Appendix A2.3**) to determine which treatments result in saplings that are significantly larger than control saplings at planting time. RCD and RGR (for both height and RCD) can be analysed in the same way.

What are the targets to aim for?

Adopt, as standard, any treatments that significantly contribute towards achieving the following targets by optimum planting-out time:

- >80% survival of saplings since pricking out;
- mean sapling heights >30 cm for fast-growing pioneer species (20 cm for *Ficus* spp.) and >50 cm for slow-growing climax tree species;
- sturdy stems, supporting mature, sun-adapted, leaves (not pale, expanding leaves) ('sturdiness quotient' can be calculated as height (cm)/RCD (mm) of <10);
- a root:shoot ratio of between 1:1 and 1:2; with an actively growing, densely branching root system that is not spiralling at the base of the container;
- no signs of pests, diseases or nutrient deficiency.

Tree seedling morphology and taxonomy

Surveys of natural forest regeneration require the identification of tree seedlings and very young saplings, but this is notoriously difficult. Plant species descriptions in floras are based primarily on reproductive structures. The morphology (particularly leaf shape) of seedlings often differs markedly from that of mature tree foliage and seedling specimens are hardly ever included in herbarium collections. Resources for the identification of tropical forest tree seedlings are almost non-existent (but see FORRU, 2000). Therefore, nurseries that are producing seedlings and saplings of known ages, from seeds collected from properly identified parent trees, provide an immensely valuable resource for studying tree seedling morphology and taxonomy.

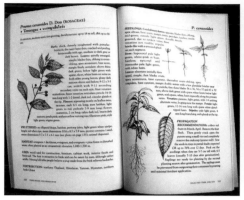

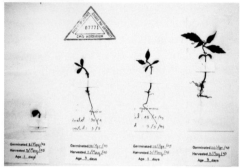

The seedlings of tropical forest trees remain largely unstudied. A tree nursery provides a unique opportunity to collect seedling specimens of known species and ages and to publish their descriptions.

Try to collect at least three specimens of seedlings or saplings at all stages of development for every species grown. Prepare them as herbarium specimens in the usual way, mounting several specimens in chronological order on a single herbarium sheet. On the herbarium label, record the age in days of each seedling or sapling specimen, and include details of the parent tree from which seeds were collected. Engage an artist to produce line drawings of the seedlings. Publish the drawings and descriptions of seedlings in an identification handbook.

Experiments with wildlings

Producing planting stock from wildings (see **Box 6.4**, p. 174) is advantageous i) when seeds are not available; ii) when seed germination and/or seedling survival and growth are problematic or slow; or iii) when the production of planting stock must be accelerated.

Experiments with wildings should address three simple questions: i) can high-quality planting stock be produced from wildings more rapidly and cost-effectively than by germinating seeds, ii) can the growing of wildings in nurseries be manipulated to achieve optimum-sized plants by planting out time; and iii) do wildings perform as well as, or better than, plants germinated from seed?

All of the seedling treatments described above can be applied to determine optimum conditions for growing-on wildings in nurseries to a plantable size. However, two additional treatments are specific for wildings: i) size when collected and ii) shoot-pruning at collection time.

Small seedlings are more delicate than larger saplings and are more easily damaged during transplantation. On the other hand, larger plants are more difficult to dig up without leaving some roots behind and can consequently suffer from transplantation shock. Group the wildings collected into three size classes (short, medium and tall).

These then become three 'treatments' in a RCBD experiment (there is no control). Collect growth and survival data, as described above, and compare mean survival and RGR among the initial size classes.

Digging up plants inevitably damages their root system, but the shoot system remains intact, and so a reduced root system must supply water to an undiminished shoot. This imbalance can cause wildings to wilt and possibly die. Pruning shoots can bring the root:shoot ratio back into balance. Apply shoot pruning treatments of varying intensities at collection time (e.g. no-pruning (control) and pruning back ⅓ or ½ of the shoot length or of the leaves). Collect growth and survival data as described above, and compare mean survival and RGR among the pruning treatments.

Continue to monitor the performance of planting stock from wildings after planting out (e.g. survival and growth rates) and then compare results with those from trees produced by germinating seeds.

FOREST RESTORATION RESEARCH UNIT

 S. 146 B4

FORRU SEEDLING PRODUCTION DATA SHEET

Use a simple data sheet to collate all information about a batch of seeds as it passes through the nursery production process, from seed collection to delivery of saplings to the restoration site.

I. COLLECTION

SPECIES: *Nyssa javonica* (Bl.) Wang   FAMILY: Cornaceae
LINKCODE: NYSSJAVA   VOUCHER NO.: 89
COLLECTION DATE: 11-Aug-06, ground   QUANTITY: 3,000 SEEDS

2. SEED GERMINATION

PRETREATMENT: seeds were soaked in water 1 night, after that sun dry 2 days
QUANTITY SOWN: 2,500 SEEDS
MEDIA/CONTAINER: Forest soil only, 8 baskets
SOWING DATE: 14-Aug-06
NUMBER GERMINATED: 2,059 SEEDS

OBSERVATION

1st germ. 26-Aug-06 to 11-Sep-06
Damping off diseases were destroyed about 12% of all germinated seedlings

3. PRICKING OUT

DATE PRICKING OUT: 3-Oct-06   QUANTITY. 1,505 SEEDLINGS
MEDIA/CONTAINER: Forest soil: Cocounut husk: Peanut husk (2:1:1) in plastic bag
NURSERY CARE:

NURSERY CARE	I	2	3	4	5
FERTILIZER	13/11/06	12/2/07	13/3/07		
PRUNING (NO)					
WEEDING	13/11/06	13/12/06	13/1/07	13/2/07	13/3/07
PEST/DISEASES CONTROL	13/1/07 Leaf eating insects				

OBSERVATION

2–3 months after pricking out, red fungus and leaf blight occurred, but all seedlings look healthy

4. HARDENING AND DESPATCH

DATE HARDENING STARTED: 17-May-07   DATE DESPATCHED: 19-Jun-07
NUMBER OF GOOD QUALITY PLANTS: 1,200 SEEDLINGS
WHERE PLANTED: MAE SA MAI, WWA PLOT

OBSERVATION

500 seedlings were planted on 30/6/07 at Ban Mae Sa Mai

Production schedules — the ultimate aim of nursery research

Growing a wide range of forest tree species is difficult to manage. Different species fruit in different months and have widely different rates of germination and seedling growth; yet all species must be ready for planting by the optimal planting time. Species production schedules make this daunting managerial task easier.

In seasonally dry tropical climates, the window of opportunity for tree planting is narrow, sometimes just a few weeks at the beginning of the rainy season. In less-seasonal climates, there may be more latitude in the timing of tree planting. In either case, species production schedules are an excellent tool to ensure that the required species of trees are ready for planting when required.

What is a production schedule?

For each tree species being grown, the production schedule is a concise description of the procedures necessary to produce planting stock of optimum size and quality from seed, wildlings or cuttings by the optimum planting-out time. It can be represented as an annotated time-line diagram that shows: i) when each operation should be performed and ii) which treatments should be applied to manipulate seed germination and seedling or sapling growth.

Information needed to prepare a production schedule

The production schedule combines all available knowledge about the reproductive ecology and cultivation of a species. It is the ultimate interpretation of the results from all the experimental procedures described above, including:

- optimum seed collection date;
- germination time or natural length of seed dormancy;
- how seed dormancy might be manipulated by pre-sowing treatments or seed storage;
- length of time required from seed sowing to pricking out;
- length of standing-down time required to grow saplings to a plantable size;
- how plant growth and standing-down time can be manipulated with fertiliser application and other treatments.

The finished product.

All of this information will become available from nursery data sheets if the procedures detailed above are followed. The production schedule is very much a working document. Draft the first version once the first batch of plants has been grown to a plantable size. This enables the identification of areas requiring further research and of appropriate treatments to be tested in subsequent experiments. As the results of experiments on each subsequent batch of plants become available, the production schedule will be gradually modified and optimised.

Box 6.6. Example production schedule (for *Cerasus cerasoides*).

Box 6.6. Example production schedule (for *Cerasus cerasoides*).

In its natural habitat, this fast-growing pioneer tree fruits in April or May. Its seeds have short dormancy and the seedlings grow rapidly during the rainy season. By December, their roots have penetrated deep enough into the soil that they are able to supply the shoot with moisture during the harsh conditions of the dry season. In the nursery, saplings that have reached a plantable size by December would have to be kept for a further 6 months before the next planting season (the following June) and would out-grow their containers.

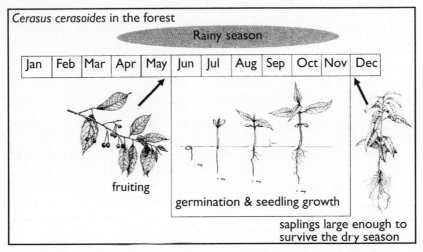

Cerasus cerasoides in the forest

Rainy season

| Jan | Feb | Mar | Apr | May | Jun | Jul | Aug | Sep | Oct | Nov | Dec |

fruiting

germination & seedling growth

saplings large enough to survive the dry season

In the nursery, the production schedule therefore involves storing the sun-dried pyrenes at 5°C until January, when they are germinated. The plants then grow to the optimum size just in time for hardening off and planting out in June. Development of this production schedule involved research on phenology, seed germination, seedling growth and seed storage.

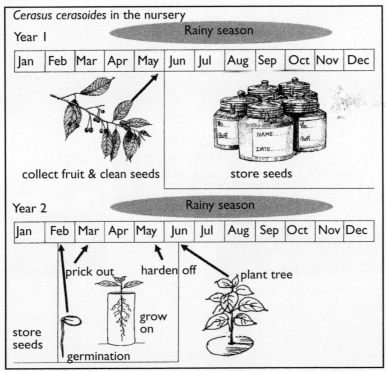

Cerasus cerasoides in the nursery

Year 1

Rainy season

| Jan | Feb | Mar | Apr | May | Jun | Jul | Aug | Sep | Oct | Nov | Dec |

collect fruit & clean seeds

store seeds

Year 2

Rainy season

| Jan | Feb | Mar | Apr | May | Jun | Jul | Aug | Sep | Oct | Nov | Dec |

prick out   harden off   plant tree

store seeds   grow on

germination

CASE STUDY 4   Doi Mae Salong: 'Treasure Tree Clubs'

Country: Thailand

Forest Type: Evergreen forest in seasonally dry tropical forestlands.

Ownership: The 'Treasure Tree Clubs' project was part of a 1,500 ha forest restoration program run as a partnership between Plant a Tree Today (PATT), the International Union for the Conservation of Nature (IUCN) and Chiang Mai University's Forest Restoration Research Unit (FORRU-CMU), working to assist Thailand's Supreme Command Office (SCO).

Management and community use: A mixture of carefully chosen cash crops and indigenous framework tree species was planted with the duel aims of alleviating poverty through sustainable agro-forestry and restoring a degraded watershed.

Level of degradation: Cleared of all but fragments of forest for agriculture.

Collecting sufficient seeds of enough tree species to restore diverse tropical forests is one of the most difficult challenges facing project managers, but it also provides an opportunity to engage entire communities in forest restoration right from the start. If many hands make light work, then ... "many eyes spot more seeds"!

At Doi Mae Salong in northern Thailand, Chiang Mai University's Forest Restoration Research Unit (FORRU-CMU) and the IUCN engaged eight village schools to start their own native tree nurseries. As part of IUCN's 'Landscapes and Livelihoods Strategy'[1] initiatives (sponsored by PATT[2]), the 'Treasure Tree Clubs' provided training to both school teachers and their pupils, increased awareness of the value of native forest trees, and provided incentives for children to collect seeds for the nurseries.

Labelling a 'treasure tree': their seeds are the treasure and children were rewarded for collecting them.

In return for collecting seeds, 'Treasure Tree Club' members accumulated stickers on their membership cards, which they could redeem for rewards.

---

[1] www.forestlandscaperestoration.org/media/uploads/File/doi_mae_salong/watershed_forest_article_6.pdf
[2] www.pattfoundation.org/what-we-do/reforestation/complete-project-list/doi-mae-salong.php

Nursery activities became part of the school curriculum, providing students with plant-growing skills that could be applied to both horticulture and forestry.

First, surviving forest trees within walking distance of the schools were identified and marked with treasure tree symbols (a diamond to imply high value with a tree seedling growing out from it) along with the local name of the tree species and the known fruiting months.

Children were issued with 'Treasure Tree Club' member cards. Any member who brought a bag of seeds from any labelled tree to the teachers in charge of the school tree nurseries received a sticker for their card. Stickers could also be gained for joining in simple tasks in the nursery, such as potting seedlings.

Tree nursery activities were included in the weekly agriculture classes and the children also applied their newly acquired arboricultural skills to growing fruit trees. For every five stickers gained, the member received a reward.

The nurseries were used to grow framework tree species for a 1,500 ha forest restoration program in the area. Saplings were sold to the program and the income used to buy materials and equipment for the schools. Thus, both individual children and the community as a whole benefited from the project. Furthermore, an element of friendly competition among the schools was introduced. Schools were judged on the basis of the species and number of tree saplings produced, as well their quality. The

Top-performing schools won trophies and all schools were awarded large packs of environmental education materials.

school children were quizzed on the tree nursery procedures they had learnt and to show that they could recognise local framework tree species. The judging process also served as the project's formal monitoring procedure. The results of the competition were revealed at a 'gala' project event, at which the top performing schools received trophies. Over one year, the project generated a total of nearly 10,000 trees of 24 species for the Supreme Command's forest restoration program, earning the schools a total of US$ 918 from tree sales.

CHAPTER 7

TREE PLANTING, MAINTENANCE AND MONITORING

Taking a tree from its container and planting it firmly into the soil is probably the archetypal image of forest restoration. It represents the culmination of months of planning, seed collection and nursery work. However, it is by no means the end of the forest restoration process. Deforested sites are harsh places: exposed, sunny and hot and often alternating between being parched and waterlogged. If trees are not provided with adequate care and protection over the first two years after planting, many will die and the effort expended in producing them will be wasted. The labour and materials required to ensure that the planted trees perform well are often under-estimated. Often, budgets run low or labour becomes unavailable, sometimes resulting in project failure and the need to start all over again. It is therefore a false economy to cut back on post-planting maintenance. Monitoring is another often-neglected task that is essential not only to provide data on tree survival and growth but also to provide an opportunity to learn from past successes and failures. Monitoring is now a required feature of all restoration projects that are funded through carbon trading.

7.1 Preparing to plant

Optimising the timing of tree planting

The optimal time for planting trees depends on soil water availability. In areas that have a seasonal climate, trees should be planted early in the rainy season, once rainfall has become regular and reliable. This gives the trees the maximum time to grow a root system that penetrates deep into the soil, allowing them to obtain sufficient water to survive the first dry season after planting. Where rainfall is more evenly spread throughout the year (i.e. no month has less than 100 mm), trees can probably be planted at any time.

Preparing the restoration site

First, take steps to protect any existing, naturally established trees, seedlings, saplings or live tree stumps. Inspect the plots thoroughly, taking care not to miss smaller tree seedlings that might be obscured by weeds. Place a brightly coloured bamboo pole next to each plant and use a hoe to dig out weeds from a circle of 1.5 m in diameter around each plant. This makes the natural sources of forest regeneration more visible to workers, so that they avoid damaging them during weeding or tree planting. Impress upon everyone working in the plots the importance of preserving these natural sources of forest regeneration.

About 1–2 weeks before the planting date, clear the entire site of herbaceous weeds so as to improve access and to reduce competition between weeds and trees (both planted and natural). The weed-pressing technique, often used for ANR, detailed in **Section 5.2**, might be adequate for sites that are dominated by soft (non-woody) grasses and herbs. Where weed pressing is ineffective, however, weeds must be dug out by the roots. First, slash the weeds down to 30 cm or so, then dig them out by their roots with a hoe and leave them to dry out on the soil surface. Make sure a first-aid kit is on hand to deal with any accidents.

Removing weed roots

Mere slashing encourages many weed species to re-sprout. As they do so, they absorb more water and nutrients from the soil than if they had never been cut in the first place. This actually intensifies root-competition with the planted trees, rather than reducing it. So, digging out the roots of weeds is essential, although the labour required to do so is considerable. Unfortunately, digging out roots also disturbs the soil, increasing the risk of soil erosion. Furthermore, there is a significant risk of accidentally slashing naturally established tree seedlings or saplings. For these reasons, and to reduce labour costs, we recommend use of glyphosate to clear plots for planting (but NOT for weeding after planting).

Herbicide use

Using a slow-acting, broad-spectrum, systemic herbicide, such as glyphosate (which is available in various formulations) can greatly increase the efficiency of weeding, reduce costs and avoid the need to disturb the soil. Such herbicides kill the entire plant, and thus prevent weeds from regenerating rapidly by vegetative growth.

Wait for the slashed weeds to begin to re-sprout before spraying them with a non-residual herbicide, such as glyphosate. Wear appropriate protective clothing as directed by the information sheet accompanying the product – usually gloves, safety glasses, rubber boots and water-proof clothing.

Slash weeds down to below knee height at least 6 weeks before the planting date. Leave the cut vegetation on site, as it will help to protect the soil from erosion and can subsequently be used as mulch around the planted trees. Wait at least 2–3 weeks for the weeds to start to sprout afresh; then spray the new shoots with glyphosate.

How does glyphosate work?

Glyphosate kills most plants, only a few species are resistant. It rapidly breaks down in the soil (i.e. it is non-residual) and so, unlike some other pesticides (e.g. DDT), it does not accumulate in the environment. The chemical is absorbed through leaves and is translocated to all other parts of the plant, including the roots. The weeds die slowly, gradually turning brown over 1–2 weeks, and the only way they can re-colonise the site is by growing from seeds. This takes much more time than re-sprouting from slashed shoots or root stocks. So, newly planted trees have about 6–8 weeks of relative freedom from weed competition. During this time, their roots can colonise soil formerly occupied fully by weed roots.

How should glyphosate be applied?

Apply herbicides on a dry windless day to prevent drift onto any naturally regenerating tree seedlings. Do not spray if rain is forecast within 24 hours after application. Within a few hours after spraying, rain and even dew render the chemical ineffective.

Large pumps mounted on pick-up trucks and long hoses that are used to spray crops might be available in agricultural communities, but they are not very accurate and their use makes it difficult to avoid spraying natural regenerants. Therefore, we recommend the use of 15 litre backpack tanks with directional spray nozzles, mounted on long wands.

Pour 150 ml of the glyphosate concentrate into a 15-litre-tank backpack sprayer and top up with clean water to the 15 litre mark. You will need to repeat this 37–50 times (using 5.6–7.5 litres of concentrate) per hectare. You should also include a wetting agent to facilitate the uptake of the chemical by the weeds.

Check the wind direction and work with the wind behind you, so that the spray is blown forwards rather than back into your face. Pump up the pressure in the backpack tank with the left hand and operate the spray wand with the right hand. Use low pressure to produce large droplets, which sink rapidly, before they can drift very far. Walk slowly across the site, spraying strips about 3 m wide by making gentle sweeps from side to side in front of you. If you accidentally spray a tree seedling or sapling, immediately tear off any leaves on which drops of the herbicide have fallen so that the chemical is not absorbed into the plant and transported to the roots. To avoid spraying the same area twice, add a dye to the glyphosate, so that you can see where you have already sprayed. If you accidentally spray the chemical onto your skin or in your eyes, wash with large amounts of water and see a doctor.

As soon as possible after spraying, take a shower and wash all clothes worn during spraying. Clean all of the equipment used (backpacks, boots and gloves) with large quantities of clean water. Make sure that the waste water does not flow into a drinking water supply; let it seep slowly into a sump pit or into the ground where there is no vegetation, far away from any water course.

Is glyphosate dangerous?

The United States Environmental Protection Agency (EPA) considers glyphosate to be relatively low in toxicity and without carcinogenic effects. It breaks down rapidly in the environment and does not accumulate in the soil. It is rated least dangerous when compared with other herbicides and pesticides. Nevertheless, if basic safety instructions are ignored, glyphosate can damage people's health and the environment, so read the instructions provided by the supplier before using it and follow them carefully. Ingestion of the concentrated solution can be lethal.

When diluted for use, glyphosate has low toxicity to mammals (including humans) but it is toxic to aquatic animals, so don't clean any contaminated equipment in streams or lakes. Research is also beginning to show that glyphosate might affect soil organisms. These minor potentially damaging effects of the chemical on the environment must, however, be weighed against the damaging long-term consequences of failing to restore forest ecosystems. Glyphosate is used only once, at the beginning of the forest restoration process. Use of the herbicides after trees have been planted is not recommended (en.wikipedia.org/wiki/Glyphosate).

Fire should not be used to clear plots

Fire kills any naturally established young trees while stimulating the re-growth of some perennial grasses and other weeds. It also kills beneficial micro-organisms such as mycorrhizal fungi and removes the possibility of using slashed weeds as mulch. If fire is used, organic matter is burnt off and soil nutrients are lost in smoke. Furthermore, fires that are intended to clear a planting plot can spread out of control and have the potential to damage nearby forest or crops.

How many saplings should be brought in?

The final combined density of planted plus naturally established trees should be about 3,100 per ha, so the required number of saplings delivered should be this figure minus the number of naturally established trees or live tree stumps estimated during the site survey (see **Section 3.2**). This results in an average spacing of about 1.8 m between planted saplings or the same distance between planted saplings and naturally established trees (or live stumps). This is much closer than the spacing used in most commercial forestry plantations because the objective is rapid canopy closure that will shade out weeds and eliminate weeding costs. Shade is the most cost-effective and environmentally friendly herbicide. Planting fewer trees results in a continued need for weeding over many years and, consequently, increases the total labour costs required to achieve canopy closure.

The tree spacing used in forest restoration is also closer than that between trees in most natural forest, so some competitive thinning will take place. This provides the restored ecosystem with an early source of dead wood, a vital resource for many forest fungi and insects. Planting at even higher densities is counter-productive as it leaves too little room for the establishment of incoming recruit tree species and therefore delays biodiversity recovery (Sinhaseni, 2008).

How many tree species should be planted?

For a plot with stage-3 degradation, count how many tree species are well represented by the sources of natural regeneration recorded in the site survey (see **Section 3.2**) and deliver enough species to top up that number to at least 30 or around 10% of the estimated species richness (if known) of the target forest type (see **Section 4.2**). For a plot with stage-4 degradation, plant as many species of the target forest type as possible. Nurse plantations can be monocultures of single species or mixtures of a few species (e.g. *Ficus* spp. + legumes; see **Section 5.5**).

Transporting saplings

Saplings are very vulnerable, particularly to wind and sun exposure, once they leave the nursery, so take care when transporting them to the site. Select the most vigorous saplings in the nursery after grading and hardening-off (see **Section 6.5**). Label the saplings that you intend to include in your monitoring program, then place all of the saplings upright (so as to prevent spillage of potting mix) in sturdy baskets. Water the saplings just before loading them into the vehicle, and transport them to the planting plot the day before planting.

If plastic bags are in use as growth containers, do not pack them so tightly that they lose their shape. Also, do not stack containers on top of each other as this will crush shoots and break stems. If an open truck is used, cover the saplings with a layer of shade netting to protect them from wind damage and dehydration. Drive slowly. Once the plots have been reached, place the saplings upright beneath any available shade and, if possible, lightly water them again. If you have enough baskets, keep the saplings in the baskets as this makes it easier to carry them around the plot on planting day.

Protect the young trees on the way to the restoration site.

SLOW DOWN! Don't throw away a year's work in the nursery on the journey to the planting site. When transporting saplings, drive with care.

Carry saplings onto the restoration site like this …

… not like this (it damages the stems) …

… and don't leave them exposed.

Planting materials and equipment

The day before planting, transport planting materials to the plots along with the saplings. These include a bamboo stake and mulching material (if required) for each sapling as well as fertiliser. Protect these materials from rain by covering them with a tarpaulin.

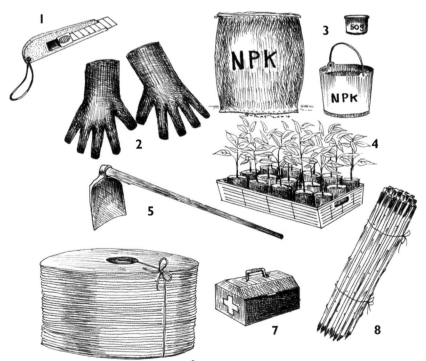

1. Knife
2. Gloves
3. Fertiliser, bucket and pre-measured cups to deliver the correct dose
4. Baskets for saplings distribution
5. Hoes for hole digging
6. Mulch mats
7. First-aid kit
8. Bamboo poles

Other preparations for the big day?

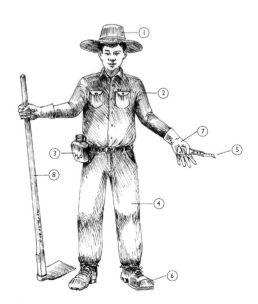

The perfectly prepared planter, with (1) a hat for sun protection; (2) a long-sleeved shirt; (3) plenty of water to drink; (4) long trousers; (5) a box cutter for slashing open plastic bags; (6) strong boots; (7) gloves and (8) a hoe for digging the planting holes.

A few days before the planting event, hold a meeting of all project organisers. Appoint a team leader for each group of planters. Make sure that all team leaders are familiar with the tree-planting techniques and that they know which area each will be responsible for planting and how many trees they must plant. Use a planting rate of 10 trees per hour to calculate the number of people required to complete the planting within the desired time limit.

Ask the team leaders to tell their team members to bring gloves, box-cutters (to slash open plastic bags), buckets, hoes or small shovels (to fill in the planting holes) and (if fertiliser is to be applied) cups. In addition, team leaders should advise the planters to carry a bottle of water and to wear a hat, sturdy footwear, a long-sleeved shirt and long trousers.

Make a final estimate of the number of people likely to participate in the planting event. Organise enough vehicles to take everyone to the plots and arrange enough food and drink to keep everyone well fed and hydrated. Make contingency plans in case of bad weather. Finally, consider whether the project and the local community might benefit from media coverage of the event and, if so, contact journalists and broadcasters.

7.2 Planting

Enthusiasm alone is not enough. A little bit of training at the start of a planting event can help to prevent costly mistakes.

Tree planting events do much more than just put trees in the ground. They provide an opportunity for ordinary people to become directly involved in improving their environment. They are also social events, helping to build community spirit. Furthermore, media coverage of a planting event can portray a positive image of the community as responsible stewards of their natural environment. Tree planting can also have an educational function. Participants can learn not only how to plant trees but also why. Take time at the beginning of the event to demonstrate the planting techniques to be used and to make sure that everyone understands the objectives of the forest restoration project. Also, take the opportunity to invite everyone to participate in follow-up operations, such as weeding, fertiliser application and fire prevention.

Spacing distances

First, mark where each tree will be planted with a 50-cm split-bamboo pole. Space the poles at 1.8 m apart or the same distance away from naturally established trees or tree stumps. Try not to position the stakes in straight rows. A random arrangement will give a more natural structure to the restored forest. Staking out the plots can be done either on planting day or a few days in advance.

Use split bamboo to space out the trees.

Planting method

Use baskets to distribute one sapling to each of the poles. Mix up the species so that saplings of the same species are not planted next to each other. This 'random' planting is known as an 'intimate mix'.

Baskets and hand carts can be used to haul the saplings to their planting positions.

Beside each bamboo pole, use a hoe to dig a hole that is at least twice the volume of the sapling's container, preferably with slanting sides (breaking up the soil around the root system will also help the roots to establish). At the same time, use the hoe to drag away dead weeds in a circle of 50–100 cm in diameter around the hole.

If the saplings are in plastic bags, slash each bag up one side with a sharp blade, taking care not to damage the root ball inside, and gently peel away the plastic bag. Try to keep the medium around the root ball intact, and expose roots to the air for no more than a few seconds if possible.

Dig holes twice the size of the container.

Carefully slice open bags and peel back the plastic.

Planting day can be enjoyed by the whole family.

Place the sapling upright in the hole and pack the space around the root ball with loose soil, making sure that the sapling's root collar is eventually positioned level with the soil surface. If the sapling has been labelled for monitoring, make sure that the label does not become buried. With the palms of your hands, press the soil around the sapling stem to make it firm. This helps to join pores in the nursery medium with those in the plot soil, thus rapidly re-establishing a supply of water and oxygen to sapling's roots. It is not usually necessary to tie the sapling to the pole for support. The poles are used merely to indicate the location where each tree should be planted.

Next, apply 50–100 g of fertiliser in a ring on the soil surface about 10–20 cm away from the sapling stem. Chemical burning can occur if fertiliser contacts the stem itself. Use pre-measured plastic cups to apply the correct fertiliser dose. Note that chemical fertilisers are usually expensive and might not be necessary for all sites.

Press soil around the stem to make it firm.

Use measuring cups to deliver the correct fertiliser dose.

Cardboard mulch mats are particularly effective on dry degraded soils. On wetter, fertile soils, they disappear too quickly.

Then (optionally) place a cardboard mulch mat of 40–50 cm in diameter around each planted sapling. Anchor the mulch mat in position by piercing it with the bamboo stake and pile dead weeds onto the cardboard mulch mat.

If there is a water supply nearby, water each planted sapling with at least 2–3 litres at the end of the planting event. A water tanker can be hired to deliver water to sites that are accessible by road but distant from natural water supplies. For inaccessible sites with no available water, schedule planting to take place when rain is forecast.

For inaccessible dry sites, plant trees when rain is forecast, but if it is possible to water the trees after planting, then do so. Pump water from a stream or truck it in by tanker.

Remove plastic bags to clean the site.

The final task is to remove all plastic bags, spare poles or cardboard mulch mats, and garbage from the site. Team leaders should personally thank all those taking part in the planting. A social event to mark the occasion is also a good way to thank participants and build support for future events.

Choosing a chemical or inorganic fertiliser

Determining that a site's soil has nutrient deficiencies requires costly chemical analyses and access to a laboratory (see **Section 5.5**). It is rarely worth the cost because, regardless of soil fertility, most tropical trees respond well to the application of a general-purpose chemical fertiliser (N:P:K 15:15:15) 3–4 times a year for 2 years after planting. Use doses of 50–100 g per tree per application. The effect is to boost growth in the first few years after planting, accelerate canopy closure, shade out herbaceous weeds and 'recapture' the site. Spreading the fertiliser in a ring around the base of the tree is more effective than placing fertiliser in the planting hole because the nutrients percolate down through the soil as the roots begin to grow into the surrounding soil.

On lowland sites with poor lateritic soils, organic fertiliser seems to be more effective than chemical fertiliser (FORRU-CMU, unpublished data), possibly because it breaks down and is leached from the soil more slowly. Thus, it delivers nutrients to the tree roots more evenly over a longer period.

The cost of chemical fertiliser fluctuates with the price of oil, and so costs have risen steeply in recent years and are likely to continue to do so. Organic fertilisers vary greatly in composition, but they are much cheaper than chemical fertilisers. Find a reliable supply of an effective local brand and stick with it, or work with local communities to start producing fertiliser from animal waste. The purchase of fertiliser from local villagers provides another way in which the community can benefit economically from a restoration project.

Mulching to reduce desiccation and weed growth

Mulch is a material placed on the ground around a sapling that can increase its survival and growth, particularly by reducing the risk of drying out immediately after planting. Mulching is particularly recommended when planting on highly degraded soils in drier areas. It has less effect when used in plots that have fertile upland soils or in the ever-wet tropics. Mulching materials vary widely from rocks and pebbles to wood chips, straw, sawdust, coconut fibre, oil palm fibre and cardboard.

Corrugated cardboard makes excellent mulch mats. It is widely available and relatively cheap. Ask your local supermarket to donate their waste cardboard or other packaging materials for making mulch mats. Cut the cardboard into 40–50 cm diameter circles. Cut a hole in the middle of the circle of about 5 cm across and make a narrow slit from the circle's perimeter to its centre. Open the circle along the slit and place the hole centrally around the tree stem. Make sure that the cardboard does not touch the stem as it could abrade them, creating wounds that can become infected by fungi. Drive a bamboo stake through the mat to keep it in place. In seasonal tropical forest, cardboard mats last one rainy season, gradually rotting down and adding organic matter to the soil. Replacing mats at the beginning of the second rainy season does not seem to result in additional benefits (FORRU-CMU data).

Most weed seeds are stimulated to germinate by light. Mulching around planted saplings blocks out light and thus prevents weeds from re-colonising the ground in the immediate vicinity of planted trees. Furthermore, mulching cools the soil, thereby reducing the evaporation of soil moisture. Soil invertebrates are attracted by the cool, moist conditions beneath the mulch. They churn up the soil around planted saplings, improving drainage and aeration.

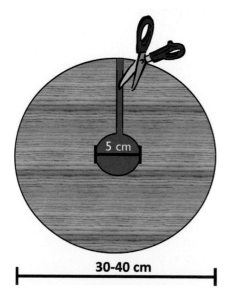

5 cm

30-40 cm

Mulch mats, cut from recycled corrugated cardboard, are cheap and effective in reducing the immediate post-planting mortality of planted trees, particularly on drought-prone sites with poor soils. They suppress weed growth and thus reduce the labour costs of weeding. Fertiliser is applied in a ring around the base of the tree. Cardboard mats last about a year if care is taken not to disturb them during weeding operations.

Polymer gel can be used to improve hydration

Water-absorbent polymer gel can help keep the roots of planted trees hydrated and reduce transplantation stress. On well-watered highland sites, it is usually unnecessary, but when used in combination with cardboard mulch mats, it can significantly reduce the immediate post-planting mortality of trees planted in dry areas on poor soils (see **Section 5.5**).

Quality control

Even when planting techniques are demonstrated at the beginning of the event, it is inevitable that some trees will not be planted properly. Once the planters have left the site, team leaders should inspect the planted trees and correct errors. Make sure that all the trees are upright, that the soil around them has been properly firmed down, and that monitoring labels have not become buried. Look for any saplings that have not been planted and either plant them or return them to the nursery. Refill any holes that have no trees in them. Clear the site of spare bamboo poles, fertiliser sacks, plastic bags and any other garbage.

Direct seeding

Direct seeding can dramatically reduce the costs of planting a forest. It is also much easier to carry out than the laborious process of planting containerised trees, but few tree species can currently be established efficiently by this technique (**Table 5.2**). At present, the method remains complementary to conventional tree planting. The pros and cons were discussed in **Section 5.3**, but the practical techniques are presented below.

Optimal timing for direct seeding

In ever-wet tropical regions, direct seeding can be implemented at any time (except during drought conditions). In seasonally dry areas, direct seeding should be carried out at the start of the rainy season (alongside conventional tree planting). This allows sufficient time for germinating seedlings to grow root systems that are capable of accessing enough soil moisture to enable the young plants to survive the first dry season after sowing. Unfortunately, the rainy season is also the peak time of year for both weed growth and the breeding of rodent seed predators, so control of these two factors is particularly important. It has been suggested that these problems would be avoided by direct seeding late in the rainy season, but recent results have confirmed that early sowing, to achieve extensive root growth before the dry season, is the overriding consideration (Tunjai, 2011).

Ensuring seed availability

Seeds must be stored from the time of fruiting until the start of the rainy season. Many tropical tree species produce recalcitrant seeds that lose viability rapidly during storage, but the storage period required for direct seeding is less than 9 months and so storage might be possible. See **Sections 6.2** and **6.6** for more information on seed storage.

Direct seeding techniques

At the beginning of a rainy season, collect seeds of the desired tree species (or remove them from storage). Apply any pre-sowing treatments that are known to accelerate the germination of the relevant species. Dig out weeds from 'seeding spots' of approximately 30 cm across, spaced about 1.5–2 m apart (or the same distance from natural regenerants if present). Dig a small hole in the soil and loosely fill it with forest soil. This ensures that beneficial symbiotic micro-organisms (e.g. mycorrhizal fungi) are present when the seed germinates. Press several seeds into each hole to a depth of about twice the diameter of the seed and cover with more forest soil. Lay mulching material, such as the pulled weeds, around the seeding spots to suppress further weed growth. During the first two rainy seasons after seeding, pull weeds by hand from the seeding spots as required. If multiple seedlings grow up at any seeding point, remove the smaller, weaker ones, so that they do not compete with the largest seedling. Carry out experiments to determine the most successful species and techniques for direct seeding at any particular site.

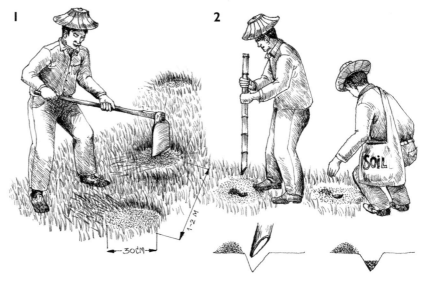

1. First, clear weeds from seeding spots.
2. Next, make small holes and add forest soil.
3. Then, press several seeds into the loose soil.
4. Finally, cover with forest soil.

7.3 Caring for planted trees

In deforested sites, planted trees are subjected to hot, dry, sunny conditions as well as to competition from fast-growing weeds. Protective measures (as described in **Section 5.1**) must be implemented to prevent fire and cattle from killing both planted trees and any natural regenerants that are present. Weeding and fertiliser application (see **Section 5.2**) are also essential for at least 18–24 months after planting in order to maximise tree growth and accelerate canopy closure. No further maintenance is necessary after canopy closure.

Fire prevention and excluding livestock

The cutting of fire breaks, organisation of fire suppression teams and exclusion of livestock from restoration sites are discussed in **Section 5.1**.

Weeding

Weeding reduces competition between planted or naturally established trees and herbaceous plants. On nearly all tropical sites, weeding is essential to prevent high tree mortality in the first two years after planting. The ring-weeding and weed-pressing methods described in **Section 5.2** can be applied equally well to planted trees as to natural regenerants.

Weeding frequency

The frequency of weeding depends on how fast the weeds grow. Visit the site frequently to observe weed growth and carry out weeding well before the weeds grow above the crowns of the planted trees. Weed growth is most rapid during rainy seasons. After planting, weed around planted trees at 4–6-week intervals for as long as the rains continue. If weed growth is slow, it may be possible to reduce the weeding frequency. It should not be necessary to weed during dry seasons.

In seasonal forests, allow some weed growth to occur before the end of the rainy season to help to shade the planted trees and thereby to prevent desiccation when the weather is hot and dry. Remember, however, that this also increases the fire risk, so only do this where fire prevention measures are effective. Where fire is particularly likely, try to keep the planted plots free of weeds at all times. The labour force required for weeding varies with weed density but, as a guide, budget for 18–24 days labour per hectare.

How long must weeding be continued?

Weeding is usually necessary for two rainy seasons after planting. In the third year after planting, the frequency of weeding can be reduced if the crowns of the planted trees begin to meet and form a forest canopy. By the fourth year, the shade of the forest canopy should be sufficient to prevent further weed growth.

Weeding is essential to keep trees alive during the first few years after planting. (A) A cardboard mulch mat can help keep weeds to a minimum immediately around the tree stem. (B) Pull out any weeds that grow near the tree base by hand (wear gloves) to avoid damaging the tree roots. Try to keep the mulch mat intact. (C) Next, use a hoe to root out weeds in a circle around the mulch mat and (D) lay the uprooted weeds on top of the mulch mat. (E) Finally, apply fertiliser (50–100 g) in a circle around the mulch mat.

Weeding techniques

The weed-pressing method described in **Section 5.2** may be used to flatten weeds growing between the trees. If the weeds are not susceptible to pressing, then use machetes or a 'weed whacker' (a mechanical hand-held weed cutter), keeping well away from both planted and natural trees to avoid accidentally slashing them.

A more delicate approach is required around the trees themselves. Wear a pair of gloves and gently pull out any weeds growing close to tree stems, including any growing through mulch. Try not to disturb the mulch too much. Use a hoe to dig out weeds close to the mulched area by their roots. Lay uprooted weeds around the trees on top of the existing mulch. This shades the soil surface and inhibits the germination of weed seeds, even as the organic mulch rots away. Try to ensure that uprooted weeds do not touch the tree stems as such contact can encourage fungal infection. Apply fertiliser around each tree immediately after weeding.

Frequency of fertiliser application

Even on fertile soils, most tree species benefit from the application of additional fertiliser during the first two years after planting. It enables the trees to grow above the weeds rapidly and to shade them out, thus reducing weeding costs. Apply 50–100 g fertiliser, at 4–6-week intervals, immediately after weeding, in a ring about 20 cm away from the tree stem. If a cardboard mulch mat has been laid, apply the fertiliser around the edge of the mulch mat. Chemical fertiliser (N:P:K 15:15:15) is recommended for upland sites, whereas organic pellets produce significantly better results on lateritic lowland soils (but see **Section 7.2**). Weeding before fertiliser application ensures that the planted trees benefit from the nutrients and not the weeds.

7.4 Monitoring progress

All tree-planting projects should be monitored, but there are many different approaches to monitoring, ranging from basic photo-monitoring and assessment of tree survival rates (described here) to complex field trial systems designed to investigate species performance, the effects of silvicultural treatments and biodiversity recovery (described in **Section 7.5**).

Why is monitoring necessary?

Funders want to know if the tree planting they pay for is successful, so monitoring results are usually an essential component of project reports. Initially, this means finding out whether or not the planted trees have survived and grown well in the first few years after planting, but the ultimate measure of success is how fast the restored forest becomes similar to the target forest ecosystem in terms of structure and function (see **Section 1.2**) and species composition (see **Section 7.5**). Interest in monitoring techniques is growing rapidly and the monitoring systems being proposed are becoming ever more complex and stringent. This is because of the value now being put on forests as carbon stores. Small errors in monitoring can result in the gain or loss of large sums of money in carbon trade. Therefore, if your project is funded by a carbon-offset scheme (e.g. REDD+), make sure that you follow the monitoring protocols stipulated by the funder and be prepared to have every aspect of your monitoring program closely scrutinised.

Simple monitoring using photography

The most simple way to assess the effects of tree planting is to take photographs before planting and at regular intervals (once per season or annually) thereafter. A neighbouring site, where no forest restoration has been implemented, can be similarly photographed so that restoration can be compared with unassisted natural regeneration. Locate points with a clear view of both the planted sites and notable landmarks. Mark the position of the points with a metal or concrete pole or paint an arrow on a large rock. Set the camera to the highest resolution and widest zoom and try to use to same camera for all shots. Frame each shot so that a landmark is positioned on the left-hand or right-hand edge of the picture and so that the horizon is aligned close to the top edge (i.e. minimise the amount of sky in the picture). Record the date, point number, location (co-ordinates if you have a GPS), and age of plot and use a compass to measure the direction in which the camera is pointed. The photos in **Section 1.3** (p. 14) are good examples. In these pictures, the large black tree stump serves as a reference point.

As soon as possible, download the photos onto a computer and back them up to a storage device or the internet. Use a logical file-naming system so that the photos can easily be arranged in chronological order and by point location (e.g. 2013_Point1_Plot141231). When you return to take more photos, take the previous ones with you, so that you can use the landmarks to position the new shots to be as similar as possible to the previous shots.

Photos are easy to take and share, and they provide an easily understood representation of the progress of restoration projects. However, funders usually require some kind of monitoring of tree survival and growth. In that case, label a subset of the trees planted and measure them at regular intervals.

Sampling trees for monitoring

The minimum requirement for adequate monitoring is a sample of 50 or more individuals of each species planted. The larger the sample is, the better. Randomly select trees to include in the sample; label them in the nursery before transporting them to the planting site. Plant them out randomly across the site, but make sure that you can find them again. Place a coloured bamboo pole by each tree to be monitored; copy the identification number from the tree label onto the bamboo pole with a weather-proof marker pen and draw a sketch map to help you find the sample trees in the future.

Labelling planted saplings

The soft metal strips used to bind electrical cables, which are available from builders' supply stores, make excellent labels for small trees. They can easily be formed into rings around tree stems. Use metal number punchers or a sharp nail to engrave an identification number on each label and bend them into a ring around the stem above

Before planting, place metal strip labels around the tree stems. Make sure they do not get buried during planting. Label numbers could include information on species, year of planting, plot number and tree number. For example, 22-48 12-3 could mean the 48th individual of species number 22, planted in plot 3 in the year 2012. Keep accurate records of your numbering system.

the lowest branch (if present). This will prevent the label from being buried when the tree is planted. Alternatively, drink cans may be cut up to make excellent tree labels. Cut off the top and bottom of the cans and slice the cans' walls into strips. Use a tough ball-point pen or nail to press identification numbers into these soft metal foil strips (on the inside surface). The strips can be formed into loose rings around saplings' stems.

Keeping the labels in position on rapidly growing trees is difficult because as the trees grow, their expanding trunks push off labels. If monitoring is carried out frequently, you will be able to re-position or replace labels before they are lost.

Once the trees have developed a girth of 10 cm or more, measured 1.3 m above the ground (girth at breast height (GBH)), more permanent labels can be nailed to the trunks, marking the girth measuring point at 1.3 m. Use 5 cm long, galvanised nails with flat heads. Hammer only about one-third of the nail length into the trunk to allow room for tree growth. Metal foil from drinks cans, cut into large squares so that the identification number can be read from a distance, makes excellent labels for larger trees.

1.30 M

Once trees have grown large, subsequent performance monitoring can be based on increase in girth at breast height (GBH).

Monitoring tree performance

To monitor tree performance, work in pairs with one partner taking measurements and the other recording data on pre-prepared record sheets. One pair can collect data on up to 400 trees per day. Prepare record sheets that include a list of the identification numbers of all labelled trees in advance (see **Section 7.5**). Take along the sketch maps made when the labelled trees were planted to help you find them. In addition, take a copy of the data collected during the previous monitoring session. This can help you sort out tree identification problems, especially for trees that might have lost their labels.

When to monitor

Measure the trees 1–2 weeks after planting to provide baseline data for growth calculations and to assess immediate mortality, which might result from transplantation shock or rough handling during the planting process. After that, monitor the trees annually; in seasonal forests, this task should be undertaken at the end of each rainy season. The most important monitoring event, however, is at the end of the second rainy season after planting (or after about 18 months), when field performance data can be used to quantify the suitability of each tree species to the prevailing site conditions (see **Section 8.5**).

What measurements should be made?

Rapid monitoring of tree performance can involve simple counts of surviving and dead trees, but recording the condition of planted trees each time they are inspected can give an early indication of whether something is going wrong. Assign a simple health score to each tree and record descriptive notes about any particular health problems observed. A simple scale of 0 to 3 is usually sufficient to record overall health. Score zero if the tree appears to be dead. For deciduous tree species, don't confuse a tree with no leaves in the dry season with a dead one. Do not stop monitoring trees just because they score zero on one occasion. Trees that appear dead above ground could still have living roots from which they might sprout new shoots. Score 1 if a tree is in poor condition (few leaves, most leaves discoloured, severe insect damage etc.). Score 2 for trees showing some signs of damage but retaining some healthy foliage. Score 3 for trees in perfect or nearly perfect health.

Measure the height of planted trees from the root collar to the highest meristem (growing point).

More detailed monitoring of tree performance involves measuring tree height and/ or girth (for calculation of growth rate) and crown width. In the first year or two after planting, tree heights can be measured with 1.5 m tape measures mounted on poles. Measure the tree height from the root collar to the highest meristem (shoot tip). For taller trees up to 10 m, telescopic measuring poles can be used. These poles are commercially available but can be home-made. Measurements of girth at breast height (GBH), rather than of height, are easier to make for taller trees and can be used to calculate growth rates.

Calculations of tree growth rates that are based on height can sometimes be unreliable as shoots can occasionally be damaged or die back, resulting in negative growth rates for small saplings even though the tree could be growing vigorously. Consequently, measurements of root collar diameter (RCD) or GBH often provide a more stable assessment of tree growth. For small trees, use callipers with a Vernier scale to measure RCD at the widest point (for use of callipers, see **Section 6.6**). Once a tree has grown tall enough to develop a GBH of 10 cm, measure both the RCD and the GBH the first time and only GBH thereafter.

Suppression of weed growth (an important framework characteristic) can also be quantified. Measuring crown width and using a scoring system for weed cover can help to determine the extent to which each tree species contributes to site 'recapture'. Use tape measures to measure the width of tree crowns at their widest point. Imagine a circle of about 1 m in diameter around the base of each tree. Score 3 if weed cover is dense over the whole circle; 2 if weed cover and leaf litter cover are both moderate; 1 if only a few weeds grow in the circle and 0 for no (or almost no) weeds. Do this before weeding is due to be carried out.

Measure crown width at the widest point to assess canopy closure and site 'recapture'.

Data analysis

For each species, calculate per cent survival at the end of the second rainy season after planting (or after 24 months) as follows:

$$\text{% Survival estimate} = \frac{\text{No. of labelled trees surviving}}{\text{No. of labelled trees planted}} \times 100$$

Use per cent survival of the labelled sample trees to estimate how many trees of each species survived across the whole site. Then determine the survival percentage of the total number of trees planted as shown in **Table 7.1**.

Table 7.1. Example calculation of species' survival rates.

Species	No. of labelled trees in sample	No. of labelled trees surviving	Estimated % survival (%S)	Total no. of trees planted (TP)	Estimated no. surviving (TP × %S/100)
S004	50	46	92	1,089	1,002
S017	50	34	68	678	461
S056	50	45	90	345	311
S123	50	48	96	567	544
S178	50	23	46	358	165
Totals				**3,037**	**2,482**
Estimated overall % survival	**81.7**				

To determine significant differences in survival between species, use a Chi-square (X^2) test. Fill in a table with the number of dead and alive trees of the two species you want to compare as follows:

Species	Alive	Dead	Total
S123	48	2	50
S178	23	27	50
Total	71	29	100

a	b	a+b
c	d	c+d
a+c	b+d	a+b+c+d

Calculate the Chi-square (X^2) statistic, using the formula:

$$(X^2) = \frac{(ad-bc)^2 \times (a+b+c+d)}{(a+b) \times (c+d) \times (b+d) \times (a+c)}$$

A significant difference in survival is indicated by a calculated Chi-square value exceeding 3.841 (with a <5% probability of error). This critical value is independent of the number of trees in the samples. In the example above, 30.35 greatly exceeds the critical value, so we can be very confident that S123 survives significantly better than S178 (for more information, go to www.math.hws.edu/javamath/ryan/ChiSquare.html). Remove species with low survival rates from future plantings and retain those with higher survival rates (see **Section 8.5**).

Calculate mean tree height and RCD for each species and calculate relative growth rates (RGR; see **Section 7.5**). To show significant differences among species, use the statistical test, ANOVA (see **Appendix A2.2**).

Monitoring other aspects of forest restoration

Detailed survey methods that are used to determine forest recovery are described in **Section 7.5**, but if you do not have the capacity or resources to implement them, then simple informal monitoring can at least provide stakeholders with the sense of achievement needed to sustain interest in the project. Make regular visits to the planted plots and record when the first flowers, fruits or birds' nests are seen on or in each of the planted tree species. Record any sighting of an animal (or their signs), especially of seed-dispersers. Once canopy closure occurs, survey the plots for naturally establishing tree seedlings or saplings and record the return of notable species. This helps to provide an impression of how quickly the restored forest begins to resemble the target forest and how fast biodiversity recovery occurs.

Monitoring for carbon accumulation

Many funders want to know how much carbon is being stored by the trees in a restoration project so that they can offset their carbon footprints or cash in on the carbon trade. Consequently, funders often require project implementers to follow international accreditation and monitoring standards, which include independent audits to verify carbon accumulation. There are many different forest carbon standards that differ in the way that they are used to survey forestry projects for carbon offset. The four that are most relevant to forest restoration projects are listed in **Table 7.2**. If your project is registered with one of these organisations, make sure you follow the monitoring protocols that they stipulate. The development of the Project Design Document (PDD), 'validation' and 'verification' of carbon storage, well as carbon credit registration can cost anywhere between US$ 2,000 to US$ 40,000. Such high charges effectively exclude small organisations from participating in these schemes, except in the case when numerous organizations bundle their projects together to obtain certification. Furthermore, community organisations often lack the expertise needed to complete the complex application and verification procedures. Our advice to small organisations is to seek sponsorship through corporate social responsibility mechanisms that are independent of carbon funding.

Table 7.2. Carbon standards organisations.

Organisation	Notes	Website
CarbonFix	Simplified, user-friendly standard that guarantees high-quality carbon credits. Adaptable to the needs of project developers and funders. Recommended for restoration projects.	www.carbonfix.info
Verified Carbon Standard (VCS)	A high-quality standard that guarantees that the carbon credits are real, verified, permanent, additional and unique. Provides detailed methodologies to quantify reduced carbon emissions.	www.v-c-s.org
Plan Vivo	Projects are allowed to develop their own methodologies in association with research institutes or universities. Objectives include a positive impact on rural communities. Quantification of carbon accumulation lacks the general rigour of other standards.	www.planvivo.org
Climate, Community & Biodiversity (CCB) standard	Quantifies the co-benefits of socio-economic and biodiversity factors, but advises use of VCS to certify carbon credits.	www.climate-standards.org

To estimate carbon accumulation in a forest that is undergoing restoration, you need to know the mass of trees per unit area. Tree trunks and roots contain most of the above-ground carbon in a forest; the amount of carbon in tree leaves and ground flora is almost negligible compared with that in the trunks and roots.

Simple measurements of the girths of the tree trunks over a known area can provide a rough approximation of most of the above-ground carbon (AGC), which is calculated using published equations (termed 'allometric' equations) that describe the relationship between a tree's diameter at breast height (and/or tree height) and its above-ground dry mass in kilograms. These equations are prepared by researchers who fell trees of widely differing girths and then dry and weigh them piece by piece. Different equations are used for different forest types and even for different tree species, so project developers need to search the literature for an equation that most closely matches the forest type that is being restored (see Brown, 1997; Chambers *et al.*, 2001; Chave *et al.*, 2005; Ketterings *et al.*, 2001; Henry *et al.*, 2011). The use of these equations involves some tricky maths, so ask a mathematician if you don't understand them.

The alternative, if allometric equations do not exist for the required forest type, is to use default values of forest type based on international, domestic or local sources. The default international values are listed in **Table 7.3**.

To sample carbon accumulation in the field, use metal poles to mark at least 10 permanent sample points across the restoration site. Use a 5-m-long piece of string to determine which trees are within 5 m of the poles and then measure their girths at breast height (1.3 m from the ground). Divide tree girth by pi (3.142) to convert to tree diameter. Then use the allometric equations to estimate the above-ground dry mass of each tree in kg from its diameter. Convert to a per-hectare value as follows:

$$\frac{\text{Sum of above-ground dry mass (kg) of all trees in all circles} \times 10{,}000}{\text{No. circles} \times 78.6}$$

Divide the result by 1,000 to convert to metric tonnes (i.e. Megagrams (Mg) in SI units) per hectare and compare your results with values typical for tropical forests (**Table 7.3**) to see how close your restored forest is to typical forest target values.

Table 7.3. Typical above-ground biomass figures for different types of tropical forests. Drier tropical forests typically contain less biomass than wetter ones (IPCC, 2006; Table 4.7).

Forest type	Continent	Above-ground biomass (tonnes dry mass per hectare)
Tropical rain forest	Africa, N. & S. America, Asia (continental), Asia (insular)	310 (130–510) 300 (120–400) 280 (120–680) 350 (280–520)
Tropical moist deciduous forest [= Seasonal tropical forest]	Africa, N. & S. America, Asia (continental), Asia (insular)	260 (160–430) 220 (210–280) 180 (10–560) 290
Tropical dry forest	Africa, N. & S. America, Asia (continental), Asia (insular)	120 (120–130) 210 (200–410) 130 (100–160) 160

To calculate the mass of tree roots multiply the above-ground biomass by 0.37 for tropical evergreen forest or by 0.56 for drier tropical forest (Table 4.4 in IPCC, 2006) or consult Cairns et al. (1997) for ratios for other forest types. When these results are added to above-ground biomass, you have an estimate for tonnes dry-mass of trees per hectare.

The carbon content of dry tropical wood varies considerably among species, but the average value is around 47% (Table 4.3 in IPCC, 2006; Martin & Thomas, 2011). Therefore, multiply the result by 0.47 to arrive at an estimate of the mass of carbon in trees per hectare.

To find out how much the carbon is worth, convert tonnes of carbon into an equivalent value of tonnes of carbon dioxide by multiplying by 3.67, then look up the value of a tonne of carbon dioxide equivalent on the carbon credit markets at: www.tgo.ot.th/english/index.php?option=com_content&view=category&id=35&Itemid=38. See also World Agroforestry Centre manual, available free from: www.worldagroforestry.org/sea/Publications/files/manual/MN0050-11/MN0050-11-1.PDF.

7.5 Research for improving tree performance

If you have sufficient resources, you may wish to consider turning your forest restoration project into a research program in which you collect more information than usually comes from the basic monitoring procedures outlined above. This requires collecting data in a systematic way, over several 'replicated' plots — a so-called 'field trial plot system' or FTPS for short. An FTPS can be used to compare the performance of planted tree species, to evaluate the effects of silvicultural treatments, to assess biodiversity recovery and carbon accumulation and to determine the optimum design and management of restoration plots. It can also become a valuable demonstration tool that can be used to teach others effective restoration techniques and how to avoid repeating expensive mistakes.

What is an FTPS?

An FTPS is a set of small plots (typically 50 × 50 m = 0.25 ha), each one planted with a different mixture of tree species and/or silvicultural treatments using the randomised complete block design described in **Chapter 6** (p. 198) and in **Appendix A2.1**. Each planting season, new plots are added to the system. In the new plots, the tree species and treatments that worked best in previous years are retained, using the selection procedure described in **Section 8.5**, while poorly performing species and unsuccessful treatments are dropped to make room for new species and treatments to be tested. If the work goes well, the younger plots out-perform the older ones because an FTPS is gradually improved in response to incoming data. Therefore, select an area for an FTPS that has plenty of unused land available for future expansion. An ideal area for planting over a 10-year period should be at least 20 ha.

Using the recommended spacing of 1.8 m between trees and a standard plot size of 50 × 50 m requires about 780 trees per plot. With a minimum acceptable sample size of 20 individuals per species, this allows for a maximum of 39 species to be tested each year.

Objectives for an FTPS

An FTPS has three main objectives: i) to generate scientific data that are used to develop 'best field practices' for effective forest restoration; ii) to test the practicability of those best practices; and iii) to provide a demonstration site for education and training in forest restoration methods.

The scientific questions addressed by the FTPS can include:-
- Which of the tree species tested meet required criteria?
- What is the optimum planting density?
- What silvicultural treatments (e.g. weeding, fertiliser application, mulching etc.) maximise the performance of the planted trees? How frequently and for how long should such treatments be applied?
- How can plantation design be optimised (e.g. how many species per plot)?
- Which species can or cannot be grown next to each other?
- How fast does biodiversity recover? How does distance to nearest forest affect biodiversity recovery?

An FTPS is also a valuable education and training facility.

Research in an FTPS should address the simpler questions (relating to species performance and silvicultural treatments) first and explore more complex issues (such as species mixes, distance to natural forest and so on) later. As all of the trees in the plots are of known age and species and most are labelled, the FTPS inevitably becomes a research resource that is much sought after by other scientists and research students.

Where should the FTPS be established?

In reality, the position of an FTPS can be determined by basic questions of land ownership and proximity to the FORRU host organisation, but where possible, try to take into account of the scientific and practical considerations below.

Scientific considerations

Uniformity — plot experiments are notoriously vulnerable to variability in site conditions. It might be difficult to separate the effects of treatments applied in different plots from the effects of differences in environmental conditions among the plots. To some extent, this problem can be compensated for by using a randomised complete block experimental design, but it helps if the FTPS is established over fairly uniform terrain in terms of elevation, slope, aspect, bedrock, soil type and so on.

Vegetation — match the restoration techniques tested in an FTPS with the initial degradation stage of the site (see **Section 3.1**).

Conservation value — FTPSs are particularly valuable when located within a protected area or its buffer zone, or wherever biodiversity conservation is the top management priority. Using an FTPS to create corridors to link forest remnants gives it extra conservation value.

Practical considerations

Accessibility and topography — reasonably convenient access, at least by 4WD vehicles, is essential, not only for planting, maintenance and monitoring the planted trees but also to facilitate visits to the plots for education purposes. Select an area within 1 or 2 hours drive of the FORRU nursery or headquarters. Obviously flatter sites are easier to work with than steep ones.

Proximity to a local community that supports the idea of forest restoration — this enables the exchange of scientific and indigenous local knowledge and access to experience of the social aspects of framework forestry. A local community can provide a source of labour and security for trial framework species plots (see **Section 8.2**). The importance of including all stakeholders in discussions about FTPS establishment was covered in **Chapter 4**. Abandoned, former agricultural land where cultivation has become too difficult or uneconomic because of deteriorating environmental conditions is ideal.

Land tenure — if the FORRU host organisation does not own the land, it must enter into agreement with the authority that controls land use in the area. This will most probably be the government department in charge of forestry or conservation or possibly a local community.

Establishing the plots

An FTPS consists of several treatment plots (T) and two kinds of control plot: 'treatment control' (TC) plots and 'non-planted control' (NPC) plots. First, decide on a standard set of procedures to follow to establish the TC plots. The standard protocol should be based on current best-known practices for growing trees in the area, which can be derived from experience and indigenous knowledge and by considering local conditions. The standard protocol can be improved year-by-year by incorporating the treatments that were most successful in the analyses of each year's field experiments. Each year the effects of new treatments, applied in the T plots, are measured by comparison with the TC plots.

Start with the following protocol and modify it to suit local conditions:
- Six to eight weeks before planting, measure out the plots; demarcate the corners with concrete posts or similar and make a map of the plots, clearly indicating plot identification numbers and which plots will receive which treatments.
- Then, slash weeds down to ground level (except in non-planted control plots), but avoid cutting any naturally established tree seedlings and saplings and coppicing shoots (mark them beforehand with coloured poles or flags).
- One month before planting, apply a non-residual herbicide (e.g. glyphosate) to kill sprouting weeds.
- Label the trees and plant at the appropriate time.
- Plant the appropriate number of candidate tree species (if possible equal numbers of all species, at least 20 trees of each species per plot) spaced, on average, 1.8 m apart. Randomly mix the species across each plot.
- If necessary, apply 50–100 g of NPK 15:15:15 fertiliser in a ring about 20 cm away from the stems of planted trees at planting time.
- During the first rainy season (or the first 6 months after planting in a wet forest) repeat the fertiliser treatment and weed around the trees (using hand tools) at least three times, at 6–8 week intervals (adjust frequency according to the rainfall and rate of weed growth).

- At the start of the first dry season after planting (in seasonal tropical forests), cut fire breaks around the plots and implement a fire prevention and suppression program.
- Repeat weeding and fertiliser application during the second rainy season after planting.
- At the beginning of the third rainy season, assess the need for further maintenance operations.

Immediately adjacent to the TC plots, establish 'treatment' (T1, T2, T3, etc.) plots, simultaneously in exactly the same way, but vary only one component of the standard protocol (e.g. fertiliser or weeding technique etc.). Tree performance in the T plots is then compared with that in the TC plots.

Experimental design

A randomised complete block design (RCBD) is recommended. Block together single replicates of each type of T plot with one TC plot and replicate the blocks in at least three locations across the study site (4–6 locations would be better). Position the blocks at least a few hundred metres apart, if possible, to take into account variability in conditions (slope, aspect and so on) across the study area. Randomly allocate treatments to each T plot within each block. Plant 'guard rows' of trees around each plot and block to prevent one treatment from influencing another and to reduce edge effects.

Next add 'non-planted control' (NPC) plots, in which no trees are planted, no treatments are applied and the vegetation is left undisturbed to undergo natural succession. The function of NPC plots is to generate baseline data on the natural rate of biodiversity recovery in the absence of forest restoration plantings and treatments. Biodiversity recovery in the restoration plots is then compared with what would have happened naturally if forest restoration had never been implemented. Associate one NPC plot with each block of TC and T plots. If the NPC plots are adjacent to the planted plots, birds that are attracted by the planted trees will 'spill over' into the NPC plots. So NPC plots should be placed at least 100 m away from planted plots.

A randomised block design, with three blocks spread across the study area. Blocks are positioned at least a few hundred metres apart and not far from remnant forest. T = treatment plot; TC = control plot and NPC = non-planted control plot.

Choice of treatments

Consider the main factors that limit tree survival and growth at the study site, and design treatments to overcome them. For example, if soil nutrients are limiting, try varying the type of fertiliser, the amount applied each time and/or the frequency of application. Alternatively, experiment with adding compost to the planting hole. If competition with weeds is the most obvious limiting factor, try varying weeding techniques (e.g. hand tools or herbicide) or the frequency of weeding, or try using dense mulch (e.g. cut weeds or corrugated cardboard) to suppress the germination of weed seeds in the immediate vicinity of planted trees. Other treatments to try include placing polymer gel or mycorrhizal inoculae in planting holes or subjecting trees to various pruning treatments before planting.

Write a field experiment plan

Prepare a working document, containing the following information:
- a sketch map of the plot system, indicating plot identification numbers and which plots receive which treatments;
- a list of species planted in the plots and the label numbers of each tree planted in each plot;
- a description of the standard planting protocol;
- a description of the treatments to be applied in each plot and a schedule for their application;
- a schedule for data collection.

Consistent application of silvicultural treatments is one of the most important and costly components of field experiments.

Make sure all FORRU staff receive a copy of the document, understand their roles in establishing, maintaining and monitoring the plots, and have been adequately trained in how to apply the specified treatments. One of the main causes of experiment failure is inadequate or inconsistent treatment application.

Monitoring field experiments

Labelling saplings

Label trees in the nursery before planting them, as described in **Section 7.4**. The minimum information on the label should be the species number and tree number. Additional information could include the plot number and year of planting, but whatever system is used, no two trees in the entire plot system should carry the same label numbers, no matter when or where they are planted.

When to monitor

As with basic monitoring (see **Section 7.4**), collect data about two weeks after planting and at the end of each growing season (i.e. rainy season), with the most important monitoring event being at the end of the second rainy season after planting. Additional monitoring at the end of each dry season can provide more detailed information about when and why trees die.

What measurements should be made?

Record survival, health, height, root collar diameter, crown width and weed score for both planted trees and natural regenerants, as for basic monitoring (see **Section 7.4**).

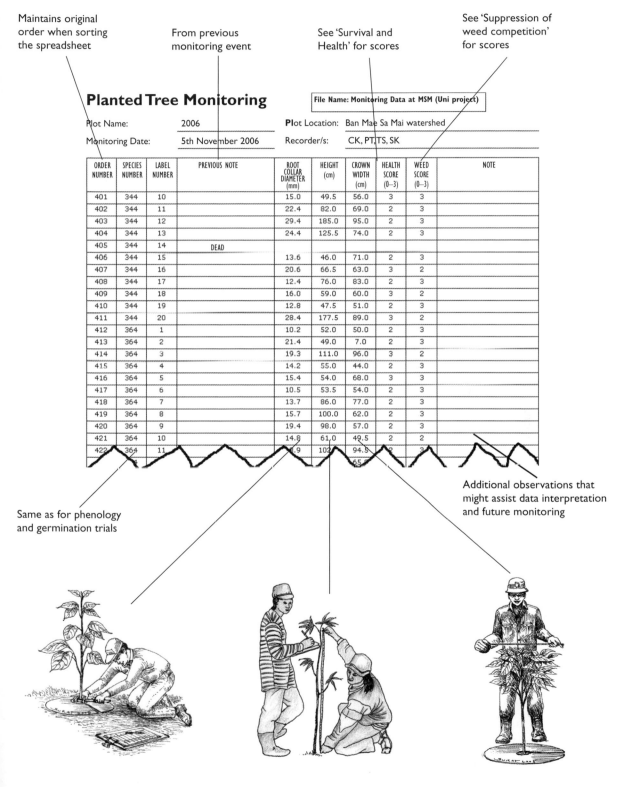

Maintains original order when sorting the spreadsheet

From previous monitoring event

See 'Survival and Health' for scores

See 'Suppression of weed competition' for scores

Planted Tree Monitoring

File Name: Monitoring Data at MSM (Uni project)

Plot Name: 2006 Plot Location: Ban Mae Sa Mai watershed

Monitoring Date: 5th November 2006 Recorder/s: CK, PT, TS, SK

ORDER NUMBER	SPECIES NUMBER	LABEL NUMBER	PREVIOUS NOTE	ROOT COLLAR DIAMETER (mm)	HEIGHT (cm)	CROWN WIDTH (cm)	HEALTH SCORE (0–3)	WEED SCORE (0–3)	NOTE
401	344	10		15.0	49.5	56.0	3	3	
402	344	11		22.4	82.0	69.0	2	3	
403	344	12		29.4	185.0	95.0	2	3	
404	344	13		24.4	125.5	74.0	2	3	
405	344	14	DEAD						
406	344	15		13.6	46.0	71.0	2	3	
407	344	16		20.6	66.5	63.0	3	2	
408	344	17		12.4	76.0	83.0	2	3	
409	344	18		16.0	59.0	60.0	3	2	
410	344	19		12.8	47.5	51.0	2	3	
411	344	20		28.4	177.5	89.0	3	2	
412	364	1		10.2	52.0	50.0	2	3	
413	364	2		21.4	49.0	7.0	2	3	
414	364	3		19.3	111.0	96.0	3	2	
415	364	4		14.2	55.0	44.0	2	3	
416	364	5		15.4	54.0	68.0	3	3	
417	364	6		10.5	53.5	54.0	2	3	
418	364	7		13.7	86.0	77.0	2	3	
419	364	8		15.7	100.0	62.0	2	3	
420	364	9		19.4	98.0	57.0	2	3	
421	364	10		14.8	61.0	49.5	2	2	
422	364	11		8.9	102	94.5	2	3	

Additional observations that might assist data interpretation and future monitoring

Same as for phenology and germination trials

Plot No.	Species No.	Tree No.	15/7/98 Health Score (0—3)	19/11/98 Health Score (0—3)	9/11/99 Health Score (0—3)	5/10/00 Health Score (0—3)	15/7/98 Height (cm)	19/11/98 Height (cm)	9/11/99 Height (cm)	5/10/0 Height (cm)
1	7	1	3	3	2	3	39	93	147	231
1	7	2	3	2	3	3	39	109	173	287
1	7	3	2	3	3	3	53	144	229	347
1	7	4	2	NF	0	0	56	NF	-	-
1	7	5	3	3	3	3	59	164	265	354
1	7	6	2.5	0	0	0	32	-	-	-
1	7	7	3	3	3	3	43	81	128	252
1	7	8	3	3	3	3	41	68	108	171
1	7	9	0.5	0	1	2	30	-	21	40
1	7	10	3	2.5	3	3	64	63	237	300
1	7	11	3	0.5	3	3	49	48	160	300
1	7	12	0.5	0	NF	0	34	-	NF	-
1	7	13	2.5	0	0	0	44	-	-	-
1	7	14	2	1.5	3	2.5	30	29	106	297
1	7	15	2	2	0	0	27	26	-	-
1	7	16	3	2.5	3	3	23	43	90	125
1	7	17	3	3	2.5	3	37	51	140	166
1	7	18	3	2.5	3	0	39	60.5	20	-
1	7	19	3	3		3	28	99	NF	341
1	7	20	2.5	2.5	1.5	3	35	46.5	53	110

Sort the data first by species number and then by tree number.

Data analysis and interpretation

Organise the spreadsheet

First, enter the field data into a computer spreadsheet. Insert new data to the right of previously collected data, so that one row represents the progress of an individual tree running chronologically from left to right. Next, sort the data by rows, first by species number and then by tree number. This groups all trees of the same species together. Insert the date on which the data were collected in the cell immediately above every column heading. Then sort the spreadsheet by column (left to right), first by column heading (row 2) and then by date (row 1). This groups the same parameters together in chronological order from left to right. The data can now be easily scanned for interesting features or anomalies and manipulated to extract the values required below for more detailed statistical analysis.

Comparing species

As in the nursery experiments, you could start by comparing survival and growth among species. To compare differences in survival, start with trees in the TC plots only: scan the spreadsheet and count the number of surviving trees in the TC plot in each block. If the same number of trees of each species was planted in every plot, simply enter the number of surviving trees into a new spreadsheet, with species as column headings and one row per block (or replicate), as shown below. If different numbers of trees of each species were planted, then calculate the percentage survival in each plot and enter those data into the new spreadsheet. Then follow the instructions in **Appendix 2** to arcsine transform the data and carry out an ANOVA. In this case, each species is the equivalent of a 'treatment' (there is no control when comparing species).

/98	19/11/98	9/11/99	5/10/00	19/11/98	9/11/99	5/10/00	9/11/99	5/10/00
CD (m)	RCD (mm)	RCD (mm)	RCD (mm)	Weed Score (0–3)	Weed Score (0–3)	Weed Score (0–3)	Width of Canopy (cm)	Width of Canopy (cm)
2	14.8	23.3	36.7	3	2.5	1	73	115
.1	17.3	27.5	45.6	2.5	2	2	86	143
.4	22.9	36.4	55.1	3	2	1	114	173
.2	NF	-	-	NF	-	-	-	-
.1	26.2	42.2	56.3	1.5	1	0.5	148	200
7	-	-	-	-	-	-	-	-
5	12.9	20.3	40.1	1	1	0.5	64	126
.5	10.8	17.1	27.2	1.5	1	1	95	150
1	-	2.1	5.4	-	-	-	-	-
7	18.2	29.6	59	1.5	1	1	150	200
1	13.4	21.6	47	1.5	1	2	103	200
3	-	NF	-	-	NF	-	NF	-
5	-	-	-	1.5	-	-	-	-
6	9.3	13	37	1.5	2	2	93	150
6	6.1	-	-	1.5	-	-	-	-
2	10.6	18	21	1.5	1.5	1	80	75
4	15.2	25	22	1.5	2	1	90	125
3	3.9	3.4	-	1.5	1.5	-	23	-
9	24	NF	54	1.5	NF	0	NF	200
6	9.2	12.8	14	1.5	0.5	2	65	108

Then sort columns by heading and then by date to group parameters together in chronological order from left to right.

The same procedure can be followed to compare the species-means of height, root collar diameter (RCD), crown width, and relative growth rates in each TC plot, although these data do not need to be arcsine transformed. In addition to the absolute size of the trees (height or RCD), it is useful to know how fast the trees are growing. This is

SPECIES

	S1	S2	S3	S4	S5	S6	S7	S8	S9	S10	S11	S12	S13	S14	S15	S16	S17	S18	S19	S20
Block 1	24	4	10	2	25	20	15	10	2	14	25	24	18	5	7	8	12	17	1	5
Block 2	22	2	11	3	25	21	16	13	3	15	24	24	13	6	8	9	13	16	2	6
Block 3	26	3	12	2	25	23	14	14	5	16	25	25	18	7	9	8	14	15	1	7
Block 4	25	4	13	3	24	22	15	13	6	13	24	28	18	8	7	7	13	17	2	6

Number of surviving trees of each species in the treatment control plot (TC) of each block at the end of the second rainy season after planting. Twenty-six trees of each species were planted in each TC plot.

especially important in forest restoration projects for carbon storage. The bigger the tree to begin with, the faster it grows, so relative growth rate (RGR) is used to compare the growth of different trees. RGR expresses the increase in the size of the plant as a percentage of the average size of the plant throughout the measurement period, and thus it can be used to compare the growth of trees that were relatively large at planting time with those that were relatively smaller. RGR can be calculated in terms of tree height as follows:

$$\frac{\ln H \ (18 \ months) - \ln H \ (at \ planting) \times 36{,}500}{No. \ days \ between \ measurements}$$

...where ln H is the natural logarithm of tree height (cm). RGR is an estimated annual percentage increase in size. It takes account of differences in the original sizes of the trees planted, so it can be used to compare trees that were larger at planting time with those that were smaller. Compare the mean values of RGR among species by ANOVA. The same formula can be used to calculate the relative growth rates of root collar diameters and crown width.

Species comparisons, based solely on field performance, are not enough to inform a definitive decision on which species to plant. See **Section 8.5** to see how field performance data can be combined with other important parameters when making the final decision on which species work best.

Comparing treatments

The effects of treatments on individual species can be determined using exactly the same analytical procedure. From the main spreadsheet, count the number of surviving trees (or calculate % survival) of a single species for each of the treatment and control plots in all blocks. Construct a new spreadsheet with treatments as the column headings (TC, T1, T2, etc.) and blocks (or replicates) as rows. Then follow the instructions in **Appendix 2** to arcsine transform the data and carry out an ANOVA.

Substitute the survival data with mean plot values for height, RCD, RGR, crown width and reduction in weed score to determine the effects of treatments on other aspects of field performance (there is no need to transform these data). Then repeat the same procedure for all other species.

Different treatments will affect different species in different ways. It is impractical to provide treatments that are optimal for each species in plots with 20 or more species, so the objective of the analysis is to determine the optimum combination of treatments that have a positive effect on most of the species planted.

Experiments with direct seeding

Direct seeding was described as a potential low-cost alternative to tree planting in **Sections 5.3** and **7.2**, but scant information is available about which tree species are suitable for this technique (**Table 5.2**). The success or failure of the direct seeding of each tree species depends on a combination of many factors, including seed structure and dormancy, attractiveness of the seeds to seed predators, susceptibility of the seeds to desiccation, soil conditions and surrounding vegetation. Therefore, experiments are necessary to determine whether a tree species establishes better by direct seeding than by planting nursery-raised saplings and to determine the cost savings achieved (if any).

Information needed for direct seeding experiments

Before a direct seeding experiment can be started, it is first necessary to know: i) what is the optimum pre-sowing treatment to accelerate seed germination; and ii) if fruiting does not occur at the optimum time for direct seeding (i.e. the start of the rainy season in seasonal tropical forests), what is the best seed-storage protocol for retaining seed viability during the period between the seed collection and the direct seeding. The nursery experiments required to answer these questions are described in **Section 6.6**. They will take at least a year to complete before direct seeding experiments can begin.

One of the main causes of failure of direct seeding is seed predation. If seeds are treated to accelerate germination before they are sown into deforested sites, the time available for seed predators to find and consume the seeds is reduced, and consequently the chances that the seed will survive long enough to germinate are increased. Treatments that accelerate germination in the nursery can, however, sometimes increase the risk of seed desiccation in the field or make seeds more attractive to ants by exposing their cotyledons. For tree species with recalcitrant seeds that are difficult to store, direct seeding is only an option for those species that fruit at the optimum direct seeding time.

Steps of direct seeding experimental design

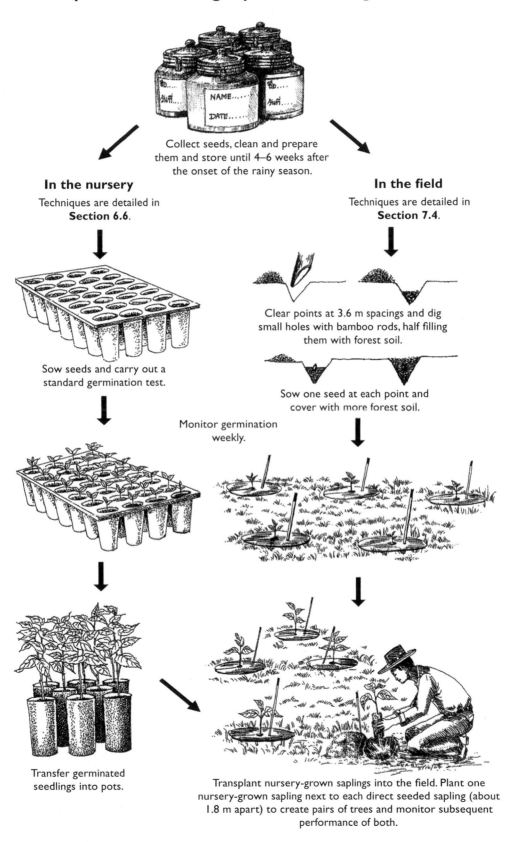

Collect seeds, clean and prepare them and store until 4–6 weeks after the onset of the rainy season.

In the nursery
Techniques are detailed in **Section 6.6**.

Sow seeds and carry out a standard germination test.

Monitor germination weekly.

Transfer germinated seedlings into pots.

In the field
Techniques are detailed in **Section 7.4**.

Clear points at 3.6 m spacings and dig small holes with bamboo rods, half filling them with forest soil.

Sow one seed at each point and cover with more forest soil.

Transplant nursery-grown saplings into the field. Plant one nursery-grown sapling next to each direct seeded sapling (about 1.8 m apart) to create pairs of trees and monitor subsequent performance of both.

Methods for direct seeding experiments

Collect seeds from several trees, combine and mix them, clean and prepare the seeds in the standard way and, if necessary, store them until planting time using the most efficient storage protocol developed from previous experiments.

In the nursery, sow seeds into modular trays and carry out a standard germination test, comparing control (non-treated) seeds with those subjected to the most efficient treatment to accelerate germination developed from previous experiments.

In the field, use the same experimental design as that used in the nursery, with the same number of treatment and control replicates and the same of seeds in each replicate, but instead of using modular germination trays, sow the seeds at direct seeding points marked with bamboo poles and spaced about 3.6 m apart across the study site. Sow one seed at each point.

Monitor seed germination weekly, both in the field and in the nursery and analyse the results using the method already described in **Section 6.6**. In the field, after germination is finished, try to dig up and inspect any non-germinated seeds. This might help to determine how many seeds were removed or damaged by seed predators and how many appear intact but simply failed to germinate.

In the nursery, once germination has ended, transfer the germinated seedlings into pots in the usual way. Use the standard protocol, developed from previous experiments, to grow the plants in the nursery. Monitor and analyse growth as described above. Monitor the plants in the same way in the field.

Once the saplings in the nursery have grown tall enough to be planted out, transplant them into the field as usual. This may be 1 or 2 years after direct seeding took place. Plant one nursery-raised sapling next to each sapling established by direct seeding (about 1.8 m apart) to create pairs of trees. Monitor the field performance of the paired trees for at least two years after planting out of the nursery-raised saplings. Use paired t-tests to compare the growth of the nursery-raised and direct-seeded trees.

Other experiments with direct seeding

There are many other treatments that can be incorporated into this basic experimental design. If burying fails to deter seed predation in the field, try experimenting with treating the seeds with chemical repellents to make the seeds unattractive to seed predators; but do not forget to test the effects of the chemical repellents on seeds germinated in the nursery in case the repellent also has an effect on germination.

Experiments that vary the maintenance procedures used around direct seeding points might also suggest how results could be improved. Try altering the weeding or mulching regime around the direct seeding points to prevent the germination of weeds in the immediate vicinity of the young seedlings, especially in the first few months after germination or sow more than one seed at each direct seeding point to overcome the effects of low germination rates.

Direct seeding certainly works for some species. Compare the direct-seeded *Sarcosperma arboreum* tree on the left with the nursery-raised one germinated from the same seed batch on the right.

Can direct seeding save money?

Since direct seeding does not require a tree nursery, it should reduce the costs of forest restoration. Direct seeding does, however, require weeding around the seeding points as the young, recently germinated seedlings are highly vulnerable to competition from weeds. The application of fertiliser and mulch around the direct seeding points in the first year also has added costs. A detailed account of all expenses must therefore be kept throughout a direct seeding experiment to determine whether this technique does actually reduce the overall costs of forest restoration.

7.6 Research on biodiversity recovery

The ultimate measure of the success of forest restoration is the extent to which biodiversity returns to the levels associated with the target forest ecosystem. The purpose of biodiversity monitoring is therefore to determine to how fast this occurs and ultimately to improve restoration methods so as to hasten biodiversity recovery.

Monitoring *all* biodiversity is not practical, so for forest restoration, biodiversity monitoring focuses on those components that relate directly to the re-establishment of natural forest regeneration mechanisms, particularly seed dispersal and the seedling establishment of recruit tree species (i.e. in-coming tree species not including those planted). Some species or groups might serve as indicators of the overall health of the forest.

Four crucial questions are:

- Do planted trees (and/or ANR techniques) produce resources (e.g. flowers, fruits and so on) at an early age that are likely to attract seed-dispersing animals?
- Are seed-dispersing animals present in the area, and if so, are they actually attracted by these resources?
- Do seeds that are brought in by those animals actually germinate, increasing the species richness of the tree seedlings or saplings naturally establishing beneath the planted trees?
- Do wind-dispersed seeds also establish naturally?

Here, we present a few techniques that can be used to answer these questions. Monitoring the performance of planted trees can show clear improvements within 2–3 years, but the recovery of biodiversity takes much longer; monitoring may continue over periods of 5–10 years, but at less frequent intervals.

The requirement for biodiversity monitoring must be considered from the beginning of field experiments during the design of a FTPS. Non-planted control plots must be included in a FTPS, and a biodiversity survey of the control plots and the plots to be subjected to restoration treatments must be carried out before the site preparation. This provides the essential baseline data against which subsequent changes in biodiversity can be compared. Biodiversity is then surveyed in both control and restoration plots and compared with that in nearby intact forest (i.e. the target forest community).

After each data collection session, two types of comparisons are performed: i) before vs. after comparisons between current data and baseline (pre-planting) data; and ii) control vs. restoration plot comparisons. In this way, the enhanced biodiversity recovery brought about by restoration actions can be distinguished from that due to natural ecological succession. Relative biodiversity recovery can then be calculated as a percentage of that recorded by the same methods in target forest.

Phenology studies

Frequent walks through the restoration plots while noting which trees are flowering and fruiting can yield most of the data needed to determine whether the trees within the restoration plots are producing resources that are likely to attract seed-dispersing animals. Establish a trail system through the centre of all plots. Walk the trails monthly, recording the following information for trees within 10 m of the trail:

- date of observation;
- block/plot identification number;
- tree number (including species number);
- presence of flowers or fruits: use the 0–4 scoring system (see **Section 6.6**);
- wildlife signs: nests, tracks, faeces and so on either on or near the trees;
- direct observations of animals using the tree for feeding, bird perching and so on.

Enter each observation, as a single row, into a spreadsheet to allow easy compilation of the data by species or date. Determine the youngest age (time since planting) at which the first individuals of a species commence flowering and fruit set. The frequency of observations (within a species) can be used as a general indication of the prevalence of flowering or fruiting at the species level. For additional detail, measure the girth at breast height (GBH) (or RCD) and the height of the flowering or fruiting trees to establish correlations between tree size and age at maturity. The flowering of some species can be inhibited if the trees are heavily shaded by adjacent tree crowns. If there is some variation in the incidence of flowering within a species, a shade score for each flowering tree can also be recorded. In addition to assessing the production of wildlife resources by planted tree species, monthly surveys can yield much additional information about the planted trees species, such as the outbreak of pests and diseases, and can provide early warning of disturbances to the plots by human activities. This kind of simple qualitative monitoring is an excellent way to involve local people in monitoring forest restoration sites as it is easily learned and requires no special skills.

Nursery stock of *Bauhinia purpurea* starts flowering and fruit set within 6 months after planting, providing food for birds and insects.

Wildlife monitoring

All re-colonising wildlife species (both plants and animals) contribute to biodiversity, but seed-dispersing animals can accelerate biodiversity recovery more than other species. Birds, fruit bats and medium-sized mammals are the major groups of interest, but of these, the bird community is the most easily studied.

Birds are an important indicator group

Birds provide a convenient indicator group for the evaluation of biodiversity because:
- they can be relatively easy to see and many are easy to identify;
- good identification guides now cover most of the tropics;
- most species are active by day;
- birds occupy most trophic levels in forest ecosystems — herbivores, insectivores, carnivores and so on — and hence a high diversity of birds usually indicates a high diversity of plants and prey species, especially insects.

What questions should be addressed?

- What bird species occurred in the area before restoration?
- What bird species are characteristic of the target forest ecosystem and do those species return to restored forest plots? If so, how soon after restoration actions?
- Which of the bird species that visit the plots are most likely to disperse the seeds of forest trees into restoration plots?
- Which bird species disappeared as a result of forest restoration activities and when?

When and where should bird surveys be carried out?

Survey the entire FTPS once it has been demarcated but before implementing any activities that are likely to alter bird habitats (i.e. before preparing the site for planting). This survey provides the baseline data against which changes are compared. Thereafter, carry out bird surveys of the same intensity in both restoration plots and control plots and also in the nearest area of target forest (see **Section 4.2**). Annual bird surveys

Bird Survey Record Sheet
Date: 17/12/05
Block number: G1
Start time: 06.30

File name: Restoration plot, 10 years old
Weather: sunny, very warm
Plot number: EG01
Finish time: 09.30
Recorders: DK, OM

Time	Species	No. of birds (sex)	Sight or song/call	Distance from point (m)	Activity	Tree species (if appropriate)
06.30	Black-crested bulbul	2	Sight	10	Feeding on fruit	*Ficus altissima*
06.30	Bar-winged flycatcher-shrike	1	"	10	Foraging for insects	*Ficus altissima*
06.30	Hill blue flycatcher	1	"	10	Fly catching	*Choerospondias axillaris*
06.40	Sooty-headed bulbul	3	"	15	Flushed from crown	*Betula alnoides*
06.45	Yellow-browed warbler	2	"	5	Moving through canopy, foraging	Many species
06.45	Pallas's warbler	1	"	5	Moving through canopy, foraging	Many species
06.45	Eurasian jay	2	Heard calls	30	Calling from nearby trees	Unknown
06.50	Magpie robin	1 male	Sight/ song	8	Foraging on forest floor, also short burst of song	
06.55	Lesser coucal	1	Sight	10	Flying through trees	
07.05	Striated yuhina	10+	"	5	Moving through canopy, feeding	Many species
07.10	Mountain bulbul	2	"	12	Feeding on fruit	*Ficus hispida*
07.22	Asian house martin	25+	"	50	Hawking insects overhead	
07.30	Scarlet-backed flowerpecker	1 male	"	5	Feeding on nectar	*Erythrina subumbrans*

are usually sufficient to detect changes in bird communities. Carry the surveys out at the same time each year as bird species richness will fluctuate according to seasonal migration patterns. Observe birds during the first 3 hours after dawn and the last 3 hours before sunset. Timetable 1-hour observation periods in each plot, alternating around the plots at hourly intervals, but ensure that, over the entire survey period, all plots are studied for the same number of hours, spread evenly among morning and evening observational periods.

Data collection

Use the 'point count' method to count birds from the centre of each plot. This method can be used to both count species and estimate bird population density (Gilbert *et al.*, 1997; Bibby *et al.*, 1998). Stand in the centre of each plot and record all bird contacts for 1 hour by both sight and song. Record the species and numbers of birds and estimated distance from the observer when birds first appear in the plot. To reduce the risk of recording the same individual birds several times, do not record the same bird species entering the plot for five minutes after first recording that species. Record the tree species (and tree number if labelled), in which birds have any activity (particularly feeding) and their position (trunk, lower canopy, upper canopy etc.).

Data analysis

Answer most of the questions listed earlier by simply scanning the species lists and counting the number of bird species that re-colonise the restoration plots and those that disappear as a result of forest restoration activities.

Use binoculars, telescopes and your ears to detect birds within 20 m of a single point in the centre of a forest restoration trial plot.

To calculate the extent of recovery in the bird community, compare the species list for pristine target forest with that for the restoration plots. Calculate the percentage of the species found in the forest that are also found in the restored plots and look at how this percentage changes over successive survey times. Next, determine which of those species are frugivorous. These are the crucial species that are most likely to disperse seeds from forest into restoration plots.

For a quantitative analysis of the species richness of bird communities, we recommend the MacKinnon list method (Mackinnon & Phillips, 1993; Bibby *et al.*, 1998), which provides a means of calculating a species recovery curve and a relative abundance index. For full step-by-step instructions and a worked example see Part 5 of FORRU, 2008 (www.forru.org/FORRUEng_Website/Pages/engpublications.htm).

Bulbuls are the 'work horses' of forest restoration in Africa and Asia. They feed on fruit in remnant forest and drop seeds of many tree species in forest restoration plots.

Mammals

Mammals can be divided into two groups of interest: i) fruit-eating species that are capable of dispersing seeds from intact forest into restored sites (e.g. large ungulates, civets, fruit bats and so on); and ii) seed predators, which could limit the seedling establishment of recruit tree species in restored sites (particularly small rodents).

Mammals are much more difficult to survey than birds as most species are nocturnal and very shy, so direct observations of mammals are usually few and far between. Opportunistic, anecdotal data (rather than systematic, quantitative data) are more commonly used to determine the recovery of mammal communities after forest restoration.

For medium-sized or larger mammals, camera trapping is a very effective way of determining the return of species to restoration sites. Digital cameras housed in camouflaged, weatherproof cases that are triggered by movement in the field of view have never been cheaper (starting at US$ 100–200). Password protected electronics means that the cameras are of no value to potential thieves. The batteries last several months and thousands of pictures can be accumulated on a single memory card (e.g. www.trailcampro.com/cameratrapsforresearchers.aspx).

Camera traps capture black and white images at night (without flash) and colour images during the day of any moving thing. The hog badger (top left) and the large Indian civet (top right) bring seeds into restoration plots. Leopard cats (lower left) help to control seed predators. Cameras can also help to detect illegal hunting (lower right).

Live trapping, using locally available rat traps, is another useful technique, particularly for small mammals such as rodents, but it is labour intensive and therefore expensive. Lay out baited traps 10–15 m apart using a 7 × 7 grid pattern. Expect capture rates of below 5%, so a great deal of effort is required for relatively few data. Expect to record a sharp decline in the populations of rodent seed predators in restoration plots by 3–4 years after planting, by which time the dense, herbaceous vegetation that provides cover for such small mammals will have been shaded out by the developing forest canopy. When handling wild animals, make sure your vaccinations against animal-borne diseases, particularly rabies, are up to date.

Most records of mammals in forest restoration plots must come from indirect observations of their tracks, feeding remains and other signs. These can be recorded during the regular phenology monitoring of planted plots and control (non-planted) plots. The frequency of observations can be used as an index of abundance and to determine whether individual mammal species are increasing or declining in numbers. Carry out a similar survey, with the same degree of sampling effort, in the nearest remnant of intact forest to determine what percentage of the original mammalian fauna re-colonises restored plots.

Sand traps make footprints clearer and easier to identify.

For a more quantitative assessment, use sand traps to record the density and frequency of mammal tracks. Clear away leaf litter from sample plots and sprinkle the soil surface with flour or sand. Mammals that walk over the sample plots will leave clear foot prints that can be measured and identified.

Last, anecdotal information can be collected from local people by interviewing. Use pictures in mammal identification hand books (rather than local names) to ask local people which mammal species they see frequently in the FTPS and remnant forest nearby and whether such species appear to be increasing or declining in abundance.

Monitoring 'recruit' tree species

In tropical forest ecosystems, most seeds are dispersed by animals. One of the main objectives of bird and mammal surveys is to determine whether restoration sites attract seed dispersers, but do the seeds that are brought in by animals actually germinate and grow into trees that contribute to overall forest structure? This question can be answered by periodic surveys to identify 'recruit' tree species (i.e. non-planted tree species that naturally re-colonise the site).

In forest ecosystems, the tree community is a good indicator of overall community biodiversity. Trees are the dominant ecosystem component, providing various habitats or niches for other organisms, such as birds and epiphytes. They are the base of the food web and account for most of the nutrients and energy in the ecosystem. The more diverse the tree community, the more likely it is that other elements of biodiversity will recover. Trees are easy to study. They are immobile, easy to find and relatively easy to identify.

What questions should be addressed?

- What tree species are present before forest restoration activities commence?
- What percentage of the tree species that comprise the target forest ecosystem re-colonise the restoration plots?
- Which forest herb species re-colonise the forest restoration plots and how soon after tree planting?

When and where should vegetation surveys be carried out?

Survey the area of the FTPS once it has been demarcated but before implementing activities that alter the vegetation (i.e. before preparing the site for planting). This provides the baseline data against which changes are compared. Thereafter, carry out vegetation surveys with the same sampling effort in both restoration plots and control plots and also in the nearest area of target forest to determine how many species from the target forest ecosystem re-colonise the restoration plots.

In seasonally dry climates, the character of the vegetation, particularly the presence or absence of annual herbs, varies dramatically with the seasons. To capture this variability, carry out vegetation surveys 2–3 times each year in the first few years after planting and subsequently at longer intervals. If you only have resources to carry out annual vegetation surveys, make sure they are always carried out at the same time of the year. Weeding in the first few years will of course disturb the vegetation. Therefore, carry out vegetation surveys just before weeding is scheduled.

Vegetation sampling methods

Establish permanent circular sampling units (SUs), across the entire study site, with equal numbers of SUs in restoration plots, controls (NPCs) and remnant target forest. Mark the centre of each sample unit with a metal or concrete (non-burnable) pole and use a 5 m piece of string to determine the perimeter of each SU. Position at least four SUs randomly in each 50 × 50 m plot. Species that are present outside SUs can also be recorded as being 'present in the environs'. Although not contributing to the diversity indices for the SUs described below, they will provide added qualitative evidence of biodiversity recovery.

Data collection

Within each SU, label every tree sapling that is taller than 50 cm. For each labelled tree, record: i) the label number; ii) whether the tree has been planted or naturally established; iii) the species name; iv) height; v) RCD (or GBH if large enough); vi) health score (see **Section 7.5**); vii) crown width; and viii) number of coppicing stems. Any tree seedlings or saplings that are shorter than 50 cm can be considered to be part of the ground flora.

A ground flora survey can be carried out at the same time, but for this survey, the radius of the SU can be reduced to 1 m. Record the names of all recognised species, including all herbs and vines and all woody trees, shrubs and climbers (shorter than 50 cm). Assign an abundance score to each species (e.g. use the Braun-Blanquet scale or the Domin scale).

When starting vegetation surveys, work with a professional botanist in the field if at all possible.

For species identifications, it is easier to work directly with an expert taxonomic botanist in the field rather than to collect voucher specimens for all of the species encountered and have them identified later at an herbarium.

Data analysis

Analyse the data for trees taller than 50 cm and the rest of the ground flora separately. Prepare a spreadsheet with the species listed in the first column (all species encountered during the entire survey in all SUs) and SU numbers in the top row. In each cell, enter the number of trees of each species in each SU (or the abundance score). The species list for the entire survey will be long and the number of species in each SU will be relatively low, so most of the values entered into the data matrix will be zero. However, the zero values must still be entered to allow calculation of indices of similarity and/or difference. Add data from each subsequent survey to the right of the current data, so that the data can be sorted into chronological order easily by column.

Begin by simply scanning the data and comparing species lists for restoration plots, non-planted controls and target forest. Which sun-tolerant pioneer species are the first to be shaded out by planted or naturally regenerating trees? Which species that are typical of the target forest type are the first to become naturally established in restoration plots? Are they wind dispersed or animal-dispersed? If the latter, which animal species are most likely to have brought their seeds into the restoration plots? Which of the planted tree species are most likely to have attracted these important seed-dispersing animals? Answers to these questions can be found without complex statistical analysis, and they will help you to decide how to improve the species mixtures and plantation design of future field trials in order to maximise biodiversity recovery rates.

One of the simplest ways to address the question of how similar the restoration plots are becoming to the target forest is to calculate a 'similarity index'. The simplest one to calculate is Sorensen's Index:

$$\frac{2C}{(RP + TF)}$$

... where RP = total number of species recorded in restoration plots, TF = total number of species recorded in target forest and C = number of species common to both habitats. When all species are found in both habitats, the value of Sorensen's index becomes 1, so biodiversity recovery can be represented by how closely the value of the index approaches 1 over time. Similarly, restoration plots can be compared with NPC plots, with the expectation that the index would decline over time as the restored forest becomes less similar to open degraded areas. In recently restored tropical forest plots, the index would be most suitable for comparing plant, bird or mammal communities.

Table 7.4. Example of how to calculate a similarity index.

	Restoration plots	Target forest
Species A	Present	Absent
Species B	Absent	Present
Species C	Present	Present
Species D	Present	Present
Species E	Absent	Present
	C	2
	RP	3
	TF	4
	Sorensen's index	**0.57**

Sorensen's index uses presence/absence data only and is easy to calculate, but it ignores the relative abundance of the species being recorded. More sophisticated 'resemblance functions', which take abundance into account, are described by Ludwig and Reynolds (1988, **Chapter 14**). These more complex calculations can be used (e.g. in cluster analysis and ordination) to classify the SUs according to how similar or different they are to each other.

CASE STUDY 5   Kaliro District

Country: Uganda

Forest type: *Albizia–Combretum* woodlands

Nature of ownership: Mainly privately owned small-scale farms.

Management and community use: Mixed farming, trees cut for charcoal production and timber, land cleared for cultivation.

Level of degradation: A substantial number of mature trees are cut down for harvesting or cleared for agriculture.

Background

This study was a part of my PhD research '*Ecology, conservation and bioactivity in food and medicinal plants in East Africa*', which investigated the seed germination and seedling growth of medicinal tree species and tested the applicability of the framework

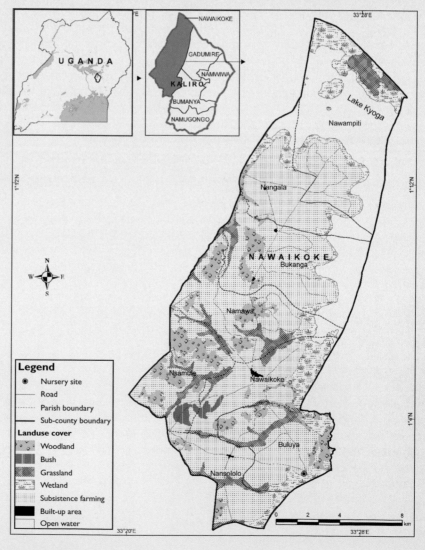

species method for conserving medicinal trees and the environment in Kaliro district, Uganda. It followed previous ethnobotanical studies to determine useful plant species, including medicinal ones (Tabuti *et al.*, 2003, Tabuti 2007).

Local traditional healers identified five woody medicinal plants as among the most important, but difficult to find: *Capparis tomentosa*, *Securidaca longipedunculata*, *Gymnosporia senegalensis*, *Sarcocephalus latifolius* and *Psorospermum febrifugum*. In a field survey, we found seeds of *C. tomentosa*, *S. longipedunculata* and *S. latifolius* and set up a direct seeding trial plot, but this method was not successful.

We therefore decided to experiment in Norway and achieved high seed germination in light and rapid early seedling growth of *Fleroya rubrostipulata* and *Sarcocephalus latifolius* (Stangeland *et al.*, 2007). We also wanted to establish new plot trials back in Uganda, but needed to find more effective field methods. Two of my colleagues, working in Thailand, told me about the framework species method used there (FORRU, 2008; www.forru.org). I adapted the technique and established a nursery according to FORRU's guidelines in March 2007. Some seeds were collected from the surrounding landscape, whereas others were procured from the National Tree Seed Centre, which also provided advisors to help with establishing the nursery.

Establishment of experimental plots

Although this study aimed to secure a supply of local medicinal plants, other useful tree species, some of them exotic, were also planted to encourage positive attitudes to tree planting: altogether 18 indigenous and 9 exotic tree species were studied (Stangeland *et al.*, 2011).

The criteria for the selection of species were as follows: i) medicinal woody species in high demand and/or becoming rare locally; ii) other useful tree species whose production might encourage a positive attitude among users (e.g. fruit trees, timber and fuel-wood trees); and iii) nitrogen-fixing tree species to improve the soil and reduce fertiliser requirements. Species selection was facilitated by previous local studies (Stangeland *et al.*, 2007; Tabuti, 2007; Tabuti *et al.*, 2009). Our objective was to test the applicability of the framework species approach in establishing multipurpose tree gardens growing products that would otherwise be harvested from woodlands.

Three groups of traditional healers provided land and took care of the seedlings after planting. One healer in each group established a multipurpose tree garden on his own land. The trees were not harvested during the first year when we monitored growth, but

Rose Akelo shows the seedlings to visitors at the inauguration of the nursery 04.08.2007 (Photo: T. Stangeland).

Nursery staff and traditional healers planting seedlings in March 2008 (Photo: T. Stangeland).

subsequently, the healers were free to cut the trees as needed. We provided seedlings and money for ploughing and fence material, while the groups of healers prepared land in March 2008, put up the fence, planted the seedlings and weeded the plots three times during the first rainy season. During the first rainy season, after tree planting in April 2008, beans were planted between the tree lines to provide some short-term benefit, increase motivation for weeding and increase soil fertility through nitrogen fixation.

How well did the FORRU methods work in Uganda?

Germination exceeded 60% for about half the species tested (48%). This contrasted with results from Thailand, where 80% of species had high germination rates (Elliott *et al.*, 2003). African tree species may thus have lower germination success or a greater

Monitoring survival and growth 13 months after planting. From the left Patrick Nzalambi, Joseph Kalule, Alexander Mbiro, Torunn Stangeland and Lucy Wanone (Photo: T. Stangeland).

need for pre-treatment than Asian species. Thirteen months after planting, seedling survival was satisfactory and comparable with the results from Thailand (Elliott *et al.*, 2003). Almost two thirds (63%) of the planted tree species achieved survival rates in excess of 70%, despite a severe drought in 2009. Height growth was also good, with one-third of species achieving excellent growth (>160 cm tall) and 30% achieving acceptable growth (>100 cm tall) 13 months after planting.

Eleven of the 27 tree species tested qualified as 'excellent' framework species (Stangeland *et al.*, 2011). Eight more species qualified as 'acceptable'. All of these species can be recommended for restoration and multipurpose tree garden planting. Eight were ranked as 'marginally acceptable'.

Potential of the framework species method in Africa

Our experience suggests that there is considerable potential for applying the framework species approach in Africa. The human populations of East African countries have more than trebled during the past 40 years, resulting in immense pressure on land for cultivation. More than 80% of people still use firewood or charcoal to cook their food, a demand met largely by plantations of exotic tree species, whilst indigenous trees have declined and become vulnerable to extinction. We found the framework species method to be practical and cost-effective. The groups of healers involved in our work have become much more interested in raising seedlings and planting trees. In fact, when we visited the site in March 2011, we found that the two groups in Nawaikoke had merged and bought land for their own nursery, building on the experience from the project.

By Torunn Stangeland

CHAPTER 8

SETTING UP A FOREST RESTORATION RESEARCH UNIT (FORRU)

Forest restoration and research go hand-in-hand. Throughout this book, we have emphasised the need to learn from restoration projects, both successful and unsuccessful, and have provided standard research protocols that will enable you to do so. In this chapter, we provide advice on setting up a dedicated Forest Restoration Research Unit (FORRU) in which to carry out the research, to organise and integrate the information derived from it and to implement education and training activities. The aim should be to place the results of the research into the hands of all those involved in forest restoration, from school children to community groups and government officials.

8.1 Organisation

Who should organise a Forest Restoration Research Unit?

The success of a Forest Restoration Research Unit (FORRU) depends on strong support from a respected institution. Without a long-term, consistent host, it is difficult to attract funding and ensure local participation in forest restoration programmes. A FORRU is best organised by a recognised institution that has established administrative procedures. This could be the national government forestry department, a university faculty or department, a botanical garden, a seed bank, a government-run research centre or a recognised NGO.

Strong institutional support is essential for establishing and maintaining good relations between the diverse organisations involved, i.e. stakeholders such as community groups, government departments, NGOs, funding agencies, international organisations, technical advisers and educational establishments. Clear and mutually acceptable arrangements governing the management of a FORRU that are laid down by the institution can both ensure its smooth running and help to prevent disputes among stakeholders.

Staffing a FORRU

An inspirational leader, a committed conservationist with experience in tropical forestry, is required to run a FORRU. In addition to having a scientific education and relevant experience, he or she should be skilled at project administration, personnel management and public relations. If a FORRU is hosted by a university, the unit leader could be a senior scientist from the faculty staff. In a government forestry research centre, a senior forestry officer could take on this role. Initially, part-time secretarial assistance might be adequate to support the leader, but as the unit grows, full-time administrative help will become necessary.

Access to a professional plant taxonomist and herbarium facilities is essential to ensure that tree species are identified accurately. Although the host organisation might not have a taxonomist on the payroll, it is essential to build a good relationship with a taxonomist who can be called upon to identify plant specimens as needed, perhaps on a part-time basis.

When starting a FORRU, two key research posts must be filled:

- a nursery manager will be needed to implement nursery research, manage data, supervise nursery staff and ultimately produce good-quality trees for field trials;
- a field officer must be employed to maintain and monitor field trials, as well as to process field data. Initially, this post might be part-time, but it will become permanent as the field-trial plot system is expanded.

Forest restoration research is not 'rocket science' and, with a little training, anyone can carry out the research protocols described in this book. So apart from the key posts described above, the rest of the unit staff can be recruited from among the local community, regardless of educational qualifications. Local people are more likely to collaborate with a FORRU if some of them are directly employed by it and if they are the first to benefit from the new knowledge and skills generated by it. Local people might be employed full time as nursery or field research assistants or on a part-time or seasonal basis, when extra work is required, such as when preparing for planting events and maintaining planted trees. Including local people in monitoring, so that they share in the project's success, is most important.

As the project proceeds, the dissemination of research results directly to those responsible for implementing forest restoration becomes increasingly important. An education and outreach programme must be designed and implemented. Educational materials must be produced, workshops and seminars organised, and someone must be available to deal with the inevitable stream of interested visitors to the unit. To begin with, the research team might be able to handle some education work, but eventually, an education manager should be recruited; otherwise, research outputs will decline as the unit's research staff are distracted from their main work.

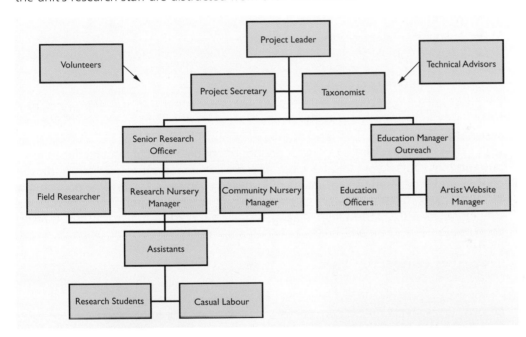

A suggested organisational structure for a FORRU. Volunteers and technical advisors can have inputs at all levels.

In addition to routine research on tree propagation and planting (carried out by the full-time staff), a FORRU provides excellent opportunities for research students to carry out thesis projects on more specialised aspects of restoration. For example, students might investigate the influence of mycorrhizas on tree growth, the best ways to control pests in the nursery, which tree species attract seed-dispersing birds or foster establishment of tree seedlings, or carbon accumulation in restored areas … to name just a few possible studies. It is important that the FORRU is freely open to students and researchers from other institutions. In this way, the unit quickly generates an impressive list of publications that can be used to encourage further funding and institutional support.

Students measuring carbon accumulation in an established forest restoration plot. A FORRU's nursery and plot system provide endless possibilities for research students.

Training requirements

It is unlikely that anyone applying to work at a FORRU will possess the full skill set necessary to develop efficient forest restoration techniques. Therefore, most new recruits will require training in at least some of the following skills:

- project management and administration, proposal writing, reporting and accounting;
- experimental design and statistics;
- tropical forest ecology;
- plant taxonomy;
- seed handling;
- nursery management and tree propagation techniques;
- managing field trials and silviculture;
- biodiversity survey techniques;
- environmental education;
- working with local communities.

Initially, the project leaders themselves must provide adequate training for all newly recruited FORRU staff, but as the levels of skill among the staff rise, nursery or field managers can begin to train assistants and casual staff. In addition to this book, the six-volume series: "Tropical Trees: Propagation and Planting Manuals" published by the Commonwealth Science Council, London, might be a helpful resource for training programmes. Outside organisations can also provide important advice or run training courses for FORRU staff. An advantage of involving overseas advisors is the opportunity to forge collaborative links, which can result in joint projects supported by international funding agencies. Opportunities may also arise for FORRU staff to attend training courses at other institutions, both locally and abroad.

Staff from the Royal Botanic Gardens, Kew train FORRU-Cambodia staff in seed-handling techniques.

Facilities

A FORRU comprises a range of facilities that are needed to the conduct the research activities described in **Sections 6.6**, **7.5** and **7.6**. These include:

- access to an area of the target forest type (see **Section 4.2**);
- a phenology trail through the target forest type (see **Section 6.6**);
- access to an herbarium;
- a research tree nursery in which tree propagation is studied and trees produced for field trials (see **Section 6.6**);
- a community tree nursery in which the feasibility of tree propagation techniques are tested by local stakeholders;
- office facilities for project administration, data handling, library and specimen storage etc.;
- a field trial plot system (see **Section 7.5**);
- an education and outreach sub-unit (see **Section 8.6**).

8.2 Working at all levels

Establishing a FORRU requires working with people from all sectors of society from high-ranking government officials to local villagers.

Contribution of FORRUs to national forest policy

To satisfy funding agencies, as well as the administrators of FORRU host institutions, it may be necessary to justify the establishment of a FORRU in terms of its contributions to:

- implementing national policies on forestry or biodiversity conservation;
- meeting the obligations of governments under international agreements.

If a government is a party to the Convention on Biological Diversity (CBD) (www.cbd.int), it is obliged to implement policies and programmes to meet the provisions of the convention; for example, it might have made commitments to:

- "rehabilitate and restore degraded ecosystems and promote the recovery of threatened species…" (Article 8 (f));
- "support local populations to develop and implement remedial action in degraded areas, where biological diversity has been reduced…" (Article 10 (d));
- "promote and encourage research which contributes to the conservation and sustainable use of biological diversity…" (Article 12 (b)).

Furthermore, under the terms of the convention, each member country must prepare a National Biodiversity Strategy and Action Plan (NBSAP). These plans usually include provisions for the restoration of forest ecosystems for biodiversity conservation, which can be used to justify the establishment of a FORRU. The full text of the CBD can be downloaded from www.biodiv.org/convention/articles.asp and NBSAPs for most countries can be found at www.cbd.int/nbsap/search/.

FORRUs can contribute towards achieving the goals of national biodiversity strategies and action plans, as required under the Convention on Biological Diversity.

If the country in which you are working is a member of the International Tropical Timber Organization (ITTO), you should consult "ITTO guidelines for the restoration, management and rehabilitation of degraded and secondary tropical forests" (www. itto.int/policypapers_guidelines). Although this document does not have the legal weight of an international convention, it does represent an international consensus of opinion that national organisations tend to respect. It includes 160 recommended actions, many of which could be supported by information generated from a FORRU.

Most countries have published national forest policies that stipulate forestry programmes and projects over periods of 5–10 years. Many of these policy statements include recommendations about the rehabilitation of degraded areas, which can be quoted to justify the establishment of a FORRU.

Finally, the UN's REDD+[1] and various other carbon-trading schemes (both voluntary and obligatory under the UN's Kyoto Protocol, e.g. the Clean Development Mechanism) aim to limit the accumulation of carbon dioxide in the atmosphere by channelling funds from carbon emitters into projects that absorb carbon or reduce emissions (see **Section 1.4**). Forestry-related carbon sequestration projects are now required to conserve biodiversity and there is therefore a growing requirement for the kind of research outputs generated by a FORRU.

[1] www.scribd.com/doc/23533826/Decoding-REDD-RESTORATION-IN-REDD-Forest-Restoration-for-Enhancing-Carbon-Stocks

Working with protected area staff

As biodiversity recovery is one of the principal aims of forest restoration, nature reserves and national parks are ideal locations for FORRU nurseries and field trials. Support from the person in charge of a protected area (PA) and his/her staff should be easier to obtain after national and local government officials have been persuaded of the value of a FORRU. A close working relationship must then be cultivated between the PA authority and FORRU staff.

The PA authority might be able to grant permission for the construction of a nursery and the establishment of field trials on PA land, provided such activities are in accordance with the area's management plan. This authority might also be able to provide staff or casual labour to assist with the activities of the unit, as well as other logistical support. When drafting funding applications, consider including the salary of one or more members of the PA staff to be seconded to the FORRU. If field trials contribute to increased forest cover, extent or quality within a PA, then the PA staff will probably want to be involved in tree planting events and in the maintenance of the planted trees. Vehicles owned by the PA might be available for transporting trees, nursery supplies and planting materials around the area. Sometimes, the full cost of providing such help can be charged to the FORRU budget, but some PAs might choose to absorb the costs into their central budget. In such cases, include a contribution to the PA overheads in funding applications.

Support from PA staff can be maintained by inviting them to attend joint workshops and training programmes at the FORRU nursery and field plots. Make sure that the Head of the PA and his/her staff are also invited to seminars and conferences at which results from the FORRU are presented and that the PA is acknowledged in all published outputs. Finally, provide the Head of the PA with regular progress reports, even if they are not requested. This will help to ensure continuity when staff changes occur at the PA headquarters.

National park officers join with local community members and FORRU-CMU staff to plant an area within Doi Suthep-Pui National Park.

The importance of working with communities

The majority of PAs are inhabited. Developing working relationships with communities is therefore essential to prevent misunderstandings about the aims of the work, and to diffuse any potential conflicts over the positioning of forest restoration plots. A good relationship with local people provides a FORRU with three important resources:

- indigenous knowledge;
- a source of labour;
- an opportunity to test the practicability of research results.

Indigenous knowledge helps with the selection of candidate framework species. Local people often know which tree species colonise abandoned cultivated areas, which attract wildlife and where suitable seed trees are located (see **Section 5.3**).

The establishment of field plots, maintenance and monitoring of planted trees, and fire prevention are labour-intensive activities. Local people should be the first to be offered such work and to benefit from payments for it. This helps to build a sense of 'stewardship' of the forest restoration plots, which increases support for the work at the community level. Thus, planted trees are more likely to be cared for and protected.

The species choices and propagation methods developed by a FORRU must be acceptable to local people. Establishing a community tree nursery, where local people can test the techniques developed by research, is therefore highly advantageous

Even the youngest members of a community can participate in restoring forests. With a long future ahead of them, children have most to gain from environmental recovery.

and provides another opportunity for local people to gain income from the project. In addition, community nurseries can produce trees close to planting sites, thereby reducing transportation costs.

Developing a close working relationship with people who live within a PA is not always easy, especially if they feel disenfranchised by the establishment of the PA. Local communities are, however, often the first to benefit from the restoration of their local environment, particularly from the re-establishment of supplies of forest products and the improvement of water supplies. A FORRU can encourage local people and PA staff to work together to establish field plots and nurseries, which can help to build closer ties between them. This benefits both local people and PA management. Stressing such benefits can help to persuade local people to participate in the activities of a FORRU.

Hold frequent meetings with the village committee to ensure that the local community is involved in all stages of a FORRU programme, particularly in the positioning of field experiments, so as not to conflict with existing land uses. Appoint someone from the local community to be the main contact person who relays information between FORRU staff and villagers. In funding applications, make provisions for the employment of local people, both in running a community tree nursery and as casual labour for the planting, maintenance and monitoring of tree planting plots and for fire prevention and suppression. Invite local people to meet visitors to the project, so that they are aware of growing interest in their work and involve them in media coverage of the project, so that they benefit from a positive public image.

Working with foreign institutes and advisors

Expertise and advice from foreign organisations can greatly accelerate the establishment of a FORRU and prevent the duplication of work that has already been done elsewhere. Foreign institutions might also be able to contribute to FORRU workshops on nursery production techniques, seed handling or other topics. Some institutions might be able to accept FORRU staff for short periods of training. Advisors could also be engaged as required to provide expertise in specialist disciplines, such as plant taxonomy.

It is unlikely that a FORRU will have the funds necessary to pay international consultancy fees to foreign experts. Consequently, it is important to build collaborative partnerships so that the costs of involving foreign advisors can be covered by their own institutions, by international funding agencies, or from collaborative project grants.

A further benefit of involving foreign institutions and their staff is that they have access to national sources of funding that are only available to projects working in partnership with the donor country. It is important to work with foreign advisors who understand the ethos of the FORRU and who do not try to change the direction of the work to suit preconceived ideas that do not accord with the ecological or socioeconomic conditions of the country where the FORRU is operating.

Box 8.1. Politics and public relations: alternative motives for participation in forest restoration.

Ban Mae Sa Mai is the largest Hmong village in northern Thailand with 190 households and a total population of more than 1,800. The Hmong is one of several ethnic minorities in northern Thailand that are collectively known as 'hill tribes'. Ban Mae Sa Mai village was originally founded at 1,300 m elevation, but was moved down the valley to its present location in 1967, after deforestation caused the village water supply to dry up. The relocation left the villagers with a strong sense of the link between deforestation and loss of water sources.

In 1981, the village and surrounding farmland were included within the boundaries of the newly declared Doi Suthep-Pui National Park. This meant that the villagers faced a legal threat of eviction as they had no formal land-ownership rights.

To avoid possible enforcement of this law, a few of the villagers formed the 'The Ban Mae Sa Mai Natural Resources Conservation Group' and built a community-wide consensus to gradually reduce the cultivation of the upper watershed and to re-plant the area with forest trees. The village committee designated a remnant of degraded primary forest above the village as 'community forest', thereby protecting three springs that supply both the village and the agricultural land below it with water.

The villagers also decided to contribute to a national project to celebrate His Majesty King Bhumibol Adulyadej's Golden Jubilee, which aimed to restore forest to more than 8,000 km² of deforested land nationwide. They agreed with the Park Authority that they would phase out cultivation of a 50-ha area in the upper watershed and replant it with forest trees; in return, they would be allowed to intensify agriculture in the lower valley. The Royal Forest Department provided eucalypt and pine trees for planting in the upper watershed, but the villagers were disappointed with the limited species choice and the results. So, when FORRU-CMU approached the village committee in 1996 with a proposal to test the framework species method in trial plots near the village, the committee enthusiastically agreed (**Case study 6**). Villagers collaborated with all aspects of the project, from planning to seed collection, growing trees in a community tree nursery, tree planting, maintenance, fire prevention and monitoring.

Village children show off the trees they have potted in the community tree nursery. Eight months later, they helped to plant them in the watershed above the village.

In 2006, questionnaires were used to evaluate the villagers' perceptions of the project and to explore their motivation for participation. Although villagers expressed general satisfaction with the project's tangible outputs, they valued the project's impact on improving relationships most highly: both relationships within the village and external relationships with the authorities and the general public.

Around 80% of respondents agreed that the project had reduced internal social conflicts over shortages in natural resources, particularly water. Interviewees stated that they had noticed an improvement in water quality and increased water quantity (particularly during the dry season), as well as a reduction in soil erosion and a better local climate.

Box 8.1. Politics and public relations

Box 8.1. continued.

The majority of villagers appreciated that the project had resulted in an improved relationship between the village and the National Park Authority, with which the villagers had previously been in conflict, and consequently they felt more secure living within the park. Villagers also highly appreciated that the project had improved their public image by attracting positive media coverage. This enabled the village to receive other forms of support, such as that from the Sub-district Administration Organisation (90% of villagers recognised this benefit) and from local units of Royal Forest Department and the National Park Authority (60% of villagers listed this as a benefit). Estimates of the amount of support attracted from these other sources varied from US$ 360 to US$ 1,070 per year.

In general, benefits that affected income were less appreciated than those that affected relationships. Nevertheless, the villagers appreciated the salaries and daily labour payments for caring for the reforestation plots and support for community development, i.e., improvements to road access, water supply, fire prevention work and religious ceremonies.

About 40% of interviewees agreed that the numbers of naturalists and/or ecotourists visiting the village had increased markedly in the previous few years, mostly because of the forest restoration programme, and that such ecotourism was generating an income of approximately US$ 350–1,250 per year, mostly through the provision of accommodation.

With regard to non-timber forest products, the villagers recognised that forest restoration had contributed to the increased production of products such as bamboo shoots and stems, banana leaves and flowers, edible leafy vegetables (mostly young leaf shoots from trees), other flowers and fruits (mostly from trees) and some mushrooms.

Tangible benefits (US$)	US$/year/household
Direct employment by the project	25.50
Attracted funds from local government	3.83
Ecotourism income	4.46
Forest products	208.93
Mean increased income per household	**242.72**

Intangible benefits	% of interviews attributing a high value
Improved relationships with:	
Forestry Department	74
NGOs	85
Others in the community	93
Improved community image	86
Improved water quality	83
Improved ability to attract funding from local government	90

Box 8.1. continued.

The marginal, steep cabbage fields above the village have largely been restored back to forest. Intensification of agriculture in the lower valley has improved the villagers' livelihoods, and was made possible by improved water quality supply from the restored watershed.

A restoration plot, established on an abandoned cabbage field, photographed 16 months after planting 30 framework tree species.

The village nursery and plots have now become vital facilities for education, attracting frequent visitors and workshops. Representatives from other communities visit the village to find out how they too, can establish successful forest restoration projects. Thus, the villagers of Ban Mae Sa Mai have converted their cabbage fields into a classroom for forest restoration, while simultaneously securing their water supply and improving both their public image and their livelihoods. Overall, this collaboration between FORRU-CMU and the Ban Mae Sa Mai community has demonstrated how scientific research and the needs of a community can be combined to create a model system for environmental education.

The village received an award from the Thai government recognising their efforts in restoring the forest around their village. An improved relationship with the authorities was a major motivational factor in this project.

8.3 Funding

Obtaining funding

If a FORRU is established within an existing, centrally funded, institution, it may be possible to make use of existing staff and facilities to initiate a research programme. As the research programme expands, however, independent funding must be found.

Funding sources for forest restoration projects have already been discussed in **Section 4.6** and all are suitable for funding a FORRU. As FORRUs are essentially academic research facilities, however, they may also draw on research grants, particularly if they are based in a university. For financial stability, it is best to maintain a varied 'portfolio' of different sources of research funding by dividing the work of the unit into clearly defined research areas (e.g. forest ecology, tree propagation and biodiversity recovery), each one supported by a different funding mechanism with different start and finish dates. In this way, the end of a single grant period does not result in staff redundancies and the collapse of the unit.

Research funding can be obtained from a wide range of different organisations. Multinational or international aid agencies (e.g. the European Union (EU) or the International Tropical Timber Organization (ITTO)) can provide large grants for large projects, but they usually impose complicated and time-consuming application and reporting procedures in order to maintain accountability and transparency to their donor governments. Therefore, only organisations with highly trained administrative staff who are capable of coping with the cumbersome bureaucratic procedures can expect to be successful in securing international funds.

Grants provided by individual foreign governments can also be very generous (e.g. under the UK's Darwin Initiative or Germany's Gesellschaft für Internationale Zusammenarbeit (GTZ)). They are usually administered through institutions in the donor country, which may also receive some support from the grant. The involvement of foreign advisors from the donor country is often a condition of the grant. This option is suitable when a good working relationship with an institution in the donor country has already been developed and the need for the involvement of foreign experts has been clearly identified. Grants from national mechanisms that support research in the project's own nation might be easier to obtain and require less bureaucracy than foreign funding, although the amounts granted are generally less.

The "CPF Sourcebook on Funding for Sustainable Forest Management", mentioned in **Section 4.6**, also covers many agencies that support forestry research (www.cpfweb.org/73034/en/).

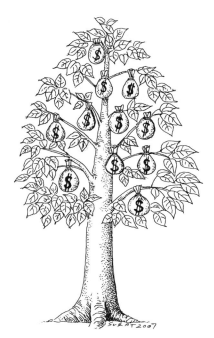

Unfortunately, money doesn't grow on trees, so fundraising, accounting and reporting are vital activities when running a FORRU. Luckily, interest in funding forest restoration, particularly to mitigate global climate change, is growing. Major funders should be interested in supporting research and in ensuring that large-scale projects are implemented using the most cost-effective methods.

8.4 Information management

Computer databases

Once established, a FORRU generates large amounts of data from diverse sources. One of the unit's most important roles must be to organise and integrate these data to generate reliable advice for practitioners. Computer databases provide the most appropriate way to i) store large diverse data sets and ii) analyse them to answer a wide range of different questions. For example, if a site at 1,300 m elevation becomes available for forest restoration, the questions asked of a database might include:

- What tree species grow at similar sites and at similar elevations?
- Of those species, which ones have fleshy fruits that attract seed-dispersing animals?
- Of those species, which ones will be fruiting in the next month so that seed collection might commence?
- Of those species, which ones have previously germinated well in the nursery?

To generate lists of species that match specified criteria, it is necessary to construct a relational database that integrates all of the data produced by a FORRU together with published data and indigenous local knowledge. Spreadsheets do not allow the sophisticated search, sort and integration facilities of dedicated database programmes, and the larger spreadsheets become, the more difficult they are to work with. Therefore, most critical data must be extracted from spreadsheets (such as those described in **Sections 6.6, 7.5** and **7.6**) and re-entered into a relational database system.

Who should set up the database

Setting up a relational database system involves intensive collaboration between the FORRU research staff, who have first-hand knowledge of the data being generated and know how they would like to analyse it, and a colleague or consultant with specific experience of working with the chosen database programme.

Database structure

Databases are like sophisticated card index systems. A 'database file' is the equivalent of one box, containing many cards. A 'record' is the equivalent of one card and a 'field' represents one of the headings on the card and the information associated with it. It is not practical to store all of the information available about a species in a single record: for some types of information, there will be a single entry (e.g. the name and characteristics of a tree species, which do not change), whereas for other types of information there may be many entries (e.g. germination trial results for each batch of seed). Therefore, the database consists of several database files, each one storing a particular category of information.

In addition, records referring to a particular species in each database file should be linkable with records referring to the same species in all other database files. Links are achieved by assigning link codes to each record; these enable records referring to the same species to be joined, regardless of which database file they are in. The most convenient link codes are the species number (S. no.) and seed-batch number (b. no.)

(see **Section 6.6**), so it is of the utmost importance that the system of species and batch numbers is maintained throughout the research process, from seed collection to planting. These identification numbers are crucial to data integration, so they should appear on all datasheets and plant labels, both in the nursery and in the field. The database system must be able to recognise these codes and group together all records that share the same codes from all database files. Thus, the database should be able to generate species reports, listing all of the recorded information on each species. It is not a good idea to use the species names (or abbreviations of them) as link codes because it may take time to identify some species correctly, and even then, taxonomists are constantly changing the scientific names of plants.

On the following pages, we suggest some record structures that contain the most basic information generated by a FORRU. This basic database structure can be expanded with new fields and database files as required. Consider adding files to hold summary data on seed storage experiments, the attractiveness of each species to wildlife, or indigenous knowledge about the uses of each tree species. But be aware that data entry is time consuming, so before embellishing the database with extra fields or files, first consider whether the data entered will actually be used to support decision-making — whether the outputs really justify the data input time.

Foresters in the Philippines learn about data management before setting up their own research tree nurseries and restoration demonstration plots at universities across the country.

Database software

Database programmes vary in terms of their sophistication and ease of use. Unfortunately, the more sophisticated the programme is, the less user-friendly it is. Microsoft Access is probably the most widely used database system, but it is expensive and several open-source database programmes are available for free (e.g. Open Office).

Whichever package you select, make sure that it supports the essential features listed below:
- the ability to link records in different database files that refer to the same species;
- searches within fields for text occurring in any position in the field (e.g. find September (i.e. "sp") occurring anywhere within a list of fruiting months.... "jl ag sp oc nv");

- the ability to generate information in one field from calculations using numbers stored in other fields, e.g. median length of dormancy could be calculated by subtracting the date of seed collection from the median date on which the seed germinated.

Also consider whether the database package supports the script of your language and/or the insertion of images (if needed). Database technology has other applications for a FORRU besides storing experimental data. Consider constructing a database that stores the names and contact details of everyone who has contact with the unit, so that you can easily organise invitations to workshops and other educational events, as well as a circulation list for the unit's newsletter. Another database could be used to catalogue books that are kept in the unit's library or photographs taken by the unit's staff.

Files, records and fields

Database file "SPECIES.DBF"

One record for each tree species. This file stores basic information about each species, which can be linked to records in other database files through the "SPECIES NUMBER:" field. Most of this information can be retrieved from a flora. Modify the list of flowering and fruiting months, as data from the phenology survey become available (see **Section 6.6**).

SPECIES NUMBER: *e.g. S71*

SCIENTIFIC NAME: *e.g. Cerasus cerasoides* **FAMILY:** *Rosaceae*

LOCAL NAME: *Nang Praya Seua Krong*

EVERGREEN/DECIDUOUS: *D*

ABUNDANCE: *e.g. 0 = Probably extirpated; 1 = Down to a few individuals, in danger of extirpation; 2 = Rare; 3 = Medium abundance; 4 = Common, but not dominant; 5 = Abundant.*

HABITAT: *develop your own codes for forest types e.g. egf = evergreen forest; species may occur in more than one forest type, list them all in any order.*

LOWER ALTITUDE: **UPPER ALTITUDE:** *from direct observations*

FLOWERING MONTHS: *ja fb mr ap my jn jl ag sp oc nv dc*

FRUITING MONTHS: *ja fb mr ap my jn jl ag sp oc nv dc*

LEAFING MONTHS: *ja fb mr ap my jn jl ag sp oc nv dc*

FRUIT TYPE: *e.g. dry/fleshy drupe/nut/samara etc.*

DISPERSAL MECHANISM: *e.g. wind/animal/water etc.*

NOTES:

ENTRY INTO DATABASE CHECKED BY: **DATE:**

Database file "SEED COLLECTION.DBF"

This database contains one record for each batch of seeds collected. Records for the different seed batches for each species are linked to a single record in "SPECIES.DBF" by the "SPECIES NUMBER:" field. Transcribe information from seed-collection data sheets (see **Section 6.6**).

SPECIES NUMBER: *e.g. S71*   **BATCH NUMBER:** *e.g. S71b1*

COLLECTION DATE: **TREE LABEL NUMBER:**   **TREE GIRTH:**

COLLECTED FROM: *e.g. ground/tree*

LOCATION: *e.g. Rusii Cave*   **GPS CO-ORDINATES:**

ELEVATION:

FOREST TYPE: *develop your own codes for forest types e.g. egf = evergreen forest.*

NO. SEEDS COLLECTED: **STORAGE/TRANSPORT DETAILS:**

SOWING DATE:

VOUCHER SPECIMEN COLLECTED: *e.g. Yes/no*

NOTES FOR HERBARIUM VOUCHER LABEL:

ENTRY INTO DATABASE CHECKED BY: **DATE:**

Database file "GERMINATION.DBF"

This database contains one record for each treatment applied to each sub-batch of seeds. Multiple records for each species or each batch, respectively, are linked to a single record in "SPECIES.DBF" by the "SPECIES NUMBER:" field and to a single record in "SEED COLLECTION.DBF" by the "BATCH NUMBER:" field. Extract data from germination data sheets (see **Section 6.6**) Use mean values from all replicates.

SPECIES NUMBER: *e.g. S71*   **BATCH NUMBER:** *e.g. S71b1*

PRE-SOWING TREATMENT: *enter only one treatment (or control) e.g. scarification.*

MEDIAN SEED GERMINATION DATE: *date on which half the seeds germinated.*

MLD: = GERMINATION.DBF/MEDIAN SEED GERMINATION DATE: *minus* **SEED COLLECTION.DBF/SOWING DATE:**

MEAN FINAL PERCENT GERMINATION:

MEAN FINAL PERCENT GERMINATED BUT DIED: *as a percentage of the number of seeds that were sown.*

ENTRY INTO DATABASE CHECKED BY: **DATE:**

Database file "SEEDLING GROWTH.DBF"

This database contains one record for each treatment applied to each batch. Multiple records for each species are linked to a single record in "SPECIES.DBF" by the "SPECIES NUMBER:" field. The record for each batch of seeds collected is linked to a single record in "SEED COLLECTION.DBF" by the "BATCH NUMBER:" field. Extract data from seedling growth data sheets (see **Section 6.6**).

SPECIES NUMBER: *e.g. S71*   **BATCH NUMBER:** *e.g. S71b1*

POTTING DATE:

TREATMENT: *enter only one treatment (or control) e.g. Osmocote once every 3 months.*

NO. OF SEEDLINGS: *total number of seedlings subjected to treatment (combined replicates).*

SURVIVAL: *as a percentage, between potting and just before planting out.*

TARGET DATE: *date on which mean seedling height reaches target value (e.g. 30 cm for fast-growing pioneers and 50 cm for slower growing climax tree species). Derived from interpolation between points on the seedling growth curve (p. 207).*

OPT. PLANTING DATE: *first optimum planting out date after the target date (usually 4–6 weeks after the first rains).*

TNT: *total nursery time =* **SEEDLING GROWTH.DBF/OPT.PLANTING OUT DATE:** *minus* **SEED COLLECTION.DBF/COLLECTION DATE:**

OST: *over storage time =* **SEEDLING GROWTH.DBF/OPT. PLANTING OUT DATE:** *minus* **SEEDLING GROWTH.DBF/TARGET DATE.** *This value is useful for identifying species for seed storage experiments.*

RGR HEIGHT: *relative growth rate, based on height measurements from just after potting to just before planting out.*

RGR RCD: *relative growth rate based on root collar diameter measurements from just after potting to just before planting out.*

ROOT/SHOOT RATIO: *from sacrificed plants just before planting out.*

NOTES ON HEALTH PROBLEMS: *descriptions of pests and diseases etc.*

ENTRY INTO DATABASE CHECKED BY: **DATE:**

Database file "FIELD PERFORMANCE.DBF"

This database contains one record for each silvicultural treatment applied to each batch. Multiple records for each species or each batch can be linked to a single record in "SPECIES.DBF" by the "SPECIES NUMBER:" field, and to records in the other database files by the "BATCH NUMBER:" field. Extract data from the field data analysis spreadsheets (see **Section 7.5**). Insert mean values for combined replicates for a single silvicultural treatment.

SPECIES NUMBER: *e.g. S71*   **BATCH NUMBER:** *e.g. S71b1*

PLANTING DATE:

FTPS LOCATION: **PLOT NUMBER(S):**

TREATMENT: *enter only one treatment (or control) e.g. cardboard mulch.*

NO. OF TREES PLANTED: *total number of trees planted and subjected to treatment (combined replicates).*

MONITORING DATE 1: *just after planting.*

SURVIVAL 1: *as a percentage.*

MEAN HEIGHT 1:   **MEAN RCD 1:**   **MEAN CANOPY:**

WIDTH 1:

MONITORING DATE 2: *after first rainy season.*

SURVIVAL 2: *as a percentage.*

MEAN HEIGHT 2:   **MEAN RCD 2:**   **MEAN CANOPY:**

WIDTH 2:

MEAN RGR HEIGHT 2:   **MEAN RGR RCD 2:**

MONITORING DATE 3: *after second rainy season.*

SURVIVAL 3: *as a percentage.*

MEAN HEIGHT 3:   **MEAN RCD 3:**   **MEAN CANOPY:**

WIDTH 3:

MEAN RGR HEIGHT 3:   **MEAN RGR RCD 3:**

MONITORING DATE 4: *add additional fields as needed for each subsequent monitoring event.*

ETC......

NOTES: *descriptions of pests and diseases etc. observed.*

ENTRY INTO DATABASE CHECKED BY: **DATE:**

8.5 Selecting suitable tree species

A relational database has many functions, but one of the most useful is to select the most suitable tree species to restore forest to any particular site. For degradation stages 3 to 5 (see **Section 3.1**), trees species should be selected according to the criteria that define framework species and/or nurse crop species (**Table 5.1** and see **Section 5.5**), combined with any other situation-specific considerations. This selection can be very subjective or involve complex analyses of the database. Therefore, we suggest two simple semi-quantitative methods to facilitate the process of species selection: the 'minimum standards' approach and a 'suitability index', which is based on a ranked scoring system. They may be used independently or in tandem, using minimum standards to create a short-list of species that is subsequently ranked by suitability index. These two methods make best use of the data available, while retaining the flexibility required to meet the various objectives of different projects.

Applying minimum acceptable standards of field performance

The most important field-performance criterion is survival rate after planting out. No matter how well a species performs in other respects (e.g. it might have rapid growth and/or be attractive to seed-dispersers), there is not much point in continuing to plant it if its survival rate after 2 years falls below 50% or so. Additional minimum acceptable standards can be applied to growth rates, canopy width, suppression of weed cover and so on, but all are subordinate to survival. The values of the minimum acceptable standards are largely subjective, although sensible values can usually be decided upon by scanning the data sets and looking for the divisions that set species apart, particularly values that contribute towards canopy closure within the desired timeframe.

Extract field data collected after 18–24 months (at the end of the second rainy season in seasonal forests) from the database into a spreadsheet with species names in the left-hand column, with data on the selected performance criteria arranged in columns to the right. Use mean values from planted control plots (see **Section 7.5**) or mean values from whichever silvicultural treatment produced the best results.

Bear in mind that whether or not a species exceeds minimum standards can depend on i) the silvicultural treatments applied, ii) climatic variability (some species may exceed the standard in one year but not the next) and iii) site conditions. So a species need not necessarily be rejected if it marginally fails to achieve the minimum standard in a single trial. Intensified site preparation or silvicultural treatments could convert a rejected species into an acceptable one.

The application of minimum standards results in three categories of species:
- category 1 species: those that fall short of most or all minimum acceptable standards (i.e. rejected species);
- category 2 species: those that exceed some minimum standards but fall short of others, or those that fall short of several standards by only a small amount (i.e. marginal species);
- category 3 species: those that greatly exceed most or all minimum standards (i.e. excellent or acceptable species).

Category 1 species are dropped from future plantings. Category 2 species could either be rejected or subjected to further experimentation to improve their performance (e.g. to improve the quality of the planting stock or to develop more intensive silvicultural treatments), while category 3 species are approved for use in future restoration work.

Example:

Three minimum standards are applied to field-performance data collected at the end of the second rainy season after planting:

- survival >50%;
- height >1 m (as seedlings should be planted when 30–50 cm tall, this represents a more than doubled height);
- crown width >90 cm (i.e. the crown has obtained more than half the width required to close canopy at a tree spacing of 1.8 m (equivalent to 3,100 trees per hectare)).

In the table below, data that fail to meet minimum standards are indicated in red.

Species	% Survival	Mean height (cm)	Mean crown width (cm)	Category	Action
S001	89	450	420	3	Accept
S009	20	62	65	1	Reject
S015	45	198	255	2	Research to increase survival
S043	38	102	20	1	Reject
S067	78	234	287	3	Accept
S072	90	506	405	3	Accept
S079	65	78	63	2	Research to increase growth
S105	48	82	77	2	Research to increase growth and survival

What if too few species exceed minimum acceptable standards?

There are several options:

- improve overall planting stock quality — review the nursery data to see whether there is anything that can be done to increase the size, health and vigour of the planting stock;
- experiment with intensified silvicultural treatments (e.g. carry out weeding or apply fertiliser more frequently), particularly if you think that site conditions could be limiting;
- try different species — review all sources of tree species information (**Table 5.2**) and start collecting seeds of species that have not already been tested.

Developing a suitability index

A semi-quantitative scoring system can be used to rank species according to a suitability index that combines a wide range of criteria. It can be applied either to refine the short-list of acceptable (or marginal) species that emerge from application of minimum standards or to all species for which data are available. Bear in mind that species with low field survival rates should always be screened out first, before calculating a suitability index.

A suitability index can take into account both easily quantifiable performance data and more subjective criteria, such as the attractiveness of each tree species to seed-dispersing animals. The simplest approach is to note whether species produce fleshy fruit or not. In older plots, this could be further refined by using the number of years to first flowering and fruiting, or the number of animal species that are attracted to a tree species.

Extract relevant data from the database and add additional information to a spreadsheet as required.

Example

Before biodiversity data are available, ability to produce fleshy fruits can be used as an indicator of 'attractiveness' to seed dispersers.

TNT = 'total nursery time' required to produce planting stock is used here to indicate ease of propagation. % germination or seedling growth rates in the nursery could also be used.

Species	% Survival	Mean height (cm)	Crown width (cm)	Fleshy fruits	TNT (years)
S001	89	450	420	Yes	<1
S015	45	198	255	Yes	<1
S067	78	234	287	Yes	1 to 2
S072	90	506	405	No	<1
S079	65	78	63	Yes	1 to 2
S105	48	82	77	Yes	>2

In this example, the species that were rejected as a result of applying minimum standards have been removed, while marginal values for some criteria remain indicated in red.

Find the species with the highest mean height. Assign a value of 100% to that maximum mean height and convert the mean heights of all other species to percentages of that maximum value to provide a height 'score' for each species. In this example, S072 has the highest mean height (506 cm) so the heights of all other species are multiplied by 100/506. Carry out the same calculation to provide scores for other chosen quantifiable criteria, including nursery performance criteria (e.g. % germination, seedling survival and so on).

Add extra weight to the criteria you feel are most important by multiplying their scores by a weighting factor (e.g. the survival has been doubled in the example below). Sum the scores and, as before, convert them into a percentage of the maximum score (adjusted score). Then rank the species in order of declining overall score.

Example

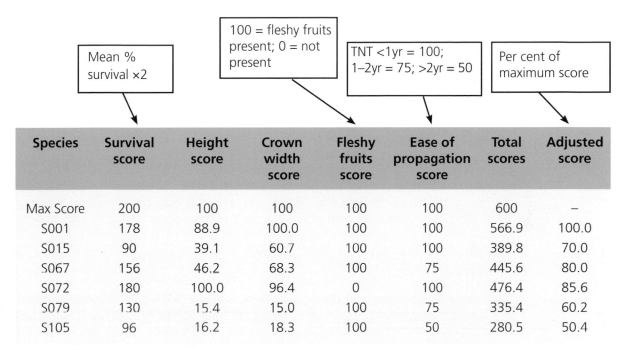

Species	Survival score	Height score	Crown width score	Fleshy fruits score	Ease of propagation score	Total scores	Adjusted score
Max Score	200	100	100	100	100	600	–
S001	178	88.9	100.0	100	100	566.9	100.0
S015	90	39.1	60.7	100	100	389.8	70.0
S067	156	46.2	68.3	100	75	445.6	80.0
S072	180	100.0	96.4	0	100	476.4	85.6
S079	130	15.4	15.0	100	75	335.4	60.2
S105	96	16.2	18.3	100	50	280.5	50.4

The boxes above the table contain:
- Mean % survival ×2
- 100 = fleshy fruits present; 0 = not present
- TNT <1yr = 100; 1–2yr = 75; >2yr = 50
- Per cent of maximum score

Based on the suitability scores above, S001, S015, S067 and S072, are the best species for planting, even though S015 would require some additional effort to increase survival. Lack of fleshy fruits in S072 is compensated for by excellent scores for other performance criteria. Rejection of both S079 and S105, which marginally failed to meet minimum standards, is confirmed as their adjusted suitability scores are only about half that of the most suitable species.

The interpretation of such a scoring system is ultimately subjective as the user must decide on which performance criteria are included, how they are quantified and how low or high the adjusted score must be to indicate the rejection or acceptance of a species.

Deciding on the species mix

One of the disadvantages of applying standards or a scoring system too rigorously is that it could result in the selection of only fast-growing pioneer species. This would create a rather uniform forest canopy (see **Section 5.3**). Planting pioneer and climax forest tree species together creates more structural diversity, even if some of the climax tree species fail to meet minimum standards or are ranked low in a scoring system.

So, when compiling the final mixture of species to be planted each year, use standards or scores to provide guidelines rather than absolute rules. Be flexible and always keep in mind the need for diversity. For example, a few slower-growing tree species could be acceptable for planting if they score highly on other criteria (e.g. early fruiting) and where most of the other species being planted are fast-growing. Similarly, a few species with narrow crowns may be desirable to add to the structural diversity of the forest canopy, provided they are planted alongside other species that score highly for canopy width. Ultimately, the species mix is selected by a subjective judgement that is modified and improved each year as a result of adaptive management.

What is adaptive management?

Ideally final species selection, as well as other management decisions, would not be made until all the data have been collected and analysed. It might, however, be many years before some of the field data are produced. Therefore, in the first few years of a FORRU, decisions are inevitably based on data that are produced early in the project, such as phenological observations or seed collection and nursery data. Tree performance data from field trials follow later, whereas data on biodiversity recovery and the establishment of recruit tree species become meaningful only after several years. Therefore, calculations of species suitability scores must be continually updated and modified as new data become available. Maintaining and updating the FORRU database is crucial to this process.

Continual re-assessment of species suitability is just one of several components of 'adaptive management', a concept central to the implementation of forest landscape restoration (see **Section 4.3**). Research results should feed into a social learning approach that is based on a process of experiential decision-making and monitoring. The database effectively acts as an archive of the outcomes of previous management trials and monitoring results, both good and bad, so that future decision-making can gradually be improved.

The process only works if all stakeholders have access to the database and can understand the outputs. Outputs must therefore be presented in user friendly formats and it is also necessary to run an education and outreach programme to ensure that all stakeholders can work with the database outputs and are thus well-equipped to participate meaningfully in management decisions. For more on adaptive management, see Chapter 4 in Rietbergen-McCracken *et al.* (2007).

8.6 Reaching out: education and extension services

Once an appreciable body of knowledge has been acquired, a FORRU should use it to provide comprehensive education and extension services that will improve the capacity of all of the stakeholders to contribute together to forest restoration initiatives. Such an outreach programme might include training courses, workshops and extension visits, supported by publications and other educational materials, each tailored to meet the different needs of each of the various stakeholder groups (e.g. government officials, NGOs, local communities, teachers, school children and so on).

Education team

To begin with, a FORRU's research staff might be called upon to provide training to interested groups as and when needed. As the project becomes more widely known, however, you should expect a rapid increase in demand for education and training services, which will begin to overwhelm the research staff, distracting them from vital research activities. It is better to recruit a team of education officers, with specialised experience of environmental education techniques, who are dedicated to providing stakeholders with the knowledge and technical support they need to implement restoration projects.

Newly recruited education staff will not be familiar with the knowledge-base acquired by the research staff. Therefore, the research team must first familiarise the education team with their research results and they must continue to provide frequent updates as the research delivers new information. The education team must then decide how to present the knowledge to stakeholders in user-friendly formats.

Education programme

Once the educators are familiar with the FORRU's knowledge-base, they must design curricula to meet the very different needs of the various stakeholders involved in forest restoration. A modular system is best, with subject material presented in different ways to match: i) the target audience and ii) the location where the module will be taught. For example, teaching forest officers about the framework species concept in a field plot requires a very different approach to teaching school children about the same concept in a classroom.

An education programme can include the following activities:

- workshops to introduce the general concepts of forest restoration and to present techniques and results; these are usually for government officers, NGOs and community groups who are considering forest restoration initiatives;
- more detailed training in forest restoration best practices for practitioners who are responsible for running nurseries and implementing planting programmes;
- extension visits to forest restoration projects that aim to provide on-site technical support directly to the people involved in implementing projects;
- hosting interested visitors to the unit such as scientists, donors, journalists and so on;
- helping with the supervision of college student thesis projects;
- presenting research results at conferences.

One of FORRU-CMU's nursery officers teaches workshop participants from the Elephant Conservation Network how to extract fig seeds. The participants subsequently set up their own FORRU in western Thailand, which is being used to restore elephant habitat (www.ecn-thailand.org/).

Special events for school children and a train-the-teachers programme (½ day to several days, for camps and teacher training) could also be undertaken as children have the most to gain from forest restoration.

Education materials

A FORRU education team should produce a wide range of educational materials to satisfy the needs of all stakeholders. Teaching aids will be needed for each module.

A video can provide a concise overview of the FORRU and its work for the opening sessions of workshops and training programmes, whereas a newsletter and a website can keep all stakeholders informed of a FORRU's outputs on a regular basis.

Publications are important educational outputs of a FORRU. Producing them can include a participatory component, involving consultations with and inputs from workshop participants. This ensures that the information provided by the FORRU is of maximum benefit for local people, and also that it makes best use of indigenous knowledge. Most of this material can easily be designed and laid out in-house with the aid of computers and desktop publishing software, particularly if someone with experience of graphic design is recruited to join the education team.

A trail through the field trials with informative sign boards turns a research facility into an educational resource of immense value.

Pamphlets and handouts

Handouts and pamphlets are one of the first outputs of a FORRU. They are useful for the unit's staff and visitors (particularly existing and potential funders). They should be both informative and help to publicise the unit. One of the first pamphlets produced could simply describe the FORRU's research programme to visitors. As the research programme develops, more technical literature should be produced, such as species data sheets and production schedules. Once this material has been written up, it can be used in other ways, for example in posters displayed in prominent places in the research unit for educational purposes.

Practical manuals

One of the first manuals produced by a FORRU should be an overview of the best practices for forest restoration, which combines the original skills and knowledge derived from the FORRU's research programme with existing knowledge and common sense. The manual serves as a text book for training both stakeholders during workshops and

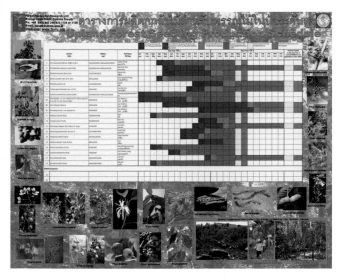

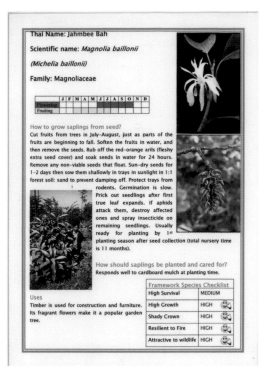

| Thai Name: Jahmbee Bah |
| Scientific name: *Magnolia baillonii* |
| (*Michelia baillonii*) |
| Family: Magnoliaceae |

How to grow saplings from seed?
Cut fruits from trees in July–August, just as parts of the fruits are beginning to fall. Soften the fruits in water, and then remove the seeds. Rub off the red-orange arils (fleshy extra seed cover) and soak seeds in water for 24 hours. Remove any non-viable seeds that float. Sun-dry seeds for 1-2 days then sow them shallowly in trays in sunlight in 1:1 forest soil: sand to prevent damping off. Protect trays from rodents. Germination is slow. Prick out seedlings after first true leaf expands. If aphids attack them, destroy affected ones and spray insecticide on remaining seedlings. Usually ready for planting by 1st planting season after seed collection (total nursery time is 11 months).

How should saplings be planted and cared for?
Responds well to cardboard mulch at planting time.

Uses
Timber is used for construction and furniture. Its fragrant flowers make it a popular garden tree.

Framework Species Checklist	
High Survival	MEDIUM
High Growth	HIGH
Shady Crown	HIGH
Resilient to Fire	HIGH
Attractive to wildlife	HIGH

A colourful production-schedule poster helps nursery staff to keep track of what seed species to collect and when to collect them.

Convert species information into user-friendly formats, such as this species profile card for *Magnolia baillonii*. Then compile information for all target species into a production-schedule poster.

extension events and newly recruited staff or visiting workers. Typically, such a manual should contain i) the basic principles and techniques of forest restoration, ii) descriptions of target forest types, and iii) descriptions and propagation methods for those tree species deemed suitable for restoration projects. It should be written in a format that is accessible to a broad readership. For an example, see FORRU-CMU's "How to Plant a Forest"[2]. This volume proved so popular that it has now been translated and adapted for use in seven Southeast Asian countries.

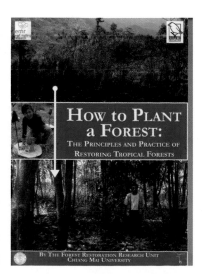

Practical manuals should be translated into the languages of neighbouring countries, to allow export of the skills and knowledge developed by a FORRU and their adaptation to different forest types and socio-economic conditions.

[2] www.forru.org/FORRUEng_Website/Pages/engpublications.htm

Research papers and an international audience

Original scientific results should be published in international journals or presented at international conferences and published in proceedings. The purpose of publications aimed at an international audience is to share research results with other people working in a similar field. Research papers also promote correspondence, discussion and exchange visits. They assist other researchers in developing their own research programmes. Furthermore, international publications enhance the status of the research unit, both at home and abroad.

Acceptance of papers by international journals and conference proceedings is important for the careers of the scientific staff (as job security in the academic world now increasingly depends on publication record) and raises the profile of the FORRU in the eyes of donor agencies. Research papers strengthen bids for funding.

Develop a communication strategy

In addition to informing and training stakeholders who are directly involved with forest restoration, the education team should also be responsible for reaching out to the broader general public by engaging the mass media. Public recognition for the work of a FORRU helps to build public acceptance of forest restoration and attracts support and funding. It also helps to establish a network of contacts with other organisations that might otherwise be unaware of the FORRU's work. So, it is worth investing some time in planning an effective communications strategy that emphasises those elements of the project that are appropriate for each of the different audiences it wishes to reach.

What questions should a communication strategy answer?

First, determine what the purpose of the communication is, what resources are available, and how to evaluate whether the message has been effectively communicated. Decide on who is the intended target audience. For example it could be the general public, land-holders, staff from government agencies, environmental organisations, teachers and students, sponsors and potential sponsors, industry organisations and so on. Be clear on what issues concern the audience, what message to communicate to them, what tools will be used, who in the FORRU will be responsible for the communication, and by when.

Writing for an audience

Develop the skills needed to present information clearly and concisely. Articles in newspapers, brochures, newsletters and on display boards will be read by people from a wide variety of backgrounds with different levels of technical expertise and language skills.

Developing a logo and promotional style

Develop a FORRU logo and a signature style (colour scheme, font style etc.) for presentations, publications, uniforms and so on. This will help audiences to recognise the FORRU 'brand'.

Photography

Good digital photographs can be used for a wide range of communication activities. Attractive, clear photos will increase the probability of having articles accepted for publication. Use a database to catalogue and organise the photo collection so as to make it easier to select the most appropriate photographs for each purpose.

FOREST RESTORATION RESEARCH UNIT

A recognisable logo helps to build a sense of unit identity and project recognition.

You can never have enough photos. Learn how to take good ones.

Communication tools

Open days, workshops and other events at the unit are all good ways to communicate with the wider public, but publicising your work at international meetings can have a broader impact. Accept invitations to speak at conferences and symposia or present posters, which can later be used around the FORRU. Keep posters short and simple with more pictures than text. Develop handouts to provide more detail.

Learn to use mass media to publicise FORRU outputs beyond the pages of scientific journals.

Use the media. Invite journalists to planting events and the opening of workshops etc. Write a press release or prepare information packs for journalists in advance so they have accurate facts and figures at their fingertips when writing articles. Ask a TV company to make a film about the unit, which can then be used as an introductory video at workshops and training events etc.

Maintain a website for regular communications with a network of interested organisations and individuals. In addition to a general description of the unit and its research and education activities, include pages with announcements of forthcoming events, a picture gallery of recent events and an interactive bulletin board. Publications and educational materials can be also posted on the website, so that anyone requesting a publication can simply be referred to the website for downloads. This saves a fortune in postage costs.

Inspiration for designing a forest restoration website can be found at:www.forru.org, www.rainforestation.ph and www.reforestation.elti.org

For those unable to access the web, a quarterly printed newsletter serves a similar function. Maintain a mailing list for the newsletter and also post copies on the website. E-mail makes it easy to communicate personally with large numbers of people, but do not allow your FORRU to gain a reputation for generating junk mail. A page on one of the web-based social media networks is a less intrusive way to keep people informed of the FORRU's activities and latest findings.

CASE STUDY 6   Chiang Mai University's Forest Restoration Research Unit (FORRU-CMU)

Country: Thailand

Forest type: Lower montane evergreen tropical forest.

Ownership: Government, national park.

Management and community use: 'Community forest' for the protection of the water supply to both Ban Mae Sa Mai village and the agricultural land below it; some harvesting of non-timber forest products.

Level of degradation: Cleared for agriculture, earlier restoration attempts had included planting of pines and eucalypts.

Like all tropical countries, Thailand has suffered from severe deforestation. Since 1961, the kingdom has lost nearly two-thirds of its forest cover (Bhumibamon, 1986), with natural forests having declined to less than 20% of the country's land area (9.8 million ha) (FAO, 1997, 2001). This has resulted in losses of biodiversity and increased rural poverty, as local people are forced to purchase, in local markets, substitutes for products formerly gathered from forests. Increases in the frequencies of landslides, droughts and flash floods have also been attributed to deforestation, while forest fires and other forms of degradation contribute approximately 30% of Thailand's total carbon emissions (Department of National Parks, Wildlife and Plant Conservation (DNP) and Royal Forest Department (RFD), 2008).

Part of the Thai government's response to these problems has been to ban logging and to attempt to conserve remaining forest in protected areas covering 24.4% of the country's land area (125,082 km^2) (Trisurat, 2007). However, many such 'protected' areas were established on former logging concessions, so large parts of them were already deforested before they were gazetted (about 20,000 km^2 (derived from Trisurat, 2007)). A 2008 report by Chiang Mai University's Academic Service Centre, found that about 14,000 km^2 of the country's forestland was "in need of urgent recovery" (Panyanuwat *et al.*, 2008).

Early attempts at reforestation involved the establishment of plantations of pines and eucalypts. For environmental protection and biodiversity conservation, forest restoration (as defined in **Section 1.2**) is more appropriate, but its implementation has been limited because of a lack of knowledge about how to grow and plant native forest tree species.

Therefore, in 1994, the Biology Department of Chiang Mai University established a Forest Restoration Research Unit (FORRU-CMU), in which to develop appropriate techniques for the restoration of tropical forest ecosystems. The unit consists of an experimental tree nursery and trial plot system in Doi Suthep-Pui National Park, which adjoins the university campus.

In 1997, FORRU-CMU began research to adapt the 'framework species' approach to restore evergreen forest in the park, having learnt about how this concept had been used in Australia (see **Box 3.1**). An herbarium collection and database of the local tree flora, established by J. F. Maxwell at the CMU Biology Department Herbarium (Maxwell & Elliott, 2001), provided an invaluable starting point, as well as a species identification service and information on the distribution of indigenous tree species.

The unit established an office and a research nursery at the park's former headquarters compound, near intact examples of the target forest types. There, a phenology study determined optimal seed collection times and provided opportunities for regular seed collection.

Experiments in the nursery developed methods for the production of containerised trees of a size suitable for planting by the optimum planting date, which is mid-June in the seasonally dry climate of northern Thailand. Germination trials (Singpetch, 2002; Kopachon, 1995), seed storage experiments and seedling growth trials (Zangkum, 1998; Jitlam, 2001) were used to develop species production schedules (see **Section 6.6**). The research facility was also used by CMU research students, who tackled more detailed research on propagation from cuttings (Vongkamjan *et al.*, 2002; see **Box 6.6**), the use of wildlings (Kuarak, 2002; see **Box 6.4**) and the role of mycorrhizas (Nandakwang *et al.*, 2008).

Every rainy season since 1997, experimental plots, ranging in size from 1.4 to 3.2 ha have been planted with varied combinations of 20–30 candidate framework tree species to: i) assess the potential of planted tree species to perform as framework species; ii) test tree species' responses to silvicultural treatments designed to maximise field performance; and iii) assess biodiversity recovery.

The plots were established in close co-operation with the people of Ban Mae Sa Mai village (see **Box 8.1**). This partnership with a local community provided FORRU-CMU with three important resources: i) a source of indigenous knowledge; ii) an opportunity for local people to test the practicability of research results; and iii) a supply of local labour. At the request of the villagers, FORRU-CMU funded the construction of a community tree nursery in the village and trained villagers in basic tree propagation methods and nursery management. The villagers now sell native forest tree seedlings to other restoration projects.

The output of the project was an effective procedure that can be used to restore rapidly lower montane evergreen in northern Thailand. The best-performing framework tree species were identified (Elliott *et al.*, 2003) and optimal silvicultural treatments determined (Elliott *et al.*, 2000; FORRU, 2006). Canopy closure can now be achieved within 3 years after planting (with a planting

Field trials tested various silvicultural treatments, including fertiliser application, weeding and mulching. Cardboard mulch mats were particularly effective in dry, degraded sites.

In experimental plots, all trees are tagged and measured 2–3 times each year: height, root-collar diameter and crown width are recorded each time. This has resulted in a large database containing information on the field performance of native forest tree species, and has allowed those that function as framework tree species to be identified.

School children from all over the world now visit FORRU-CMU's nursery and field plots to learn forest restoration techniques.

density of 3,100 trees per hectare). Rapid biodiversity recovery was also achieved. Sinhaseni (2008) reported that 73 non-planted trees species re-colonised the plots within 8–9 years. When combined with the 57 planted framework tree species, the total tree species richness in the sampled plots amounted to 130 (85% of the tree flora of the target evergreen forest). The species richness of the bird community increased from about 30 before planting to 88 after 6 years, including 54% of the species found in the target forest (Toktang, 2005).

The techniques developed were published in a user-friendly, practitioner's guide entitled "How to Plant a Forest", in both Thai and English (FORRU, 2006), and subsequently translated into five other regional languages. The project also resulted in a set of protocols that could be applied by researchers in other tropical regions to develop techniques for the restoration of any tropical forest type, taking into account the indigenous tree flora and local climatic and socio-economic conditions. These were published in a manual for researchers, entitled "Research for Restoring Tropical Forest Ecosystems" (FORRU, 2008), also in several languages. Both books can be downloaded free from www.forru.org. These manuals have subsequently been used to replicate the FORRU concept in the restoration of other forest types, largely with the support of the UK's Darwin Initiative: in southern Thailand (http://darwin.defra.gov.uk/project/13030/), China (http://darwin.defra.gov.uk/project/14010/) and Cambodia (http://darwin.defra.gov.uk/project/EIDPO026/).

The most important output of the project was a set of techniques for the restoration of evergreen tropical forest on abandoned agricultural fields at altitudes greater than 1,000 m above sea level. Eight and a half years after planting 29 framework species, weeds were eliminated, humus had accumulated, a multi-level canopy had developed and biodiversity recovery was well underway.

APPENDIX 1: TEMPLATES FOR DATA COLLECTION SHEETS

A1.1 Rapid site assessment

A1.2 Phenology

A1.3 Seed collection

A1.4 Germination

A1.5 Seedling growth

A1.6 Nursery production record

A1.7 Field performance of planted trees

A1.8a Vegetation survey — trees

A1.8b Vegetation survey — ground flora

A1.9 Bird survey

A1.1 Rapid site assessment

Circle	Livestock signs	Fire signs	Soil (% exposed, condition, erosion)	Weeds (% cover and mean height, +/- tree seedlings)	No. trees >1 m tall (<30 cm gbh)	No. live tree stumps	No. trees >30 cm gbh	Total no. regenerants
1								
2								
3								
4								
5								
6								
7								
8								
9								
10								

Total	
Mean	
Average/ha	(= mean × 10,000/78)
No. of trees to plant per ha	(= 3100 — Average/ha)

Location GPS			
Recorder			
Date			
Total species of regenerants:	Pioneers		Climax

A1.2 Phenology

Order	Label	Species no.	Species	GBH	FB	FL	FT	BA	YL	ML	SL	Tree position/Notes

RECORDERS: DATE: LOCATION:

A1.3 Seed collection

Date collected:	Species no.:	Batch no.:
Family:		Common name:
Botanical name:		
Location:		
GPS co-ordinates:		Elevation:
Forest type:		
Collected from:	Ground [ ]	Tree [ ]
Tree label no.:	Tree girth cm	Tree height m
Collector:		Date seed sown:
Notes		
		Voucher collected? [ ]

✂ -

FOREST RESTORATION RESEARCH UNIT
VOUCHER SPECIMEN HERBARIUM LABEL
NOTE: all dates are day/month/year

FAMILY:	COMMON NAME:
BOTANICAL NAME:	DATE:
PROVINCE:	DISTRICT:
LOCATION:	
BPS CO-ORDINATES:	ELEVATION:
HABITAT:	
DESCRIPTIVE NOTES: TREE GIRTH: cm	TREE HEIGHT: m

Bark

Fruit

Seed

Leaf

COLLECTED BY:	SPECIMEN IDENTIFICATION NO.:	NO. OF DUPLICATES:

A1.4 Germination

Species name:

Date seeds collected:

Species number:

Family:

Date seeds sown:

Batch number:

No. seeds sown per replicate:

TREATMENT DESCRIPTIONS

C	
T1	
T2	
T3	
T4	
ETC.	

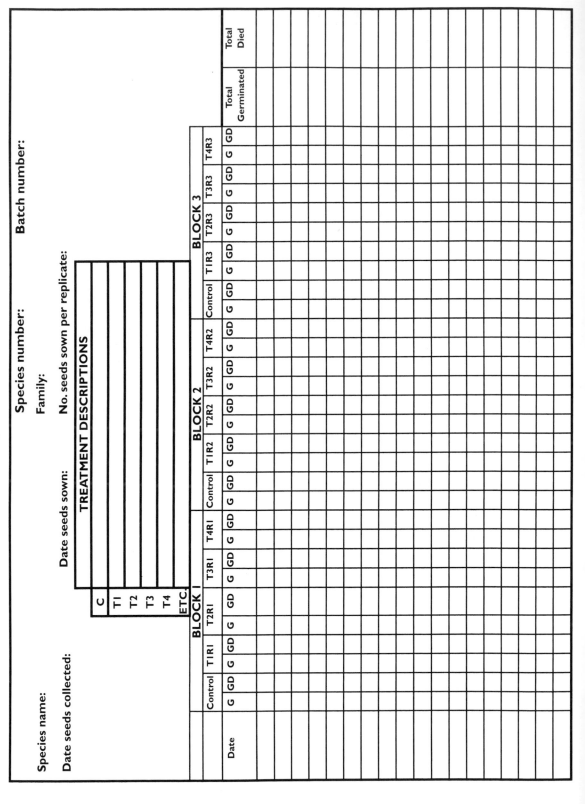

BLOCK 1

Date	Control		T1R1		T2R1		T3R1		T4R1	
	G	GD	G	GD	G	GD	G	GD	G	GD

BLOCK 2

Control		T1R2		T2R2		T3R2		T4R2	
G	GD	G	GD	G	GD	G	GD	G	GD

BLOCK 3

Control		T1R3		T2R3		T3R3		T4R3		Total Germinated	Total Died
G	GD	G	GD	G	GD	G	GD	G	GD		

A1.5 Seedling growth

Species:		S. no.:		Batch no.::	

Seeds sown date:	Pricking out date:

HEIGHT (cm)

Date	Days	SEEDLING NUMBER															AVG	RGR
		1	2	3	4	5	6	7	8	9	10	11	12	13	14	15		

ROOT COLLAR WIDTH (mm)

Date	Days	SEEDLING NUMBER															AVG	RGR
		1	2	3	4	5	6	7	8	9	10	11	12	13	14	15		

HEALTH SCORE (0–3)

Date	Days	SEEDLING NUMBER															AVG
		1	2	3	4	5	6	7	8	9	10	11	12	13	14	15	

A1.6 Nursery production record

Year..............   No................

SPECIES	SPECIES NO.		BATCH NO.	
I. SEED GERMINATION				
Pre-sowing seeding treatment:				
Media and tray type:				
	Date:		Quantity:	
Seed collection:				
Seed sowing:				
First germination date:				
OBSERVATIONS:				
2. PRICKING OUT				
Media:				
Container type:				
	Date:		Quantity:	
Pricking out:				
OBSERVATIONS (condition of seedlings):				

3. NURSERY CARE	Dates			
	1	2	3	4
Fertilizer:				
Pruning:				
Weeding:				
Pest/disease control measures:				

4. HARDENING AND DISPATCH				
Hardening date started	No. of seedlings:			
	Date	Quantity	Where	
Dispatched				
Dispatched				
Dispatched				
OBSERVATIONS;				

A1.7 Field performance of planted trees

PLOT NAME:		PLOT LOCATION:	
MONITORING DATE:		RECORDER(S):	

Order no.	Sp. no.	Label no.	Previous notes	RCD (mm)	Height (cm)	Crown width (cm)	Weed score (0–3)	Health score (0–3)	Notes

A1.8a Vegetation survey — trees

DATE:	RECORDER:		PLOT NO.:		CIRCLE NO.:
LOCATION:	FOREST TYPE:			GPS	

Trees >50 cm tall within 5 m radius circle								

Tree species	Planted/ natural	Label no.	RCD (mm)	GBH (cm) (>6.3)	Health score (0–3)	Crown width (cm)	No. coppicing stems	Notes

A1.8b Vegetation survey — ground flora

DATE:	RECORDER:		PLOT NO.:	CIRCLE NO.:
LOCATION:	FOREST TYPE:			GPS:

Ground flora species within 1 m radius circle			SCORE
			SCORE
Leaf litter			
Bare soil			
Rocks			

SPECIES						SCORE	Tree seedling? Y/N

Score <5% = 0.5, 5–9% = 1, 10–24% = 2, 25–49% = 3, 50–79% = 4, and 80%+ = 5

A1.9 Bird survey

Recorders:		File name:	
Date:		Weather:	
Block number:		Plot number:	
Start time:		Finish time:	

Time	Species	No. of birds (sex)	Sight or song/ call	Distance from point (m)	Activity	Tree species (if appropriate)

APPENDIX 2: EXPERIMENTAL DESIGN AND STATISTICAL TESTS

A2.1 Randomised complete block design experiments

All ecological experiments generate highly variable results. Therefore, experiments must be repeated or 'replicated' several times, and the results must be presented as mean values followed by a measure of variation among replicates that are subjected to the same treatment (e.g. variance or standard deviation). Luckily, most of the experiments required for forest restoration research (e.g. germination tests, seedling growth experiments and field trials) can all be set up using the same basic experimental design and the same method of statistical analysis: a 'randomised complete block design' (RCBD), with the results analysed by a two-way analysis of variance (ANOVA) followed by pair-wise comparisons.

What is a randomised complete block design?

Each of the replicated 'blocks' within an RCBD consisting of one replicate of the control, plus one replicate of each of the treatments being tested. Each treatment and the control are represented equally in every block (i.e. by using the same number of seeds, plants etc.). In each block, the positions of the control and the treatments are allocated randomly. The replicate blocks are placed randomly across the study area (or nursery).

Why use RCBD?

An RCBD separates the effects that are due to environmental variability from those of the treatments being tested. Each block may be exposed to slightly different environmental conditions (light, temperature, moisture etc.). This creates variability in the data that can obscure the effects of applied treatments; but as a control replicate and treatment replicates are grouped together in each block, all germination trays or plots within a block are exposed to similar conditions. Consequently, the effects of variable external conditions can be accounted for and the effects of the treatments applied (or the absence of effects) revealed by a two-way ANOVA (see **Section A2.2**).

How many blocks and treatments?

Ideally, the combined number of blocks and treatments used should result in at least 12 'residual degrees of freedom' (rdf) according to the equation below…

$$rdf = (t-1) \times (b-1)$$

…where t is the number of treatments (including the control) and b is the number of blocks. In reality, it is often very difficult to achieve an rdf of more than 12 in nursery or field experiments because of shortages in the availability of seeds, trees, land or labour. An rdf of <12 can still yield robust results if you ensure as much uniformity among the blocks as possible. Otherwise, you could use a simpler experimental design (e.g. paired experiments, which compare a single treatment with a control) and simpler analytical methods (e.g. Chi-square for germination or survival data (see **Section 7.4**)).

A2.2 Analysis of variance (ANOVA)

Data from RCBD experiments can be analysed by a rigorous standard statistical test called analysis of variance (ANOVA). There are several forms of this test. The one used to analyse RCBD experiments is a 'two-way ANOVA (without replication)'. The 'without replication' part is confusing because treatments are replicated across the blocks, but in statistical jargon, it means that there is only one value for each treatment in each block; for example, for germination experiments, there is one value for the number of seeds germinating in each replicate germination tray.

The simplest way to perform an ANOVA is to use the Analysis ToolPak that comes bundled with Microsoft Excel, so first make sure that you have the Analysis ToolPak installed on your computer.

If you are using Windows XP, open Excel and click on 'Tools' in the toolbar and then click on 'Add-Ins…'. Make sure that the box next to 'Analysis ToolPak' has a tick in it. If the tick box does not appear, you must re-run Excel set-up and install the Analysis ToolPak add-in.

If using Vista or Windows 7, click on the Microsoft Office button (top left), then on the Excel Options button (bottom right of the dropdown menu), then on 'Add Ins' and finally on the 'Go' button next to 'Manage Excel Add Ins'. Tick the box labelled 'Analysis ToolPak'.

The experiments described in **Chapters 6** and **7** generate two kinds of data: i) binomial data, which describe variables that have only two states, e.g. germination (i.e. germinated or not germinated) and survival (i.e. alive or dead); and ii) continuous data (which can have any value), e.g. seedling height, root collar diameter, crown width or relative growth rate. If you are analysing binomial data, you should first arcsine transform the data, for statistical reasons, before carrying out the analysis of variance. If you are analysing continuous data, you can skip the next section and move straight to ANOVA.

Preparing binomal data for ANOVA

Enter your data (e.g. number of germinated seeds or number of surviving trees) in a table as shown below (original data), with blocks as rows and treatments as columns.

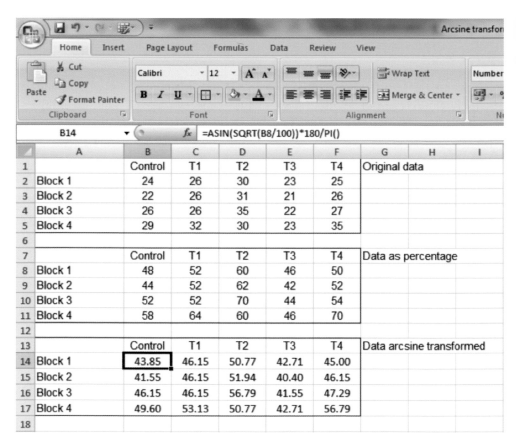

In this example, the original data are the number of seeds germinated (out of 50) in each of 4 blocks for each of 5 pre-sowing treatments: for example, T1 = soaking in hot water for 1 hour, T2 = scarification with sand paper, T3 = soaking in acid for 1 minute, and T4 = soaking in cold water overnight.

Next, construct another table to calculate percentage values: e.g. for the control in block 1, 24 seeds germinated out of 50 sown, so the percentage germinating = 24/50 × 100 = 48%.

Then set up a third table below, to calculate the arcsine-transformed percentages; for example, for the control in block 1 (located in cell B8), type the following formula into the third table:

$$=ASIN(SQRT(\mathbf{B8/100}))*180/PI().$$

Then, copy the formula into the other cells of the third table. To make sure you have entered the formula correctly, entering 90 in the percentage table. An arcsine-transformed value of 71.57 should be returned in the third table.

Now carry out the ANOVA as described below, using the arcsine-transformed percentages.

ANOVA

In this example, we are using the mean height of trees (cm) 18 months after planting in a field trial plot system (see **Section 7.5**), subjected to different fertiliser treatments. Open a new spreadsheet and type in your data with blocks as rows and treatments as columns, as shown below.

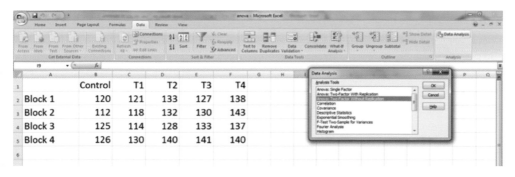

	Control	T1	T2	T3	T4
Block 1	120	121	133	127	138
Block 2	112	118	132	130	143
Block 3	125	114	128	133	137
Block 4	126	130	140	141	140

In this example, the data show tree height (cm). Different fertiliser doses were applied to the trees at planting time and three times in the rainy season: T1 = 25g fertiliser, T2 = 50g, T3 = 75g and T4 = 100g.

Next, if using Windows XP, click on 'Tools' and then on 'Data Analysis…'. With Vista or Windows 7 click on the 'Data' tab at the top of the screen and then on 'Data Analysis' (top right). A dialogue box, containing a list of various statistical tests, will appear. Click on 'ANOVA: Two-Factor Without Replication' and then click 'OK'.

Another dialogue box will appear. Click on the square button to the right of the 'Input Data' box ('Input Range' in Windows 7). Then, using the mouse, drag the cursor across the data table to select the entire data set, including column and row headings. Click on the square button again to get back to the dialogue box, then make sure there is a tick in the 'Labels' box and that the value in the 'Alpha' box is 0.05. Click on the circular button, 'Output Range:' and then on the square button to the right of the output range box. In the spreadsheet, move the cursor to a cell immediately below your data table and click. Then go back to the dialogue box and click 'OK'. Two tables of output results will appear below your data table. The upper one summarises mean values for each treatment and for each block, along with a measure of variability (i.e. variance). The lower one will tell you if there are significant differences among the treatments.

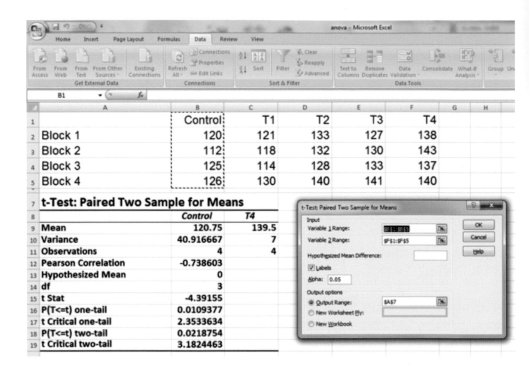

Full results from the upper table of output results are as follows:

ANOVA: Two-Factor Without Replication.

Summary	Count	Sum	Average	Variance
Block 1	5	639	127.8	59.7
Block 2	5	635	127.0	149.0
Block 3	5	637	127.4	77.3
Block 4	5	677	135.4	47.8
Control	4	483	120.75	40.92
T1	4	483	120.75	46.25
T2	4	533	133.25	24.92
T3	4	531	132.75	36.25
T4	4	558	139.50	7.00

In this example, variances within blocks (among treatments) are generally higher than variances within treatments (among blocks), suggesting that the effects of the treatments are stronger than random variations resulting from differences in conditions among the blocks. It looks like treatments 2, 3 and 4 increase germination compared with the control, whereas treatment 1 has no effect. But are these results significant? The lower table answers this question.

ANOVA						
Source of Variation	**SS**	**df**	**MS**	**F**	**P-value**	**F crit**
Rows	241.6	3	80.5333	4.3066	0.02799	3.49029
Columns	1110.8	4	277.7	14.8503	0.00014	3.25917
Error	224.4	12	18.7			
Total	1576.8	19				

In this table, 'rows' refers to blocks and 'columns' refers to treatments. ANOVA tests the 'null hypothesis' that there are no real differences among the control and the treatments tested and that any variation among the mean values is just due to chance. Consequently, if large differences among the mean values for treatments and blocks are found, then the assumption will be false, and at least one of the treatments has had a significant effect. The important values to look at are the P-values, which quantify the probability that the null hypothesis (i.e. no differences) is valid. The table, therefore, shows that there is only a 0.00014 in 1, or 0.014% probability that differences among treatments do not exist (and hence a 99.986% probability that they do). Similarly, real differences among the blocks are highly probable (97.2% likely). The significant differences among blocks show that a randomised block design was necessary in order to remove a substantial amount of variation associated with differences in the micro-environments that affect each block. Although this ANOVA shows significant differences among treatments, it does not say which of the differences are significant. In order to determine that, it is necessary to perform a pair-wise comparison. For further information about ANOVA and for a wider choice of analytical techniques, please refer to Dytham (2011) and Bailey (1995).

A2.3 Paired t-tests

If significant differences among mean values are confirmed by ANOVA, pair-wise comparisons are needed to determine which differences are significant. Statistical tests that determine whether the difference between two means is significant include Fisher's Least Significant Difference (LSD) test, Tukey's Honestly Significant Difference (HSD) test and the Newman Keuls test. These tests can be performed using statistical software, such as Minitab or SPSS, trial versions of which can be downloaded from the internet [1].

[1] spss.en.softonic.com

In Excel, you can perform a paired t-test using the analysis ToolPak. It is not statistically valid to use this test to compare all means with all other means automatically. Adopt the so-called *a priori* approach, i.e. decide on the questions you want to answer beforehand and only carry out only those tests that answer those questions. In this case, the main question is "do treatments significantly increase or reduce performance compared with the control?"

In 'Data Analysis', click on 't-test: Paired Two Sample for Means' and then click 'OK'. In the dialogue box, click on the square button, to the right of the 'Variable 1 Range' box. Then, using the mouse, drag the cursor down the table to select the data set for 'control', including the column heading. Repeat for 'Variable 2 Range' by selecting the data set for whichever treatment you have decided to test (the screen print below shows the results for 'control' compared with 'T4'). Back in the dialogue box, select a 'Hypothesized Mean Difference' of '0' (the null hypothesis being that there is no significant difference between the treatment data). Make sure there is a tick in the 'Labels' box and that the value in the 'Alpha' box is 0.05. Click on the circular radio button, 'Output Range:' and then on the square button to the right of the output range box. In the spreadsheet, move the cursor to a cell immediately adjacent to your data table and click. Then go back to the dialogue box and click 'OK'. A table of output results will appear adjacent to your data table. Repeat the process for all pair-wise comparisons that you decide will be useful.

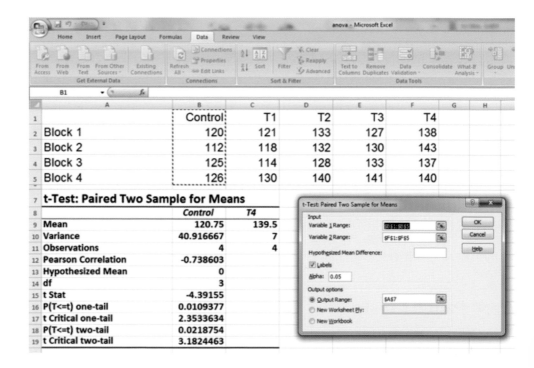

t-test: Paired Two Sample for Means.

	Control	T2	Control	T3	Control	T4
Mean	120.75	133.25	120.75	132.75	120.75	139.5
Variance	40.91667	24.91667	40.91667	36.25	40.91667	7
Observations	4	4	4	4	4	4
Pearson Correlation	0.25316		0.629662		-0.7386	
Hypothesized Mean Difference	0		0		0	
df	3		3		3	
t Stat	-3.54738		-4.48252		-4.39155	
P(T<=t) one-tail	0.019081		0.010353		0.010938	
t Critical one-tail	2.353363		2.353363		2.353363	
P(T<=t) two-tail	0.038162		0.020706	sig	0.021875	sig
t Critical two-tail	3.182446		3.182446		3.182446	

The ANOVA results table for tree height in **Section A2.2** above showed higher mean values for treatments 2, 3 and 4 and a similar mean value for treatment 1. For these differences to be significant, the value of 't Stat' must be greater than a critical value determined from the number of degrees of freedom and the acceptable value of P (usually 5%). The significance of the differences is therefore determined by looking at the value for 'P(T<=t) two-tail'. If the value is **0.05 or less**, the difference is significant. It means that there is only a **5% or lower** probability that the null hypothesis (i.e. that the difference between the means is zero) is correct. In the t-test example above, treatments T2, T3 and T4 all satisfy this condition. So the result is that applying 50–100g fertiliser most probably increased tree height compared with the control from around 121 cm to around 133 to 140 cm, depending on the amount of fertiliser used. Applying 25g fertiliser most probably had no effect. You can ignore the other data shown in the t-test table, such as the values for one-tail tests, unless you are confident about interpreting them.

GLOSSARY

Agro-forestry: a plantation design that increases and diversifies the economic benefits from forestry by adding crops and/or livestock to the system.

Accelerated (assisted) natural regeneration (ANR): management actions to enhance the natural processes of forest restoration, focusing on encouraging the natural establishment and subsequent growth of indigenous forest trees, while preventing any factors that might harm them.

Analogue forestry: forestry that uses a combination of domesticated and indigenous forest tree species and other plants to re-establish a forest structure similar to that of climax forest.

Biodiversity: the variety of life encompassing genes, species and ecosystems.

Biodiversity offset: payments made by agencies whose actions destroy or diminish biodiversity in one place that are used to restore biodiversity in another place, thereby achieving no net loss of biodiversity.

Candidate framework species: local tree species undergoing nursery and field performance testing to determine their suitability as framework species.

Carbon credits: payments by carbon emitters (companies, governments or individuals) that are used to finance projects that aim to absorb carbon dioxide from the atmosphere, leading to zero net increase in atmospheric carbon dioxide.

Climax forest: the final stage of forest succession, a relatively stable forest ecosystem having attained the maximum development in terms of biomass, structural complexity and biodiversity that can be sustained within the limits imposed by the soil and prevailing climatic conditions.

Climax tree species: tree species that comprise climax forest.

Community forest: a forest that is managed collectively by local people, usually with the extraction of timber and non-timber forest products.

Conservation: the preservation, management, and care of natural and cultural resources.

Damping off: fungal diseases that attack the stems of young seedlings.

DBH (diameter at breast height): diameter of the tree trunk at 1.3 m above ground level.

Deciduous: shedding leaves annually or periodically; not evergreen.

Deforestation: conversion of forest into other land uses with less than 10% tree cover, e.g. arable land, pasture, urban uses, logged areas, or wasteland.

Degradation: disturbance leading to decrease forest quality and impeded ecological functioning of the forest ecosystem.

Direct seeding: the establishment of trees on deforested sites by sowing seeds rather than by planting nursery-raised saplings.

Dormancy: a period during which viable seeds delay germination, despite having conditions (moisture, light, temperature etc.) that are normally favourable for the later stages of germination and seedling establishment.

Ecotourism: low impact, nature-based tourism that produces positive benefits for the conservation of biodiversity.

Ectomycorrhiza: an association between vascular plant roots and fungi that forms a fungal sheath on root surfaces and between root cortical cells.

Endemic: indigenous to and confined to a particular area.

Enrichment planting: planting trees to i) increase the population density of existing tree species or ii) increase tree species richness by adding tree species to degraded forest; also used to mean restocking logged-over or otherwise degraded forest with economic species.

Epiphyte: a plant growing on (but not penetrating) another plant, e.g. orchids growing on the branch of a tree.

Evergreen: a plant that retains green foliage throughout the year.

Exotic: of species – introduced, not native.

Extinction: the complete loss of a species globally; when no more individuals of a species exist.

Extirpation: the disappearance of a species from a particular area, while it survives elsewhere.

Extractive reserve: designated conservation areas in which natural-resource extraction is carried out complementary to the objective of conserving biological diversity and the natural resource base.

Forest landscape restoration (FLR): integrated management of all landscape functions in deforested or degraded areas to regain ecological integrity and enhance human well-being; usually including some forest restoration.

Forest restoration: actions to re-instate ecological processes that the accelerate recovery of forest structure, ecological functioning and biodiversity levels towards those typical of climax forest.

Forest Restoration Research Unit (FORRU): established to develop methods to harness and accelerate the natural processes of forest regeneration, so that biodiversity-rich forest ecosystems, similar to climax forest, can be re-established.

Foster ecosystem: tree plantations of not necessarily indigenous species used to facilitate the natural regeneration of native species.

Framework species method (or Framework forestry): planting the minimum number of indigenous tree species required to re-instate the natural processes of forest regeneration and recover biodiversity. It combines the planting of 20–30 key tree species with various ANR techniques to enhance natural regeneration, creating a self-sustained forest ecosystem from a single planting event.

Framework tree species: indigenous, non-domesticated, forest tree species, which, when planted on deforested sites, rapidly re-establish forest structure and ecological functioning, while attracting seed-dispersing wildlife.

Frugivorous: fruit-eating.

FTPS (field trial plot system): a set of small plots, each one planted with a mixture of different tree species and/or silvicultural treatments using the randomised complete block design (RCBD).

GBH (girth at breast height): circumference of the tree trunk at 1.3 m above ground level.

Gross domestic product (GDP): the total value of all goods and services bought or sold in an economy.

Genetic diversity: diversity within a species.

Geographic positioning system (GPS): a handheld or vehicle-mounted system that uses satellite communications to determine geographical position and other navigational information.

Germination: the growth of seeds or spores after a period of dormancy; emergence of an embryonic root through the seed coverings.

Geographical information system (GIS): Computerised manipulation of maps and other geographical information, useful for the planning of forest restoration projects.

Herbarium: a repository for easy accessible collections of dried, preserved and well-labelled specimens of plants and fungi.

Hypha: a long, branching filamentous cell of a fungus; the main mode of vegetative growth of a fungus; collectively called 'mycelium'.

Indigenous: native to an area, not introduced; the opposite of exotic.

Intermediate seeds: seeds that can be dried to low moisture contents, approaching those of orthodox seed, but are sensitive to chilling when dried.

Keystone tree species: species that flower or fruit at times when other food resources for animals are in short supply.

Maximum diversity/Miyawaki methods of forest restoration: restoring as much of the tree species richness of the original forest as possible without relying on natural seed dispersal.

MLD (median length of dormancy): the time taken from seed sowing of a batch of seeds to germination of half of the seeds that finally germinate; for example, if 10 seeds germinate out of a batch of 100 sown, it is the time to germination of the 5th seed.

Mycorrhiza: symbiotic (occasionally weakly pathogenic) association between a fungus and the roots of a plant.

Natural regeneration: the recovery of forest following disturbance in the absence of human intervention, resulting in increasing ecosystem functionality, vegetation species diversity, structural complexity, habitat availability and so on.

Non-governmental organisation (NGO): a legally constituted organisation created by private persons or organisations with no participation or representation of any government.

Non-timber forest products (NTFPs): broadly includes all non-timber vegetation in forests and agro-forestry environments that have commercial value. They include plants, parts of plants, fungi and other biological materials harvested from natural, manipulated, or disturbed forests. NTFPs can be classified into four major product categories: culinary, floral and decorative, wood-based, and medicinal and dietary supplements.

Nurse tree species: extremely hardy, usually fast-growing pioneer tree species planted specifically to restore environmental and soil conditions that are favourable for the establishment of a broader range of indigenous forest tree species.

Orthodox seeds: seeds that are easy to store for many months or even years.

Payments for environmental services (PES): compensating those involved in forest restoration or conservation for carbon storage, watershed protection, conservation of biodiversity and all the other environmental services provided by restored or conserved forest.

Phenology: the study of the responses of living organisms to seasonal cycles in environmental conditions, e.g. the periodic flowering and fruiting of trees.

Pioneer tree species: early-successional species that germinate only in full sun or the largest gaps. They exhibit high photosynthetic and growth rates, have simple branching patterns, and require high temperature and/or high light intensity for germination. These species are usually short-lived and are characteristic of pioneer forest.

Primary forest: climax forest that has not been substantially disturbed in recent history.

Production schedule: a concise description of the procedures for producing planting stock of optimum size and quality from seed (or wildlings) by the optimum planting out time. This timetable combines all available knowledge about the reproductive ecology and cultivation of a species.

Protected area: an area of land and/or sea that is especially dedicated to the protection and maintenance of biological diversity and of natural and associated cultural resources that is managed through legal or other effective means.

Rainforestation: a forest restoration technique, developed in the Philippines, that uses indigenous tree species to restore ecological integrity and biodiversity while also producing a diverse range of timbers and other forest products for local people.

RCBD (randomised complete block design): experimental design where single replicates of each treatment and the control are randomly positioned within a 'block', with each block being replicated in at least 3 locations across the study site.

RCD (root collar diameter): diameter of the plant stem at the root collar, just above the level of the soil.

Recalcitrant seeds: seeds that are sensitive to drying and chilling.

Recruit species: additional (non-planted) tree species that establish naturally in forest restoration sites.

Reforestation: planting trees to re-establish tree cover of any kind; includes plantation forestry, agroforestry, community forestry and forest restoration.

Remnant forest: small areas of forest that survive in a landscape following large-scale deforestation.

RGR (relative growth rate): a measurement of plant growth rate that takes into account the plant's initial size.

Root collar: the point at which the above-ground parts of a plant meet the tap-root.

Secondary forest: a forest or woodland area that has re-grown after a major disturbance but is not yet at the end point of succession (climax forest), usually distinguished by differences in ecosystem functionality, vegetation species diversity, structural complexity and so on.

Seed bank: all of the seeds, often in a dormant state, that are stored within the soil of many terrestrial ecosystems. A seed bank can also refer to the storage of collected seeds as a source for forest restoration activities.

Seed rain: the movement of seed into an area through natural processes. This can occur through various mechanisms of dispersal, including wind and animal dispersal.

Senescent leaves: leaves that are losing their chlorophyll (and hence green colour) just before leaf fall.

Silviculture: controlling the establishment, growth, composition, health, and quality of forests to meet the diverse needs and values of landowners.

Site recapture: elimination of herbaceous vegetation by the shading effects of planted trees or by ANR.

Stakeholder: anyone affected by or involved in a forest restoration project.

Target forest: a forest ecosystem that defines the goals of a forest restoration program in terms of tree species composition, structure, and biodiversity levels and so on; usually the nearest surviving patch of climax forest that remains in the landscape at a similar elevation, slope, aspect etc. to those of the restoration site.

Vesicular arbuscular mycorrhizas (VAM): mycorrhizal fungi that grow into the root cortex of the host plant and penetrate root cells, forming two kinds of specialised structures: arbuscules and vesicles. Also known as arbuscular mycorrhizas.

Voucher specimens: dried specimens of tree leaves, flowers and fruits etc. that are kept for confirmation of species names (from phenology-study trees, seed-collection trees etc.)

Wildings: seedlings or saplings growing naturally in native forest that are dug up to be grown on in a nursery.

Wildlife: all non-domesticated plant and animal species living in natural habitats.

REFERENCES

Aide, T. M., M. C. Ruiz-Jaen and H. R. Grau, 2011. What is the state of tropical montane cloud forest restoration? In Bruijnzeel, A., F. N. Scatena and L. S. Hamilton (eds.), Tropical Montane Cloud Forests: Science for Conservation. Cambridge University Press, Cambridge, pp 101–110.

Alvarez-Aquino, C., G. Williams-Linera and A. C. Newton, 2004. Experimental native tree seedling establishment for the restoration of a Mexican cloud forest. Restor. Ecol. 12(3): 412–418.

Anderson, J. A. R., 1961. The Ecology and Forest Types of the Peat Swamp Forests of Sarawak and Brunei in Relation to their Silviculture. PhD thesis, Edinburgh University, UK.

Aronson, J., D. Valluri, T. Jaffré and P. P. Lowry, 2005. Restoring Dry tropical forests. In: Mansourian, S., D. Vallauri and N. Dudley (eds.) (in co-operation with WWF International), Forest Restoration in Landscapes: Beyond Planting Trees. Springer, New York, pp 285–290.

Ashton, M. S., C. V. S. Gunatilleke, B. M. P. Singhakurmara, I. A. U. N. Gunatilleke, 2001. Restoration pathways for rain forest in southwest Sri Lanka: a review of concepts and models. For. Ecol. Manage. 154: 409–430.

Asia Forest Network, 2002. Participatory Rural Appraisal for Community Forest Management: Tools and Techniques. Asia Forest Network. www.communityforestryinternational.org/publications/field_methods_manual/pra_manual_tools_and_techniques.pdf

Assembly of Life Sciences (U.S.A.), 1982. Ecological Aspects of Development in the Humid Tropics. National Academy Press, Washington, D.C.

Bagong Pagasa Foundation, 2009. Cost comparison analysis ANR vs. conventional reforestation. Paper presented at the concluding seminar of FAO-assisted project TCP/PHI/3010 (A), Advancing the Application of Assisted Natural Regeneration (ANR) For Effective, Low-Cost Forest Restoration.

Bailey, N. T. J., 1995. Statistical Methods in Biology (3rd edition). Cambridge University Press, Cambridge.

Barlow, J. and C. A. Peres, 2007. Fire-mediated dieback and compositional cascade in an Amazonian forest. Phil. Trans. R. Soc. B, doi:10.1098/rstb.2007.0013. www.tropicalforestresearch.org/Content/people/jbarlow/Barlow%20and%20Peres%20PTRS%202008.pdf

Baskin, C. and J. Baskin, 2005. Seed dormancy in trees of climax tropical vegetation types. Trop. Ecol. 46(1): 17–28.

Bennett, A. F., 2003. Linkages in the Landscape: the Role of Corridors and Connectivity in Wildlife Conservation. IUCN, Gland and Cambridge.

Bertenshaw, V. and J. Adams, 2009a. Low-cost monitors of seed moisture status. Millennium Seedbank Technical Information Sheet No. 7. www.kew.org/msbp/scitech/publications/07-Low-cost%20moisture%20monitors.pdf

Bertenshaw, V. and J. Adams, 2009b. Small-scale seed drying methods. Millennium Seedbank Technical Information Sheet No. 8. www.kew.org/msbp/scitech/publications/08-Low-cost%20drying%20methods.pdf

Bhumibamon, S., 1986. The Environmental and Socio-economic Aspects of Tropical Deforestation: a Case Study of Thailand. Department of Silviculture, Faculty of Forestry, Kasetsart University, Thailand.

Bibby, C., M. Jones and S. Marsden, 1998. Expedition Field Techniques: Bird Surveys. The Expedition Advisory Centre, Royal Geographical Society, London.

Bone, R., M. Lawrence and Z. Magombo, 1997. The effect of *Eucalyptus camaldulensis* (Dehn) plantation on native woodland recovery on Ulumba Mountain, southern Malawi. For. Ecol. Manage. 99: 83–99.

Bonilla-Moheno, M. and Holl, K. D., 2010. Direct seeding to restore tropical mature-forest species in areas of slash-and-burn agriculture. Restor. Ecol. 18: 438–445.

Borchert, R., S. A. Meyer, R. S. Felger and L. Porter-Bolland, 2004. Environmental control of flowering periodicity in Costa Rican and Mexican tropical dry forests. Global Ecol. Biogeogr. 13: 409–425.

Boucher, D., 2008. Out of the Woods: A realistic role for tropical forests in curbing global warming. Union of Concerned Scientists, Cambridge, Massachusettes. www.ucsusa.org/assets/documents/global_warming/UCS-REDD-Boucher-report.pdf

Bradshaw, A. D., 1987. Restoration as an acid test for ecology. In: Jordan W. R., M. Gilpin and J. D. Aber (eds.), Restoration Ecology. Cambridge University Press, Cambridge, pp 23–29.

Broadhurst, L., A. Lowe, D. J. Coates, S. A. Cunningham, M. McDonald, P. A. Vesk and C. Yates, 2008. Seed supply for broad-scale restoration: maximizing evolutionary potential. Evol. Appl. 1: 587–597.

Brown, S., 1997. Estimating Biomass and Biomass Change of Tropical Forests: a Primer. FAO Forest. Pap. 134, Food and Agriculture Organization, Rome.

Bruijnzeel, L. A., 2004. Hydrological functions of tropical forests: not seeing the soil for the trees? Agric. Ecosyst. Environ. 104: 185–228. www.asb.cgiar.org/pdfwebdocs/AGEE_special_Bruijnzeel_Hydrological_functions.pdf

Brundrett, M., N. Bougher, B. Dell, T. Grove and N. Malajczuk, 1996. Working with Mycorrhizas in Forestry and Agriculture. ACIAR Monograph 32, ACIAR, Canberra.

Butler, R. A., 2009. Changing drivers of deforestation provide new opportunities for conservation. http://news.mongabay.com/2009/1208-drivers_of_deforestation.html

Cairns, M. A., S. Brown, E. Helmer and G. A. Baumgardner, 1997. Root biomass allocation in the world's upland forests. Oecologia 111: 1–11.

Calle, Z., B. O. Schlumpberger, L. Piedrahita, A. Leftin, S. A. Hammer, A. Tye and R. Borchert, 2010. Seasonal variation in daily insolation induces synchronous bud break and flowering in the tropics. Trees 24: 865–877.

Cambodia Tree Seed Project, 2004. Direct seeding. Project report, Forestry Administration, Phnom Penh, Cambodia. http://treeseedfa.org/uploaddocuments/DirectseedingEnglish.pdf

Carmago, J. L. C., Ferraz I. D. K. and Imakawa A. M., 2002. Rehabilitation of degraded areas of central Amazonia using direct sowing of forest tree seeds. Restor. Ecol. 10: 636–644.

Castillo, A., 1986. An Analysis of Selected Reforestation Projects in the Philippines. PhD thesis, University of the Philippines, Los Banos.

Chambers, J. Q., L. Santos, R. J. Ribeiro and N. Higuchi, 2001. Tree damage, allometric relationships, and above-ground net primary production in a tropical forest. For. Ecol. Manage. 152: 73–84.

Chave, J., C. Andalo, S. Brown, M. A. Cairns, J. Q. Chambers, D. Eamus, H. Folster, F. Fromard, N. Higuchi, T. Kira, J. P. Lescure, B. W. Nelson, H. Ogawa, H. Puig, B. Riera and E .T. Yamakura, 2005. Tree allometry and improved estimation of carbon stocks and balance in tropical forests. Oecologia 145: 87–99.

Clark, J. S., 1998. Why trees migrate so fast: confronting theory with dispersal biology and the paleorecord. Amer. Naturalist 152 (2): 204–224.

Cochrane, M. A., 2003. Fire science for rain forests. Nature 421: 913–919.

Cole, R. J., K. D. Holl, C. L. Keene and R. A. Zahawi, 2011. Direct seeding of late-successional trees to restore tropical montane forest. For. Ecol. Manage. 261 (10): 1590–1597.

Coley, P. D. and J. A. Barone, 1996. Herbivory and plant defenses in tropical forests. Annual Rev. Ecol. Syst. 27: 305–35.

Cropper, M., J. Puri and C. Griffiths, 2001. Predicting the location of deforestation: the role of roads and protected areas in north Thailand. Land Economics 77 (2): 172–186.

Dalmacio, M. V., 1989. Assisted natural regeneration: a strategy for cheap, fast, and effective regeneration of denuded forest lands. Manuscript, Philippines Department of Environment and Natural Resources Regional Office, Tacloban City, Philippines.

Danaiya Usher, A., 2009. Thai Forestry: A Critical History. Silkworm Books, Bangkok.

Davis, A. P., T. W. Gole, S. Baena and J. Moat, 2012. The impact of climate change on indigenous Arabica coffee (*Coffea arabica*): predicting future trends and identifying priorities. PLoS ONE 7(11): e47981. doi:10.1371/journal.pone.0047981

Department of National Parks, Wildlife and Plant Conservation (DNP) and Royal Forest Department (RFD), 2008. Reducing Emissions from Deforestation and Forest Degradation in The Tenasserim Biodiversity Corridor (BCI Pilot Site) and National Capacity Building for Benchmarking and Monitoring (REDD Readiness Plan). www.forestcarbonpartnership.org/fcp/sites/forestcarbonpartnership.org/files/Documents/PDF/Thailand_R-PIN_Annex.pdf

Diamond, J. M., 1975. The island dilemma: lessons of modern biogeographic studies for the design of natural reserves. Biological Conservation 7: 129–46.

Douglas, I., 1996. The impact of land-use changes, especially logging, shifting cultivation, mining and urbanization on sediment yields in humid tropical southeast Asia: a review with special reference to Borneo. Int. Assoc. Hydrol. Sci. Publ. 236: 463–471.

Doust, S. J., P. D. Erskine and D. Lamb, 2006. Direct seeding to restore rainforest species: Microsite effects on the early establishment and growth of rainforest tree seedlings on degraded land in the wet tropics of Australia. For. Ecol. Manage. 234: 333–343.

Doust, S. J., P. D. Erskine and D. Lamb, 2008. Restoring rainforest species by direct seeding: tree seedling establishment and growth performance on degraded land in the wet tropics of Australia. For. Ecol. Manage. 256: 1178–1188.

Dugan, P., 2000. Assisted natural regeneration: methods, results and issues relevant to sustained participation by communities. In: Elliott, S., J. Kerby, D. Blakesley, K. Hardwick, K. Woods and V. Anusarnsunthorn (eds.), Forest Restoration for Wildlife Conservation. Chiang Mai University, pp 195–199.

Dytham, C., 2011. Choosing and Using Statistics: a Biologist's Guide (3rd edition). Wiley-Blackwell, Oxford.

Elliott, S., 2000. Defining forest restoration for wildlife conservation. In: Elliott, S., J. Kerby, D. Blakesley, K. Hardwick, K. Woods and V. Anusarnsunthorn (eds.), Forest Restoration for Wildlife Conservation, Chiang Mai University, pp 13–17.

Elliott, S., J. F. Maxwell and O. Prakobvitayakit, 1989. A transect survey of monsoon forest in Doi Suthep-Pui National Park. Nat. Hist. Bull. Siam Soc. 37 (2): 137–171.

Elliott, S., P. Navakitbumrung, C. Kuarak, S. Zangkum, V. Anusarnsunthorn and D. Blakesley, 2003. Selecting framework tree species for restoring seasonally dry tropical forests in northern Thailand based on field performance. For. Ecol. Manage. 184: 177–191.

Elliott, S., P. Navakitbumrung, S. Zangkum, C. Kuarak, J. Kerby, D. Blakesley and V. Anusarnsunthorn, 2000. Performance of six native tree species, planted to restore degraded forestland in northern Thailand and their response to fertiliser. In: Elliott, S., J. Kerby, D. Blakesley, K. Hardwick, K. Woods and V. Anusarnsunthorn (eds.), Forest Restoration for Wildlife Conservation. Chiang Mai University, pp 244–255.

Elliott, S., S. Promkutkaew and J. F. Maxwell, 1994. The phenology of flowering and seed production of dry tropical forest trees in northern Thailand. Proc. Int. Symp. on Genetic Conservation and Production of Tropical Forest Tree Seed, ASEAN-Canada Forest Tree Seed Project, pp 52–62. www.forru.org/FORRUEng_Website/Pages/engscientificpapers.htm

Elster, C., 2000. Reasons for reforestation success and failure with three mangrove species in Colombia. For. Ecol. Manage. 131: 201–214.

Engel, V. L. and J. Parrotta, 2001. An evaluation of direct seeding for reforestation of degraded lands in central Sao Paulo state, Brazil. For. Ecol. Manage. 152: 169–181.

Environmental Investigation Agency, 2008. Demanding Deforestation. EIA Briefing. www.eia-international.org/files/reports175-1.pdf

Erwin, T. L., 1982. Tropical forests: their richness in *Coleoptera* and other arthropod species. Coleop. Bull. 36: 74–75.

Fandey, H. M., 2009. The Impact of Fire on Soil Seed Bank: a Case Study in the Tanzania Miombo Woodlands. MSc thesis, University of Sussex, UK.

Ferguson, B. G., 2007. Dispersal of Neotropical tree seeds by cattle as a tool for eco-agricultural restoration. Paper presentation at the Joint ESA/SER Joint Meeting on Ecological Restoration in a Changing World. http://eco.confex.com/eco/2007/techprogram/P2428.htm.

Food and Agriculture Organization of the United Nations, 1981. Tropical Forest Resource Assessment Project United Nations Food and Agriculture Organization, Rome.

Food and Agriculture Organization of the United Nations, 1997. State of the World's Forests 1997. UN FAO, Rome.

Food and Agriculture Organization of the United Nations, 2001. State of the World's Forests 2001. UN FAO, Rome.

Food and Agriculture Organization of the United Nations, 2006. Global Forest Resources Assessment 2005 – Progress towards sustainable forest management. FAO Forest. Pap. 147, UN FAO, Rome.

Food and Agriculture Organization of the United Nations, 2009. State of the World's Forests 2009. UN FAO, Rome.

Forget, P., T. Millerton and F. Feer, 1998. Patterns in post-dispersal seed removal by neotropical rodents and seed fate in relation to seed size. In: Newbery, D., H. Prins and N. Brown (eds.), Dynamics of Tropical Communities. Blackwell Science, Cambridge, pp 25–49.

FORRU (Forest Restoration Research Unit), 2000. Tree Seeds and Seedlings for Restoring Forests in Northern Thailand. Biology Department, Science Faculty, Chiang Mai University, Thailand. www.forru.org

FORRU, 2006. How to Plant a Forest: the Principles and Practice of Restoring Tropical Forests. Biology Department, Science Faculty, Chiang Mai University, Thailand. www.forru.org

FORRU, 2008. Research for Restoring Tropical Forest Ecosystems: A Practical Guide. Biology Department, Science Faculty, Chiang Mai University, Thailand. www.forru.org/FORRUEng_Website/Pages/engpublications.htm

Gamez, L., undated. Internalization of watershed environmental benefits in water utilities in Heredia, Costa Rica. http://moderncms.ecosystemmarketplace.com/repository/moderncms_documents/ESPH_Heredia_Costa_Rica.pdf

Gardner, T. A., J. Barlow, L. W. Parry and C. A. Peres, 2007. Predicting the uncertain future of tropical forest species in a data vacuum. Biotropica 39(1): 25–30.

Garwood, N., 1983. Seed germination in a seasonal tropical forest in Panama: a community study. Ecol. Monogr. 53 (2): 159–181.

Gentry, A. H., 1995. Diversity and floristic composition of neotropical dry forests. In: Bullock, S. H., H. A. Mooney and E. Medina (eds.), Seasonally Dry Tropical Forests. Cambridge University Press, Cambridge.

Ghimire, K. P., 2005. Community forestry and its impact on watershed condition and productivity in Nepal. In: Zoebisch, M., K. M. Cho, S. Hein and R. Mowla (eds.), Integrated Watershed Management: Studies and Experiences from Asia. AIT, Bangkok.

Gilbert, L. E., 1980. Food web organization and the conservation of neotropical diversity. In: Soule, M. E. and B. A. Wilcox (eds.), Conservation Biology: An Evolutionary-Ecological Perspective. Sinauer Associates, Sunderland, Massachusetts, pp 11–33.

Gilbert G., D. W. Gibbons and J. Evans, 1998. Bird Monitoring Methods: a Manual of Techniques for Key UK Species. RSPB, Sandy, Bedfordshire, UK.

Goosem, S. and N. I. J. Tucker, 1995. Repairing the Rainforest. Wet Tropics Management Authority, Cairns, Australia. www.wettropics.gov.au/media/med_landholders.html

Grainger, A., 2008. Difficulties in tracking the long-term global trend in tropical forest area. Proc. Natl. Acad. Sci. USA 105 (2): 818–823.

Hardwick, K. A., 1999. Tree Colonization of Abandoned Agricultural Clearings in Seasonal Tropical Montane Forest in Northern Thailand. PhD thesis, University of Wales, Bangor, UK.

Hardwick, K., J. R. Healey and D. Blakesley, 2000. Research needs for the ecology of natural regeneration of seasonally dry tropical forests in Southeast Asia. In: Elliott, S., J. Kerby, D. Blakesley, K. Hardwick, K. Woods and V. Anusarnsunthorn (eds.), Forest Restoration for Wildlife Conservation. Chiang Mai University, pp 165–180.

Harvey, C. A., 2000. Colonization of agricultural wind-breaks by forest trees: effects of connectivity and remnant trees. Ecol. Appl. 10: 1762–1773.

Hau, C. H., 1997. Tree seed predation on degraded hillsides in Hong Kong. For. Ecol. Manage. 99: 215–221.

Hau, C. H., 1999. The Establishment and Survival of Native Trees on Degraded Hillsides in Hong Kong. PhD thesis, University of Hong Kong.

Heng, R. K. J., N. M. Abd. Majid, S. Gandaseca, O. H. Ahmed, S. Jemat and M. K. K. Kin, 2011. Forest structure assessment of a rehabilitated forest. American Journal of Agricultural and Biological Sciences 6 (2): 256–260.

Henry, M., N. Picard, C. Trotta, R. J. Manlay, R. Valentini, M. Bernoux and L. Saint-André, 2011. Estimating tree biomass of sub-Saharan African forests: a review of available allometric equations. Silva Fenn. 45 (3B): 477–569. www.metla.fi/silvafennica/full/sf45/sf453477.pdf

Hodgson, B. and P. McGhee, 1992. Development of aerial seeding for the regeneration of Tasmanian Eucalypt forests. Tasforests, July 1992.

Hoffmann, W. A., R. Adasme, M. Haridasan, M. T. deCarvalho, E. L. Geiger, M. A. B. Pereira, S. G. Gotsch and A. C. Franco, 2009. Tree topkill, not mortality, governs the dynamics of savanna–forest boundaries under frequent fire in central Brazil. Ecology 90: 1326–1337.

Holl, K., 1998. Effects of above- and below-ground competition of shrubs and grass on *Calophyllum brasiliense* (Camb.) seedling growth in abandoned tropical pasture. For. Ecol. Manage. 109: 187–195.

Holl, K. D., M. E. Loik, E. H. V. Lin and I. A. Samuels, 2000. Tropical montane forest restoration in Costa Rica: overcoming barriers to dispersal and establishment. Restor. Ecol. 8 (4): 330–349.

IPCC (Intergovernmental Panel on Climate Change), 2000. Land Use, Land-Use Change and Forestry. Watson, R. T., I. R. Noble, B. Bolin, N. H. Ravindranath, D. J. Verardo and D. J. Dokken (eds.), Cambridge University Press, Cambridge.

IPCC, 2006. 2006 IPCC Guidelines for National Greenhouse Gas Inventories. Prepared by the National Greenhouse Gas Inventories Programme, Eggleston H. S., L. Buendia, K. Miwa, T. Ngara and K. Tanabe (eds.), Institute for Global Environmental Strategies (IGES), Japan. www.ipcc-nggip.iges.or.jp/public/2006gl/vol4.html

IPCC, 2007. Climate Change 2007: the Fourth Assessment Report (AR4) of the United Nations Intergovernmental Panel on Climate Change (IPCC). www.ipcc.ch/pdf/assessment-report/ar4/wg1/ar4-wg1-ts.pdf.

Janzen, D. H., 1981. *Enterolobium cyclocarpum* seed passage rate and survival in horses, Costa Rican Pleistocene seed-dispersal agents. Ecology 62: 593–601.

Janzen, D. H., 1988. Dry tropical forests. The most endangered major tropical ecosystem. In: Wilson, E. O. (ed.), Biodiversity. National Academy of Sciences/Smithsonian Institution, Washington DC, pp 130–137.

Janzen, D. H., 2000. Costa Rica's Area de Conservación Guanacaste: a long march to survival through non-damaging biodevelopment. Biodiversity 1 (2): 7–20.

Janzen, D. H., 2002. Tropical dry forest: Area de Conservación Guanacaste, northwestern Costa Rica. In: Perrow, M. R., and A. J. Davy (eds.), Handbook of Ecological Restoration, Vol. 2, Restoration in Practice. Cambridge University Press, Cambridge, pp 559–583.

Jitlam, N., 2001. Effects of Container Type, Air Pruning and Fertilizer on the Propagation of Tree Seedlings for Forest Restoration. MSc thesis, Chiang Mai University, Thailand.

Kafle, S. K., 1997. Effects of Forest Fire Protection on Plant Diversity, Tree Phenology and Soil Nutrients in a Deciduous Dipterocarp-Oak Forest in Doi Suthep-Pui National Park. MSc thesis, Chiang Mai University, Thailand.

Kappelle, M. and J. J. A. M. Wilms, 1998. Seed-dispersal by birds and successional change in a tropical montane cloud forest. Acta Bot. Neerl. 47: 155–156.

Ketterings, Q. M., R. Coe, M. van Noordwijk, Y. Ambagau, Y. and C. A. Palm, 2001. Reducing uncertainty in the use of allometric biomass equations for predicting above-ground tree biomass in mixed secondary forests. For. Ecol. Manage. 146, 199–209.

Knowles, O. H. and J. A. Parrotta, 1995. Amazon forest restoration: an innovative system for native species selection based on phonological data and field performance indices. Commonwealth Forestry Review 74: 230–243.

Kodandapani, N. M. Cochrane and R. Sukumar, 2008. A comparative analysis of spatial, temporal, and ecological characteristics of forest fires in seasonally dry tropical ecosystems in the Western Ghats, India. For. Ecol. Manage. 256: 607–617.

Koelmeyer, K. O., 1959. The periodicity of leaf change and flowering in the principal forest communities of Ceylon. Ceylon Forest. 4: 157–189, 308–364.

Kopachon, S. 1995. Effects of Heat Treatment (60-70°C) on Seed Germination of some Native Trees on Doi Suthep. MSc thesis, Chiang Mai University, Thailand.

Kuarak, C., 2002. Factors Affecting Growth of Wildlings in the Forest and Nurturing Methods in the Nursery. MSc thesis, Chiang Mai University, Thailand. www.forru.org/FORRUEng_Website/Pages/engstudentabstracts.htm

Kuaraksa, C. and S. Elliott, 2012. The use of Asian *Ficus* species for restoring tropical forest ecosystems. Restor. Ecol. 21; 86–95.

Lamb, D., 2011. Regreening the Bare Hills. Springer, Dordecht.

Lamb, D., J. Parrotta, R. Keenan and N. I. J. Tucker, 1997. Rejoining habitat remnants: restoring degraded rainforest lands. In: Laurence W. F. and R. O. Bierrgaard Jr. (eds.), Tropical Forest Remnants: Ecology, Management and Conservation of Fragmented Communities. University of Chicago Press, Chicago, pp 366–385.

Laurance, S. G. and W. F. Laurance, 1999. Tropical wildlife corridors: use of linear rainforest remnants by arboreal mammals. Biol. Conserv. 91: 231–239.

Lewis, L. S., G. Lopez-Gonzalez, B. Sonké, K. Affum-Baffoe, T. R. Baker, L. O. Ojo, O. L. Phillips, J. M. Reitsma, L. White, J. A. Comiskey, K. M.-N. Djuikouo, C. E. N. Ewango, T. R. Feldpausch, A. C. Hamilton, M. Gloor, T. Hart, A. Hladik, J. Lloyd, J. C. Lovett, J.-R. Makana, Y. Malhi, F. M. Mbago, H. J. Ndangalasi, J. Peacock, K. S.-H. Peh, D. Sheil, T. Sunderland, M. D. Swaine, J. Taplin, D. Taylor, S. C. Thomas, R. Votere and H. Woll, 2009. Increasing carbon storage in intact African tropical forests. Nature 457: 1003–1007.

Lewis, S. L., P. M. Brando, O. L. Phillips, G. M. F. van der Herijden and D. Nepstad, 2011. The 2010 Amazon drought. Science 331: 554.

Longman, K. A. and R. H. F. Wilson, 1993. Tropical Trees: Propagation and Planting Manuals. Vol. 1. Rooting Cuttings of Tropical Trees. Commonwealth Science Council, London.

Lowe, A. J., 2010. Composite provenancing of seed for restoration: progressing the 'local is best' paradigm for seed sourcing. The State of Australia's Birds 2009: restoring woodland habitats for birds. Compiled by David Paton and James O'Conner. Supplement to Wingspan Newsletter 20(1) (March). www.birdlife.org.au/documents/SOAB-2009.pdf

Lucas, R. M., M. Honzak, P. J. Curran, G. M. Foody, R. Milnes, T. Brown and S. Amaral, 2000. Mapping the regional extent of tropical forest regeneration stages in the Brazilian legal Amazon using NOAA AVHRR data. Int. J. Remote Sens. 21 (15): 2855–2881.

Ludwig, J. A. and J. E. Reynolds, 1988. Statistical Ecology. Chapter 14. John Wiley & Sons, New York.

Maia, J. and M. R. Scotti, 2010. Growth of *Inga vera* Willd. subsp. *affinis* under *Rhizobia* inoculation. Nutr. Veg. 10 (2): 139–149.

Malhi, Y., L. E. O. C. Aragão, D. Galbraith, C. Huntingford, R. Fisher, P. Zelazowski, S. Sitche, C. McSweeney and P. Meir, 2009. Exploring the likelihood and mechanism of a climate-change-induced dieback of the Amazon rainforest. Proc. Natl. Acad. Sci. USA 106 (49): 20610–20615.

Mansourian, S., D. Vallauri, and N. Dudley (eds.) (in co-operation with WWF International), 2005. Forest Restoration in Landscapes: Beyond Planting Trees. Springer, New York.

Marland, G., T. A. Boden and R. J. Andres, 2006. Global, regional, and national CARBON DIOXIDE emissions. In: Trends: a Compendium of Data on Global Change. Carbon Dioxide Information Analysis Center, Oak Ridge National Laboratory, U.S. Department of Energy, Oak Ridge, TN. http://cdiac.esd.ornl.gov/trends/emis/tre_glob.htm.

Martin, A. R and S. C. Thomas, 2011. A reassessment of carbon content in tropical trees. PLoS ONE 6(8): e23533. doi:10.1371/journal.pone.0023533

Martin, G. J., 1995. Ethnobotany: a Methods Manual. Chapman and Hall, London.

Maxwell, J. F. and S. Elliott, 2001. Vegetation and Vascular Flora of Doi Sutep–Pui National Park, Chiang Mai Province, Thailand. Thai Studies in Biodiversity 5. Biodiversity Research and Training Programme, Bangkok.

McKinnon, J. and K. Phillips, 1993. A Field Guide to the Birds of Borneo, Sumatra, Java and Bali. Oxford University Press, Oxford.

McLaren, K. P. and M. A. McDonald, 2003. The effects of moisture and shade on seed germination and seedling survival in a tropical dry forest in Jamaica. For. Ecol. Manage. 183: 61–75.

Mendoza, E. and R. Dirzo, 2007. Seed size variation determines inter-specific differential predation by mammals in a neotropical rain forest. Oikos 116: 1841–1852.

Meng, M., 1997. Effects of Forest Fire Protection on Seed-dispersal, Seed Bank and Tree Seedling Establishment in a Deciduous Dipterocarp-Oak Forest in Doi Suthep-Pui National Park. MSc thesis, Chiang Mai University, Thailand.

Midgley, J. J., M. J. Lawes and S. Chamaillé-Jammes, 2010. Savanna woody plant dynamics: the role of fire and herbivory, separately and synergistically. Turner Review No.19, Austral. J. Bot. 58: 1–11.

Milan, P., M. Ceniza, E. Fernando, M. Bande, P. Noriel-Labastilla, J. Pogosa, H. Mondal, R. Omega, A. Fernandez and D. Posas, undated. Rainforestation Training Manual. Environmental Leadership and Training Initiative (ELTI), Singapore.

Miyawaki, A., 1993. Restoration of native forests from Japan to Malaysia. In: Lieth, H. and M. Lohmann (eds.), Restoration of Tropical Forest Ecosystems, Kluwer Academic Publishers, Dordrecht, The Netherlands, pp 5–24.

Miyawaki, A. and S. Abe, 2004. Public awareness generation for the reforestation in Amazon tropical lowland region. Trop. Ecol. 45 (1): 59–65.

Montagnini, F. and C. F. Jordan, 2005. Tropical Forest Ecology – The Basis for Conservation and Management. Springer, Berlin.

Muhanguzi, H. D. R., J. Obua, H. Oreym-Origa and O. R. Vetaas, 2005. Forest site disturbances and seedling emergence in Kalinzu Forest, Uganda. Trop. Ecol. 46 (1): 91–98.

Myers, N., 1992. Primary Source: Tropical Forests and Our Future (Updated for the Nineties). W. W. Norton and Co., London.

Nair, J. K. P., and C. R. Babu, 1994. Development of an inexpensive legume-*Rhizobium* inoculation technology which may be used in aerial seeding. J. Basic Microbiol. 34: 231–243.

Nandakwang, P. S. Elliott, S. Youpensuk, B. Dell, N. Teaumroong and S. Lumyong, 2008. Arbuscular mycorrhizal status of indigenous tree species used to restore seasonally dry tropical forest in northern Thailand. Res. J. Microbiol. 3 (2): 51–61.

Negreros, C. P. and R. B. Hall, 1996. First-year results of partial overstory removal and direct seeding of mahogany (*Swietenia macrophylla*) in Quintana Roo, Mexico. J. Sustain. For. 3: 65–76.

Nepstad, D. C., 2007. The Amazon's Vicious Cycles: Drought and Fire in the Greenhouse. WWF International, Gland. http://assets.wwf.org.uk/downloads/amazonas_vicious_cycles.pdf

Nepstad, D., G. Carvalho, A. C., Barros, A. Alencar, J. P. Capobianco, J. Bishop, P. Mountinho, P. Lefebre, U. Lopes Silva and E. Prins, 2001. Road paving, fire regime feedbacks and the future of Amazon forests. For. Ecol. Manage. 154: 395–407.

Nepstad, D.C., C. Uhl, C. A. Pereira and J. M. C. da Silva, 1996. A comparative study of tree establishment in abandoned pastures and mature forest of eastern Amazonia. Oikos 76 (1): 25–39.

Newmark, W. D., 1991. Tropical forest fragmentation and the local extinction of understorey birds in the Eastern Usambara Mountains, Tanzania. Conserv. Biol. 5: 67–78.

Newmark, W. D., 1993. The role and design of wildlife corridors with examples from Tanzania. Ambio 22: 500–504.

Ng, F. S. P., 1980. Germination ecology of Malaysian woody plants. Malaysian Forester 43: 406–437.

Nuyun, L. and Z. Jingchun, 1995. China aerial seeding achievement and development. Forestry and Society Newsletter, November 1995, 3 (2): 9–11.

Ødegaard, F., 2008. How many species of arthropods? Erwin's estimate revised. Biol. J. Linn. Soc. 71 (4) 583–597.

Paetkau, D., E. Vazquez-Dominguez, N. I. J. Tucker and C. Moritz, 2009. Monitoring movement into and through a newly restored rainforest corridor using genetic analysis of natal origin. Ecol. Manag. & Restn. 10 (3): 210–216.

Pagano, M. C., 2008. Rhizobia associated with neotropical tree *Centrolobium tomentosum* used in riparian restoration. Plant Soil Environ. 54 (11): 498–508.

Page, S., A. Hosciło, H. Wösten, J. Jauhiainen, M. Silvius, J. Rieley, H. Ritzema, K. Tansey, L. Graham, H. Vasander and S. Limin, 2009. Restoration ecology of lowland tropical peatlands in Southeast Asia: current knowledge and future research directions. Ecosystems 12: 888–905.

Panyanuwat, A., T. Chiengchee, U. Panyo, C. Mikled, S. Sangawongse, T. Jetiyanukornkun, S. Ratchusanti, C. Rueangdetnarong, T. Saowaphak, J. Prangkoaw, C. Malumpong, S. Tovicchakchaikul, B. Sairorkhom and O. Chaiya, 2008. The Evaluation Project of the Forestation Plantation and Water Source Check Dam Construction. The University Academic Service Center, Chiang Mai University, Thailand (in Thai).

Parrotta, J. A., 1993. Secondary forest regeneration on degraded tropical lands: the role of plantations as "foster ecosystems." In Lieth, H. and M. Lohmann (eds.). Restoration of Tropical Forest Ecosystems. Kluwer Academic Publishers, Dordrecht, The Netherlands, pp 63–73.

Parrotta, J. A., 2000. Catalyzing natural forest restoration on degraded tropical landscapes. In: Elliott S., J. Kerby, D. Blakesley, K. Hardwick, K. Woods and V. Anusarnsunthorn (eds.), Forest Restoration for Wildlife Conservation. Chiang Mai University, pp 45–56.

Parrotta, J. A., J. W. Turnbull and N. Jones, 1997a. Catalyzing native forest regeneration on degraded tropical lands. For. Ecol. Manage. 99: 1–7.

Parrotta, J. A., O. H. Knowles and J. N. Wunderle, 1997b. Development of floristic diversity in 10-year old restoration forests on a bauxite mine in Amazonia. For. Ecol. Manage. 99: 21–42.

Pearson, T. R. H., D. F. R. P. Burslem, C. E. Mullins and J. W. Dalling, 2003. Functional significance of photoblastic germination in neotropical pioneer trees: a seed's eye view. Funct. Ecol. 17 (3): 394–404.

Pena-Claros, M. and H. De Boo, 2002. The effect of successional stage on seed removal of tropical rainforest tree species. J. Trop. Ecol. 18: 261–274.

Pennington, T. D. and E. C. M. Fernandes, 1998. Genus *Inga*; Utilization. Royal Botanic Gardens, Kew.

Pfund, J. and P. Robinson (eds.), 2005. Non-Timber Forest Products: Between Poverty Alleviation and Market Forces. Special publication of Inter Cooperation, and the editorial team of the Working Group "Trees and Forests in Development Cooperation", Switzerland. http://frameweb.org/adl/en-US/2427/file/274/NTFP-between-poverty-alleviation-and-market-forces.pdf

Philachanh, B., 2003. Effects of Presowing Seed Treatments and Mycorrhizae on Germination and Seedling Growth of Native Tree Species for Forest Restoration. MSc thesis, Chiang Mai University, Thailand. www.forru.org/FORRUEng_Website/Pages/engstudentabstracts.htm

Posada, J. M., T. M. Aide, and J. Cavelier, 2000. Livestock and weedy shrubs as restoration tools of tropical montane rainforest. Restor. Ecol. 8: 361–370.

Putz, F. E., P. Sist, T. Fredericksen and D. Dykstra, 2008. Reduced-impact logging: challenges and opportunities, For. Ecol. Manage. 256: 1427–1433.

Reitbergen-McCraken, J., S. Maginnis and A. Sarre, 2007. The Forest Landscape Restoration Handbook. Earthscan, London.

Richards, P. W., 1996. The Tropical Rain Forest (2nd Edition). Cambridge University Press, Cambridge.

Rodríguez, J. M. (ed.), 2005. The Environmental Services Program: A Success Story of Sustainable Development Implementation in Costa Rica. National Forestry Fund (FONAFIFO), San José.

Ros-Tonen, M. A. F. and K. F. Wiersum, 2003. The Importance of Non-Timber Forest Products for Forest-Based Rural Livelihoods: an Evolving Research Agenda. Amsterdam AGIDS/UvA. http://pdf.wri.org/ref/shackleton_04_the_importance.pdf

Sanchez-Cordero, V. and R. Martínez-Gallardo, 1998. Post-dispersal fruit and seed removal by forest-dwelling rodents in a lowland rain forest in Mexico. J. Trop. Ecol. 14: 139–151.

Sansevero, J. B. B., P. V. Prieto, L. F. D. de Moraes and P. J. P. Rodrigues, 2011. Natural regeneration in plantations of native trees in lowland Brazilian Atlantic forest: community structure, diversity, and dispersal syndromes. Restor. Ecol. 19: 379–389.

Scatena, F. N., L. A. Bruijnzeel, P. Bubb and S. Das, 2010. Setting the stage. In: Bruijnzeel, L. A., F. N. Scatena and L. S. Hamilton (eds.), Tropical Montane Cloud Forests: Science for Conservation and Management. Cambridge University Press, Cambridge, pp 3–13.

Schmidt, L., 2000. A Guide to Handling Tropical and Subtropical Forest Seed. DANIDA Forest Seed Centre, Denmark.

Schulte, A., 2002. Rainforestation Farming: Option for Rural Development and Biodiversity Conservation in the Humid Tropics of Southeast Asia. Shaker Verlag, Aachen.

Scott, R., P. Pattanakaew, J. F. Maxwell, S. Elliott and G. Gale, 2000. The effect of artificial perches and local vegetation on bird-dispersed seed deposition into regenerating sites. In: Elliott, S., J. Kerby, D. Blakesley, K. Hardwick, K. Woods and V. Anusarnsunthorn (eds.), Forest Restoration for Wildlife Conservation. Chiang Mai University, pp 326–337.

Sekercioglu, C. H., 2009. Tropical ecology: riparian corridors connect fragmented forest bird populations. Current Biology 19: 210–213.

Sgró, C.M., A. J. Lowe and A. A. Hoffmann, 2011. Building evolutionary resilience for conserving biodiversity under climate change. Evol. Appl. 4 (2): 326–337.

Shiels, A. and L. Walker, 2003. Bird perches increase forest seeds on Puerto Rican landslides. Restor. Ecol. 11 (4): 457–465.

Shono, K., E. A. Cadaweng and P. B. Durst, 2007. Application of Assisted Natural Regeneration to restore degraded tropical forestlands. Restor. Ecol. 15 (4): 620–626.

Siddique, I., V. L. Engel, J. A. Parrotta, D. Lamb, G. B. Nardoto, J. P. H. B. Ometto, L. A. Martinelli and S. Schmidt, 2008. Dominance of legume trees alters nutrient relations in mixed species forest restoration plantings within seven years. Biogeochem. 88: 89–101.

Silk, J. W. F., 2005. Assessing tropical lowland forest disturbance using plant morphology and ecological attributes. For. Ecol. Manage. 205: 241–250.

Singh, A. and P. Raizada, 2010. Seed germination of selected dry deciduous trees in response to fire and smoke. J. Trop. Forest Sci. 22 (4): 465–468.

Singpetch, S., 2002. Propagation and Growth of Potential Framework Tree Species for Forest Restoration. MSc thesis, Chiang Mai University, Thailand.

Sinhaseni, K., 2008. Natural Establishment of Tree Seedlings in Forest Restoration Trials at Ban Mae Sa Mai, Chiang Mai Province. MSc thesis, Chiang Mai University, Thailand.

Slik, J. W. F., F. C. Breman, C. Bernard, M. van Beek, C. H. Cannon, K. A. O. Eichhorn and K. Sidiyasa, 2010. Fire as a selective force in a Bornean tropical everwet forest. Oecologia 164: 841–849.

Soule, M. E. and J. Terborgh, 1999. The policy and science of regional conservation. In: Soule, M. E. and J. Terborgh (eds.), Continental Conservation: Scientific Foundations of Regional Reserve Networks. Island Press, New York, pp 1–17.

Stangeland, T., J. R. S. Tabuti and K. A. Lye, 2007. The influence of light and temperature on the germination of two Ugandan medicinal trees. Afr. J. Ecol. 46: 565–571.

Stangeland, T., J. R. S. Tabuti and K. A. Lye, 2011. The framework tree species approach to conserve medicinal trees in Uganda. Agrofor. Syst. 82 (3): 275–284.

Stokes, E. J., 2010. Improving effectiveness of protection efforts in tiger source sites: developing a framework for law enforcement monitoring using MIST. Integrative Zoology 5: 363–377.

Stoner, E. and J. Lambert, 2007. The role of mammals in creating and modifying seed shadows in tropical forests and some possible consequences of their elimination. Biotropica 39 (3): 316–327.

Stouffer, P. C. and R. O. Bierregaard, 1995. Use of Amazonian forest fragments by understorey insectivorous birds. Ecology 76: 2429–2445.

Tabuti, J. R. S., 2007. The uses, local perceptions and ecological status of 16 woody species of Gadumire Sub-county, Uganda. Biodivers. Conserv. 16: 1901-1915.

Tabuti, J. R. S., K. A. Lye and S. S. Dhillion, 2003. Traditional herbal drugs of Bulamogi, Uganda: plants, use and administration. J. Ethnopharmacol. 88, 19–44.

Tabuti, J. R. S., T. Ticktin, M. Z. Arinaitwe and V. B. Muwanika, 2009. Community attitudes and preferences towards woody species and their implications for conservation in Nawaikoke Sub-county, Uganda. Oryx 43 (3): 393–402.

TEEB, 2009. TEEB Climate Issues Update. September 2009. www.teebweb.org/teeb-study-and-reports/additional-reports/climate-issues-update/

Thira, O. and O. Sopheary, 2004. The Integration of Participatory Land Use Planning Tools (PLUP) in the Community Forestry Establishment Process: a Case Study, Tuol Sambo Village, Trapeang Pring Commune, Damer District, Kompong Cham Province, Cambodia. CBNRM Learning Institute, Phnom Penh, Cambodia. www.learninginstitute.org/files/publications/Catalogues/Final_Publication_Catalogue.pdf

Toktang, T., 2005. The Effects of Forest Restoration on the Species Diversity and Composition of a Bird Community in Doi Suthep-Pui National Park Thailand from 2002–2003. MSc thesis, Chiang Mai University, Thailand.

Traveset, A., 1998. Effect of seed passage through vertebrate frugivores' guts on germination: a review. Perspect. Plant Ecol. Evol. Syst. 1 (2): 151–190.

Trisurat, Y., 2007. Applying gap analysis and a comparison index to evaluate protected areas in Thailand. Eviron. Manage. 39: 235–245.

Tucker, N., 2000. Wildlife colonisation on restored tropical lands: what can it do, how can we hasten it and what can we expect? In Elliott, S., J. Kerby, D. Blakesley, K. Hardwick, K. Woods and V. Anusarnsunthorn (eds.), Forest Restoration for Wildlife Conservation. Chiang Mai University, pp 278–295.

Tucker, N. and T. Murphy, 1997. The effects of ecological rehabilitation on vegetation recruitment: some observations from the Wet Tropics of North Queensland. For. Ecol. Manage. 99: 133–152.

Tucker, N. I. J. and T. Simmons, 2009. Restoring a rainforest habitat linkage in north Queensland: Donaghy's Corridor. Ecol. Manage. Restn. 10 (2): 98–112.

Tunjai, P., 2005. Appropriate Tree Species and Techniques for Direct Seeding for Forest Restoration in Chiang Mai and Lamphun Provinces. MSc thesis, Chiang Mai University, Thailand.

Tunjai, P., 2011. Direct Seeding For Restoring Tropical Lowland Forest Ecosystems In Southern Thailand. PhD thesis, Walailak University, Thailand.

Tunjai, P., 2012. Effects of seed traits on the success of direct seeding for restoring southern Thailand's lowland evergreen forest ecosystem. New Forests 43 (3), 319–333.

Turkelboom, F., 1999. On-farm Diagnosis of Steepland Erosion in Northern Thailand. PhD thesis, KU Leuven, The Netherlands.

UNEP-WCMC, 2000. Global Distribution of Current Forests, United Nations Environment Programme – World Conservation Monitoring Centre (UNEP-WCMC). www.unepwcmc.org/forest/global_map.htm.

Union of Concerned Scientists, 2009. Scientists and NGOs: Deforestation and Degradation Responsible for Approximately 15 Percent of Global Warming Emissions. www.ucsusa.org/news/press_release/scientists-and-ngos-0302.html

United Nations, 2001. World Population Monitoring – 2001. UN Department of Economic and Social Affairs, Population Division, New York. www.un.org/esa/population/publications/wpm/wpm2001.pdf

United Nations, 2009. World Population Prospects – The 2008 Revision – Highlights. UN Department of Economic and Social Affairs – Population Division. www.un.org/esa/population/publications/wpp2008/wpp2008_highlights.pdf.

Van Nieuwstadt, M. G. L. and D. Sheil, 2005. Drought, fire and tree survival in a Borneo rain forest, East Kalimantan, Indonesia. J. Ecol. 93: 191–201.

Vanthomme, H., B. Belle and P. Forget, 2010. Bushmeat hunting alters recruitment of large-seeded plant species in central Africa. Biotropica 42 (6): 672–679.

Vasconcellos, H. L. and J. M. Cherret, 1995. Changes in leaf-cutting ant populations (Formicidae: Attini) after clearing of mature forest in Brazilian Amazonia. Studies on Neotropical Fauna and Environment 30: 107–113.

Vicente, R., R. Martins, J. J. Zocche and B. Harter-Marques, 2010. Seed dispersal by birds on artificial perches in reclaimed areas after surface coal mining in Siderópolis municipality, Santa Catarina State, Brazil. R. Bras. Bioci., Porto Alegre 8 (1): 14–23.

Vieira, D. L. M. and A. Scariot, 2006. Principles of natural regeneration of dry tropical forests for restoration. Restor. Ecol. 14 (1): 11–20.

Vongkamjan, S., 2003. Propagation of Native Forest Tree Species for Forest Restoration in Doi Suthep-Pui National Park. PhD thesis, Chiang Mai University, Thailand. www.forru.org/FORRUEng_Website/Pages/engstudentabstracts.htm

Vongkamjan, S., S. Elliott, V. Anusarnsunthorn and J. F. Maxwell, 2002. Propagation of native forest tree species for forest restoration in northern Thailand. In: Chien, C. and R. Rose (eds.), The Art and Practice of Conservation Planting. Taiwan Forestry Research Institute, Taipei, pp 175–183.

Whitmore, T. C., 1998. An Introduction to Tropical Rain Forests (2nd edition). Oxford University Press, Oxford.

Wiersum, K. F., 1984. Surface erosion under various tropical agroforestry systems. In: O'Loughlin, C. L. and A. J. Pearce (eds.), Effects of Forest Land Use on Erosion and Slope Stability. IUFRO, Vienna, pp 231–239.

Wilson, E. O., 1992. The Diversity of Life. Harvard University Press, Cambridge, Massachusetts.

Woods, K. and S. Elliott, 2004. Direct seeding for forest restoration on abandoned agricultural land in northern Thailand. J. Trop. Forest Sci. 16 (2): 248–259.

Wright, S. J. and H. C. Muller-Landau, 2006. The future of tropical forest species. Biotropica 38: 287–301.

Zangkum, S., 1998. Growing Tree Seedlings to Restore Forests: Effects of Container Type and Media on Seedling Growth and Morphology. MSc thesis, Chiang Mai University, Thailand.

Zappi, D., D. Sasaki, W. Milliken, J. Piva, G. S. Henicka, N. Biggs and S. Frisby, 2011. Plantas vasculares da região do Parque Estadual Cristalino, norte de Mato Grosso, Brasil. Acta Amazonica 41 (1): 29–38.

Zelazowski, P., Y. Malhi, C. Huntingford, S. Sitch and J. B. Fisher, 2011. Changes in the potential distribution of humid tropical forests on a warmer planet. Phil. Trans. R. Soc. A 369: 137–160.

INDEX

camera trapping, 256
Cameroon, 5, 30, 59
canker (disease), 183
canopy, 29
    closure, 126, 138, 140, 144, 235, 294
    width, 282–5
Capparis tomentosa, 261
Captan, 169–70, 178, 184
carbon,
    carbon dioxide, 10, 17, 61
    carbon standards organisations, 237–8
    credits, **17–8**, 108, 238–9
        compliance credits, 17
        markets, 17–8, 238–9
        REDD+, 18, 101, 231, 268
        trading, 17, 91, 102, 147, 216
        voluntary credits, 17, 108
    emissions, 17–8
    foot prints, 17
    monitoring carbon accumulation, 237–9
    offset schemes, 231, 237, 268
    sinks, 17, 108
    storage, 10, 13, **17–8**, 42, 62, 108, 146, 237–9, 239,
        247
carbon-based global economy, 17
CarbonFix, 238
cardamon, 8
cardboard (see also 'mulch'), 224–6, 230
Cardwellia sublimis, 80
Casuarius casuarius johnsonii, 96
cattle, 3, 10, 20, 50, 58, **117**, 149
    domestic, 58, **117**, 149
    ranching, 3, 10, 20, 117, 149
    wild, 50, 58
CBD (Convention on Biological Diversity), 109, **267–8**
CDM (Clean Development Mechanism), 17, 268
Cecropia, 22, 46
Ceiba, 46
Cerasus cerasoides, 213
Cercopithecus diana, 8
Chairuangsri, Sutthathorn, 81
charcoal,
    deposits in soil profile, 59
    for seed drying, 165
    making, 4, 260
Chiang Mai University, 14, 293
Chico Mendes Reserve, 114
children (involving), 214–5
China, 36, 134
Chi-square test, 236, 237
Chromolaena, 57

odorata, 56
civets, 51, 256
Clean Development Mechanism (CDM), 17, 268
climate change, 2, 7, 10, 33, 59, 60, **61–3**
    adaptation to, 90, 159
    mitigation of, 17–18
climate regulation, 19
Climate, Community & Biodiversity (CCB) standard,
        238
climates,
    everwet, 25, 54, 63
    seasonally dry, 25, 54, 63, 124, 157, 212
climax forest, 12–3, **45–8**, 63, 88–9, 123
climax species, **45–7**, 49, 53, 60, 119–20, 125–7, 131,
        135, 138, 285
cloud stripping, 40
CO_2, 10, 17, 61
coconut husk (see also 'growth media'), 166, 204
Collaborative Partnership of Forests (CPF), 109, 275
Columbia, 117
communication strategy, 290–2
community forestry, 112, 272–4, 293
Compositae (Asteraceae), 56–7
composite provenancing, 159–60
conflict resolution, 91
Congo Basin, 24, 43
conifers, 182
connectivity, 90
conservation (see also 'biodiversity'), 136, 139, **267–8**
containers, 171, 176, 204
coppicing, **49**, 116, 122, 124
Corporate Social Responsibility, see 'CSR'
corporate sponsorship, 108, 141
corridors, 90, **95–8**, 100, 136
Costa Rica, 18, 95, 117, 146, 149–51
costs and benefits of forest restoration (see also 'forest
        restoration'), **146–8**, 273–4
cotyledons, 156
Cristalino Ecological Foundation (FEC), 20
Cristalino State Park, 20–2
cross-pollination, 158
crown density method, see 'phenology, scoring'
crows, 50
CSR (Corporate Social Responsibility) schemes, 108,
        237
cultivation, 45
cultural services, 19
cut and carry, 117
cuttings, 170, 178–9